Statistical Process Control

Statistical Process Control

A practical guide

Second edition

John S. Oakland

PhD, CChem, MRSC, FIQA, FSS, MASQC, FAQMC, MBIM
Exxon Chemical Professor of Total Quality Management
and

Roy F. Followell

BSc, PhD, CPhys, MInstP, MInstPS, FIQA, FAQMC
University of Bradford, Management Centre

HEINEMANN NEWNES

Heinemann Newnes
An imprint of Heinemann Professional Publishing Ltd
Halley Court, Jordan Hill, Oxford OX2 8EJ

OXFORD LONDON MELBOURNE AUCKLAND SINGAPORE
IBADAN NAIROBI GABORONE KINGSTON

First published 1986
Reprinted 1986, 1987, 1989
First published as a paperback edition 1987
Reprinted 1988
Second edition 1990
Reprinted 1990

British Library Cataloguing in Publication Data
Oakland, John S.
    Statistical process control. – 2nd ed
    1. Industries. Quality control. Statistical **methods**
    I. Title II. Followell, Roy F.
    658.562015195

ISBN 0 434 91484 3

Printed and bound in Great Britain
by Billing & Sons Ltd, Worcester

To Susan and Doris

Contents

Preface   xi

1   Quality, processes and control   1

1.1   The basic concepts   1
1.2   Design, conformance and costs   5
1.3   The SPC system   11
1.4   Some basic tools   14
Chapter highlights   14

2   Understanding the process   18

2.1   Information about the process   18
2.2   Flowcharting   21
2.3   Process examination   25
2.4   Development of the process   28
Chapter highlights   30

3   Data collection and presentation   32

3.1   The systematic approach   32
3.2   Data collection   33
3.3   Bar charts and histograms   36
3.4   Graphs and other pictures   46
3.5   Conclusions   49
Chapter highlights   50

4   Process problem solving   51

4.1   Introduction   51
4.2   Pareto analysis   52
4.3   Cause and effect analysis   62
4.4   Scatter diagrams   70

	4.5	Control charts	74
		Chapter highlights	78

5   Variables and process variation — 80

	5.1	Causes of process variability	80
	5.2	Measures of location of central values (accuracy)	88
	5.3	Measures of spread of values (precision)	92
	5.4	Understanding variation – the normal distribution	94
	5.5	Sampling and averages	98
		Chapter highlights	105

6   Process control using variables — 106

	6.1	Means, ranges and charts	106
	6.2	Are we in control?	118
	6.3	Do we continue to be in control?	120
	6.4	International control charts for variables	124
	6.5	Choice of sample size and frequency of sampling	125
	6.6	Summary of SPC for variables using $\overline{X}$ and R charts	127
		Chapter highlights	127

7   Process capability for variables and its measurement — 129

	7.1	Will it meet the requirements?	129
	7.2	Process capability indices	133
	7.3	Interpreting capability indices	135
	7.4	The use of control chart and process capability data	141
	7.5	Modified or relaxed control charts	147
		Chapter highlights	148

8   Other types of control charts for variables — 151

	8.1	Life beyond the mean and range chart	151
	8.2	Median, mid-range and multi-vari charts	157
	8.3	Moving mean and moving range charts	161
	8.4	Control charts for standard deviation	174
		Chapter highlights	180

9   Random variation and its management — 182

	9.1	Introduction	182
	9.2	The components of capability	183
	9.3	The components of variability	184
	9.4	The addition of independent components of variation	185

9.5   The addition of dependent components of variation   191
9.6   Blending and mixing   196
Chapter highlights   199

10   Managing out-of-control processes   201

10.1   Introduction – process types and variability   201
10.2   The evaluation of actual process control procedures   203
10.3   The analysis of existing data – some examples   206
10.4   The use of control charts for trouble shooting   212
10.5   Assignable causes   219
Chapter highlights   221

11   Process control by attributes   223

11.1   Underlying concepts   223
11.2   Charts for number of defectives or non-conforming
       units (np)   225
11.3   Charts for proportion defective or non-conforming (p) 237
11.4   Charts for number of defects or non-conformities (c) 241
11.5   Charts for number of defects or non-conformities
       per unit (u)   246
11.6   Managing specifications based on subjective
       assessments   250
Chapter highlights   254

12   Cumulative sum charts   256

12.1   Introduction   256
12.2   The detection of trends and runs – attributes   256
12.3   Cusum charts – variables   262
12.4   Decision procedures   268
12.5   The design and use of V-masks – attributes   269
12.6   The design and use of V-masks – variables   272
12.7   Shewhart charts and cusums in combination   275
12.8   Some examples of cusum and Shewhart charts   278
Chapter highlights   282

13   Designing the process control system   284

13.1   SPC and the quality system   284
13.2   Teamwork and process control   287
13.3   Improvements in the process   290
13.4   Taguchi methods   297
Chapter highlights   303

14 SPC in non-manufacturing   **305**

    14.1   The process and the data   305
    14.2   Process capability analysis in a bank   306
    14.3   Profits on sales   308
    14.4   Forecasting income   311
    14.5   Ranking in managing product range   312
    14.6   Activity sampling   315
    14.7   Absenteeism   316
    14.8   Errors on invoices   318
    14.9   Injury data   320
    14.10 Summarizing SPC in non-manufacturing   322
    Chapter highlights   322

15 The implementation of statistical process control   **324**

    15.1   Introduction   324
    15.2   Successful users of SPC and the benefits derived   326
    15.3   The initial barriers to the implementation of SPC 328
    15.4   A proposed methodology for SPC implementation   329
    15.5   How to start the implementation of SPC   333
    Chapter highlights   336

Appendices   **338**

A   The normal distribution and non-normality   338
B   Constants used in the design of control charts for mean   348
C   Constants used in the design of control charts for range 349
D   Constants used in the design of control charts for
    median and range   350
E   Constants used in the design of control charts for
    standard deviaton   351
F   Cumulative Poisson probability tables   352
G   Confidence limits and tests of significance   363
H   OC curves and ARL curves for $\bar{X}$ and R charts   373
I   Autocorrelation   378
J   Approximations to assist in process control of attributes 380
K   Glossary of terms and symbols   385
L   Problems for the reader to solve   392
M   Some worked examples and answers to numerical
    problems in Appendix L   407
N   Further reading   419

Index   421

Preface

In the first edition of this book the author recalled a a telephone call from a worried Managing Director of a medium-sized engineering firm who was desperate for help. Approximately 15 per cent of his workforce were employed as finished product inspectors which had helped tremendously to reduce the number of customer complaints and warranty claims. However, it had created a huge pile of scrap material and products for rework, the sight of which preyed on his mind and kept him awake every night. It also prevented him from getting his car into his reserved space every morning!

Since that time the business and commercial world has changed quite a lot, and pressure from companies supplying directly to the consumer has forced those in charge of production and service operations to think more about preventing problems than how to find and fix them.

This has created even greater demands than before for a 'tool kit' to supplement good quality management systems and teamwork in the search for customer satisfaction. The second edition of this book is essentially about that tool kit, although it includes more than the first edition on the 'philosophy' behind SPC, and the management approaches necessary to make the tools work effectively.

To be successful in today's economic climate, any organization and its suppliers must be dedicated to continuing improvements in quality. More efficient ways to produce goods and services that consistently meet the needs of the customer must be sought. There is now ever-increasing pressure on those in the supply chains of manufacturing companies to do so.

To achieve this, several things are required: senior management commitment to improvements in quality; a good quality system; teamwork; and the use of effective methods. The second edition of this best-selling book sets down some of the basic principles of quality to provide a platform for improvement and describes the techniques used in process quality control. As before, the book covers the subject of

Statistical Process Control (SPC) in a basic but comprehensive manner, with the emphasis on a practical approach throughout. Again a special feature is the use of real-life examples from a number of industries, and these have been extended in several ways in this edition.

John Oakland has been joined in authorship by his friend and colleague Roy Followell. In addition to working together in quality management since the early 1980s at the University of Bradford Management Centre, they have formed one of the best-known international consultancies in the area, O & F Quality Management Consultants Ltd, based at Shipley in West Yorkshire, UK.

The wisdom gained by John and Roy, and their colleagues at the Management Centre and in the consultancy, in helping literally thousands of organizations to implement total quality management, good quality systems and SPC has been incorporated, where possible, into this edition. The book now provides possibly the most comprehensive guide written on how to actually use SPC 'in anger'. Numerous facets of the implementation process, gleaned from many man-years' work in a variety of industries, have been threaded through the book, as the individual techniques are covered.

SPC never has been and never will be simply a 'tool kit' and in this edition the authors hope to provide not only the instruction guide for the tools, but communicate the philosophy of never-ending improvement, which has become so vital to success in business throughout the world.

This book has been completely rewritten and there have been many additions and changes to the first edition. Specifically, new areas covered or extended include flowcharting, the measurement and control of variability, explanations of the use and pitfalls of process capability indices, Taguchi methods, the use of SPC in trouble shooting, and SPC in non-manufacturing, which is covered throughout the text and specifically in one chapter.

Neither the first nor the second edition of this book were written for the professional statistician or mathematician. As before, attempts have been made to eliminate much of the mathematical jargon that often causes distress. Those interested in pursuing the theoretical aspects will find references to books and papers for further study.

Where possible, further procedures have been included to simplify the application of 'statistical tools' to provide powerful techniques with which to improve product and service quality and reduce costs.

The book is written to meet the requirements of students in universities, polytechnics, and colleges engaged in courses on science, technology, engineering, and management subjects, including quality assurance. It also serves as a textbook for self or group instruction of production/operations managers, supervisors, engineers, scientists, and technologists. The text offers clear guidances and help to those unfamiliar with either quality control or statistical applications.

The authors would like to acknowledge the contributions of both their colleagues in the European Centre for Total Quality Management at the Management Centre of the University of Bradford and their associate consultants in O & F Quality Management Consultants Limited. This collaboration, both in an academic environment and in a vast array of public and private organizations, has resulted in their understanding of the part to be played by the use of SPC techniques and the recommendations of how to implement them. They would also like to thank Barbara Shutt who has fought courageously with the manuscript, Roy's handwriting, and the word processor. In the first edition, the sole author claimed credit for errors and omissions. In the present work any imperfections are the fault of the other author, and good old 100 per cent inspection.

John S. Oakland and Roy F. Followell

1   Quality, processes and control

1.1   The basic concepts

Organizations, whatever their nature, compete on three issues; quality, delivery, and price. There cannot be many people in the Western World who remain to be convinced that the reputation attached to an organization for the quality of its products and services is a key to its success and the future of its employees.

What is quality?

The word 'quality' is often used to signify 'excellence' of a product or service – we talk about 'Rolls Royce quality' and 'top quality'. In some manufacturing companies quality may be used to indicate that a product conforms to certain physical characteristics set down with a particularly 'tight' specification. But if we are to manage quality it must be defined in a way which recognizes the true requirements of the 'customer'.

Quality is defined simply as *meeting the requirements of the customer* and this has been expressed in many ways by other authors:

'fitness for purpose or use' – Juran

'the totality of features and characteristics of a product or service that bear on its ability to satisfy stated or implied needs – BS 4778: Part 1: 1987 (ISO 8402 – 1986)

'the total composite product and service characteristics of marketing, engineering, manufacture, and maintenance through which the product and service in use will meet the expectation by the customer' – Feigenbaum.

The ability to meet the customer requirements is vital, not only between two separate organizations, but within the same organization. There exists in every factory, every department, every office, a series of suppliers and customers. The typist is a supplier to the boss – is the typist meeting the requirements? Does the boss receive error-free typing set out as he wants it, when he wants it? If so, then we have a quality typing service. Does the factory receive from its supplier defect-free parts which conform to the requirements of the assembly process? If so, then we have a quality supplier.

For industrial and commercial organizations, which are viable only if they provide satisfaction to the consumer, competitiveness in quality is not only central to profitability, but crucial to business survival. The consumer should not be required to make a choice between price and quality, and for manufacturing or service organizations to continue to exist they must learn how to manage quality. In today's tough and challenging business environment, the development and implementation of a comprehensive quality policy is not merely desirable – it is essential.

Every day two people who work in a certain factory scrutinize together the results of the examination of the previous day's production, and commence the ritual battle over whether the material is suitable for despatch to the customer. One is called the Production Manager, the other the Quality Control Manager. They argue and debate the evidence before them, the rights and wrongs of the specification, and each tries to convince the other of the validity of their argument. Sometimes they nearly break into fighting.

This ritual is associated with trying to answer the question:

'*Have we done* the job correctly?'

'correctly' being a flexible word depending on the interpretation given to the specification on that particular day. This is not quality *control*, it is postproduction *detection*, wasteful detection of bad product before it hits the customer. There is a belief in some quarters that to achieve quality we much check, test, inspect or measure – the ritual pouring on of quality at the end of the process – and that quality, therefore, is expensive. This is nonsense, but it is frequently encountered. In the office one finds staff checking other people's work before it goes out, validating computer input data, checking invoices, typing, etc. There is also quite a lot of looking for things, chasing things that are late, apologizing to customers for non-delivery, and so on, waste, waste, and more waste.

The problems are only a symptom of the real, underlying cause of this type of behaviour, the lack of understanding of quality management. The concentration of inspection effort at the final product or service stage merely shifts the failures and their associated costs from outside the company to inside. To reduce the total costs of quality, control must be at the point of manufacture or operation; quality cannot be inspected into an item or service after it has been produced. It is essential for cost-effective control to ensure that articles are manufactured, documents are typed, or that services are generated correctly the first time. The aim of process control is the *prevention* of the manufacture of defective products and the generation of errors and waste in non-manufacturing areas.

To get away from the natural tendency to rush into the detection mode, it is necessary to ask different questions in the first place. We should not ask whether the job has been done correctly, we should ask first:

'*Can we do* the job correctly?'

This has wide implications and this book aims to provide some of the tools which must be used to ensure that the answer is 'Yes'. However, we should realize straight away that such an answer will only be obtained using satisfactory methods, materials, equipment, skills and instruction, and a satisfactory or capable 'process'.

What is a process?

A process is the transformation of a set of inputs, which can include materials, actions, methods and operations, into desired outputs, in the form of products, information, services or – generally – results. In each area or function of an organization there will be many processes taking place. Each process in every department or functional area can be analysed by an examination of the inputs and outputs. This will determine the action necessary to improve quality.

The output from a process is that which is transferred to somewhere or to someone – the customer. Clearly, to produce an output which meets the requirements of the customer, it is necessary to define, monitor and control the inputs to the process, which in turn may have been supplied as output from an earlier process. At every supplier–customer interface there resides a transformation process and every single task throughout an organization must be viewed as a process in this way.

To begin to monitor and analyse any process, it is necessary to first of all identify what the process is, and what the inputs and outputs are. Many processes are easily understood and relate to known procedures, e.g. drilling a hole, compressing tablets, filling cans with paint, polymerizing a chemical. Others are less easily identified, e.g. servicing a customer, delivering a lecture, storing a product, inputting to a computer. In some situations it can be difficult to define the process. For example, if the process is making a sales call, it is vital to know if the scope of the process includes obtaining access to the potential customer or client. Defining the scope of a process is vital, since it will determine both the required inputs and the resultant outputs.

Once the process is specified, the inputs and suppliers, outputs and customers can also be defined, together with the requirements at each of the interfaces. Often the most difficult areas in which to do this are in non-manufacturing organizations or non-manufacturing parts of manufacturing organizations, but careful use of appropriate questioning methods can release the necessary information. Sometimes this difficulty stems from the previous absence of a precise definition of the requirements and possibilities. Inputs to processes include: equipment, tools, or plant required; materials – including paper; information – including the specification for the outputs; methods or procedures – including instructions; people (and the inputs they provide, such as skills, training, knowledge, etc.), records and the environment (Figure 1.1).

Prevention of failure in any transformation is possible only if the process definition, inputs, and outputs are properly documented and agreed. The documentation of procedures will allow reliable data about the process itself to be collected, analysis to be performed, and action to be taken to improve the process and prevent failure or non-conformance with the requirements. The target in the operation of any process is the total avoidance of failure. If the objective of no failures or error-free

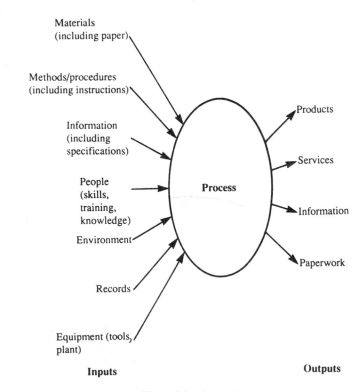

Figure 1.1   *A process*

work is not adopted, at least as a target, then certainly it will never be achieved.

What is control?

All processes can be monitored and brought 'under control' by gathering and using data. This refers to measurements of the performance of the process and the feedback required for corrective action, where necessary. Once we have established that our process is 'in control' and capable of meeting the requirements, we can address the next question:

'*Are we doing* the job correctly?'

which brings a requirement to monitor the process and the controls on it. Managers are in control only when they have created a system and climate in which their subordinates can exercise control over their own processes – in other words, the operator of the process has been given the 'tools' to control it.

If we now re-examine the first question: '*Have we done* it correctly?' we can see that, if we have been able to answer both of the questions: '*Can we do* it correctly?' (*capability*) and '*Are we doing* it correctly?' (*control*) with a 'yes', we must have done the job correctly – any other outcome would be illogical. By asking the questions in the right order, we have removed the need to ask the 'inspection' question and replaced a strategy of *detection* with one of *prevention*. This concentrates all the attention on the front end of any process – the inputs – and changes the emphasis to making sure the inputs are capable of meeting the requirements of the process. This is a managerial responsibility and these ideas apply to every transformation process, which must be subjected to the same scrutiny of the methods, the people, the skills, the equipment and so on to make sure they are correct for the job.

The control of quality clearly can take place only at the point of transformation of the inputs into the outputs, the point of operation or production, where the letter is typed or the artefact made. The act of product inspection is not quality control. When the answer to 'Have we done it correctly?' is given indirectly by answering the questions on capability and control, then we have assured quality and the activity of checking becomes one of *quality assurance* – making sure that the product or service represents the output from an effective system which ensures capability and control.

1.2   Design, conformance and costs

Before any discussion on quality can take place it is necessary to be clear about the purpose of the product or service, in other words, what the customer requirements are. The customer may be inside or outside the organization and his/her satisfaction must be the first and most important ingredient in any plan for success. Clearly, the customer's perception of quality changes with time and an organization's attitude to quality must, therefore, change with this perception. The skills and attitudes of the people in the organization are also subject to change, and failure to monitor such changes will inevitably lead to dissatisfied customers. Quality, like all other corporate matters, must be continually reviewed in the light of current circumstances.

The quality of a product or service has two distinct but interrelated aspects:

- Quality of design
- Quality of conformance to design

Quality of design

This is a measure of how well the product or service is designed to achieve its stated purpose. If the quality of design is low, either the service or product will not meet the requirements, or it will only meet the requirement at a low level.

A major feature of the design is the specification. This describes and defines the product or service and should be a comprehensive statement of all aspects which must be present to meet the customer's requirements.

A precise specification is vital in the purchase of materials and services for use in any conversion process. All too frequently, the terms 'as previously supplied', or 'as agreed with your representative', are to be found on purchasing orders for bought-out goods and services. The importance of obtaining materials and services of the appropriate quality cannot be overemphasized and it cannot be achieved without proper specifications. Published standards should be incorporated into purchasing documents wherever possible.

There must be a corporate understanding of the company's quality position in the marketplace. It is not sufficient that the marketing department specifies a product or service, 'because that is what the customer wants'. There must also be an agreement that the producing departments can produce to the specification. Should 'production' or 'operations' be incapable of achieving this, then one of two things must happen: either the company finds a different position in the marketplace, or substantially changes the operational facilities.

Quality of conformance to design

This is the extent to which the product or service achieves the specified design. What the customer actually receives should conform to the design, and operating costs are tied firmly to the level of conformance achieved. The customer satisfaction must be designed into the production system. A high level of inspection or checking at the end is often indicative of attempts to inspect in quality. This will achieve nothing but spiralling costs and decreasing viability. Conformance to a design is concerned largely with the quality performance of the actual operations. The recording and analysis of information and data play a major role in this aspect of quality and this is where statistical methods must be applied for effective interpretation.

The costs of quality

Obtaining a quality product or service is not enough. The cost of achieving it must be carefully managed so that the long-term effect of 'quality costs' on the business is a desirable one. These costs are a true measure of the quality effort. A competitive product or service based on a balance between quality and cost factors is the principal goal of responsible production/operations management and operators. This objective is best accomplished with the aid of a competent analysis of the costs of quality.

The analysis of quality costs is a significant management tool which provides:

- A method of assessing and monitoring the overall effectiveness of the management of quality
- A means of determining problem areas and action priorities.

The costs of quality are no different from any other costs in that, like the costs of maintenance, design, sales, distribution, promotion, production, and other activities, they can be budgeted, monitored and analysed.

Having specified the quality of design, the producing or operating units have the task of making a product or service which matches the requirement. To do this they add value by incurring costs. These costs include quality-related costs such as prevention costs, appraisal costs, and failure costs. Failure costs can be further split into those resulting from internal and external failure.

Prevention costs
These are associated with the design, implementation and maintenance of the quality management system. Prevention costs are planned and are incurred prior to production or operation. Prevention includes:

Product or service requirements   The determination of the requirements and the setting of corresponding specifications, which also take account of capability, for incoming materials, processes, intermediates, finished products and services.

Quality planning   The creation of quality, reliability, production, supervision, process control, inspection and other special plans (e.g. preproduction trials) required to acheive the quality objective.

Quality assurance   The creation and maintenance of the overall quality system.

Inspection equipment   The design, development and/or purchase of equipment for use in inspection work.

Training   The development, preparation and maintenance of quality training programmes for operators, supervisors and managers to both achieve and maintain capability.

Miscellaneous   Clerical, travel, supply, shipping, communications and other general office management activities associated with quality.

Resources devoted to prevention give rise to the 'costs of getting it right the first time'.

Appraisal costs

These costs are associated with the supplier's and customers' evaluation of purchased materials, processes, intermediates, products and services to assure conformance with the specified requirements. Appraisal includes:

Verification   Of incoming material, process set-up, first-offs, running processes, intermediates and final products or services, and includes product or service performance appraisal against agreed specifications.

Quality audits   To check that the quality management system is functioning satisfactorily.

Inspection equipment   The calibration and maintenance of equipment used in all inspection activities.

Vendor rating   The assessment and approval of all suppliers – of both products and services.

Appraisal activities result in the 'cost of checking it is right'.

Internal failure costs

These costs occur when products or services fail to reach designed standards and are detected before transfer to the consumer takes place. Internal failure includes:

Scrap   Defective product which cannot be repaired, used or sold.

Rework or rectification   The correction of defective material or errors to meet the requirements.

Reinspection   The re-examination of products or work which has been rectified.

Downgrading Product which is usable but does not meet specifications and may be sold as 'second quality' at a low price.

Waste The activities associated with doing unnecessary work or holding stocks as the result of errors, poor organization, the wrong materials, exceptional as well as generally accepted losses, etc.

Failure analysis The activity required to establish the causes of internal product or service failure.

External failure costs

These costs occur when products or services fail to reach design quality standards and are not detected until after transfer to the consumer. External failure includes:

Repair and servicing Either of returned products or those in the field.

Warranty claims Failed products which are replaced or services redone under guarantee.

Complaints All work and costs associated with the servicing of customers' complaints.

Returns The handling and investigation of rejected products, including transport costs.

Liability The result of product liability litigation and other claims, which may include change of contract.

Loss of goodwill The impact on reputation and image which impinges directly on future prospects for sales.

External and internal failures produce the 'costs of getting it wrong'.

The relationship between these so-called direct costs of prevention, appraisal and failure costs, and the ability of the organization to meet the customer requirements is shown in Figure 1.2. Where the ability to produce a quality product or service acceptable to the customer is low, the total direct quality costs are high and the failure costs predominate. As ability is improved by modest investment in prevention, the failure costs and total cost drop very steeply. It is possible to envisage the combination of failure (declining), appraisal (declining less rapidly) and prevention costs (increasing) as leading to a minimum in the combined costs. Such a minimum does not exist because, as it is approached, the requirements become more exacting. Frank Price, the author of *Right First Time* also refutes the minimum and calls it 'the mathematics of mediocrity'.

So far little has been said about the often intractable indirect quality costs associated with customer dissatisfaction, and loss of reputation or goodwill. These costs reflect the customer attitude toward an organization and many be both considerable and elusive in estimation but

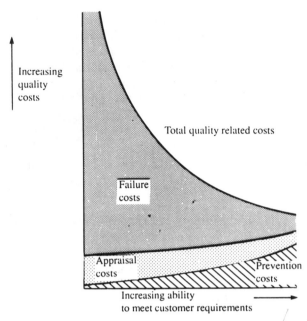

Figure 1.2   *Relationship between costs of quality and organization capability*

not in fact. A full discussion of the measurement and management of the costs of quality is outside the scope of this book, but may be found in *Total Quality Management* by John Oakland published by Heinemann Professional Publishing.

Total direct quality costs, and their division between the categories of prevention, appraisal, internal failure and external failure, vary considerably from industry to industry and from site to site. A figure for quality-related costs of less than 10 per cent of sales turnover is seldom quoted when perfection is the goal. This means that in an average organization there exists a 'hidden plant' or 'hidden operation', amounting to perhaps one-tenth of productive capacity. This hidden plant is devoted to producing scrap, rework, correcting errors, replacing defective goods, services and so on. Thus, a direct link exists between quality and productivity and there is no better way to improve productivity than to convert this hidden resource to truly productive use. A systematic approach to the control of processes provides the only way to accomplish this.

Technologies and market conditions vary between different industries and markets, but the basic concepts of quality management and the financial implications are of general validity. The objective should be to produce, at an acceptable cost, goods and services which conform to the requirements of the customer. The way to accomplish this is to use a systematic approach in the operating departments of: design, manufacturing, quality assurance, purchasing, sales, personnel, administration and all others – nobody is exempt. The statistical approach to quality control is not a separate science or a unique theory of quality control – rather a set of valuable tools which becomes an integral part of the 'total' quality approach.

Two of the original and most famous authors on the subject of statistical methods applied to quality management are Shewhart and Deming. In their book, *Statistical Method from the Viewpoint of Quality Control* they wrote:

> The long-range contribution of statistics depends not so much upon getting a lot of highly trained statisticians into industry as it does on creating a statistically minded generation of physicists, chemists, engineers and others who will in any way have a hand in developing and directing production processes of tomorrow.

This was written in 1939. It is as true today as it was then.

1.3   The SPC system

Statistical Process Control (SPC) methods, backed by management commitment and good organization, provide objective means of controlling quality in any transformation process, whether used in the manufacture of artefacts, the provision of services, or the transfer of information.

SPC is not only a tool kit. It is a strategy for reducing variability, the cause of most quality problems; variation in products, in times of deliveries, in ways of doing things, in materials, in people's attitudes, in equipment and its use, in maintenance practices, in everything. Control by itself is not sufficient. Total Quality Management (TQM)[1], like SPC, requires that the process should be improved continually by reducing its variability. This is brought about by studying all aspects of the process using the basic question:

'*Could we* do the job more consistently and on target (i.e. better)?'

the answering of which drives the search for improvements. This significant feature of SPC means that it is not constrained to measuring conformance, and that it is intended to lead to action on processes which are operating within the 'specification' to minimize variability. There must be a willingness to implement changes, even in the ways in which an organization does business, in order to achieve continuous improvement. Innovation and resources will be required to satisfy the long-term

---

[1] See Oakland, John S., *Total Quality Management*, Heinemann Professional Publishing, 1989.

requirements of the customer and the organization, and these must be placed before or alongside short-term profitability.

⊬ Process control is vital and SPC should form a vital part of the overall corporate strategy. Incapable and inconsistent processes render the best designs impotent and make supplier quality assurance irrelevant. Whatever process is being operated, it must be reliable and consistent. SPC can be used to achieve this objective.

In the application of SPC there is often an emphasis on techniques rather than on the implied wider managerial strategies. SPC is not about plotting charts and pinning them to the walls of a plant or office, it must be a component part of a company-wide adoption of 'total quality' and act as the focal point of never-ending improvement. Changing an organization's environment into one in which SPC can operate properly may take several years rather than months. For many companies SPC will bring a new approach, a new 'philosophy', but the importance of the statistical techniques should not be disguised. Simple presentation of data using diagrams, graphs and charts should become the means of communication concerning the state of control of processes.

Many companies, particularly those in the motor industry or its suppliers, have adopted the Deming philosophy and approach to quality. Dr Deming is a statistician who gained fame by helping Japanese companies to improve quality after the Second World War. His basic philosophy is that quality and productivity increase as variability decreases and, because all things vary, statistical methods of quality control must be used to measure and gain understanding of the causes of the variation. In these companies, attention has been focused on quality improvement through the use of quality management systems and SPC.

The responsibility for quality in any tranformation process must lie with the operators of that process – the producers. To fulfil this responsibility, however, people must be provided with the tools necessary to:

- Know whether the process is capable of meeting the requirements;
- Know whether the process is meeting the requirements at any point in time;
- Correct or adjust the process or its inputs when it is not meeting the requirements.

The success of this approach has caused messages to cascade through the supplier chains and companies in all industries, including those in the process and service industries which have become aware of the enormous potential of SPC, in terms of cost savings, improvements in quality, productivity and market share. As the authors know from experience, this has created a massive demand for knowledge, education, and understanding of SPC and its applications.

A quality management system, based on the fact that many functions will share the responsibility for any particular process, provides an effective method of acquiring and maintaining desired quality standards. The 'Quality Department' should not assume direct responsibility for quality but should support, advise and audit the work of the other functions, in much the same way as a financial auditor performs his duty without assuming responsibility for the profitability of the company.

A systematic study of a process through answering the questions:

Can we do the job correctly? (capability)
Are we doing the job correctly? (control)
Have we done the job correctly? (quality assurance)
Could we do the job better? (improvement)[1]

provides knowledge of the process capability and the sources of non-conforming outputs. This information can then be fed back quickly to marketing, design, and the 'technology' functions. Knowledge of the current state of a process also enables a more balanced judgement of equipment, both with regard to the tasks within its capability and its rational utilization.

It is worth repeating that statistical process control procedures exist because there is variation in the characteristics of materials, articles, services and people. The inherent variability in every transformation process causes the output from it to vary over a period of time. If this variability is considerable, it may be impossible to predict the value of a characteristic of any single item or at any point in time. Using statistical methods, however, it is possible to take meagre knowledge of the output and turn it into meaningful statements which may then be used to describe the process itself. Hence, statistically-based process control procedures are designed to divert attention from individual pieces of data and focus it on the process as a whole. SPC techniques may be used to measure and control the degree of variation of any purchased materials, services, processes, and products and to compare this, if required, to previously agreed specifications.

---

[1]   This system for process capability and control is based on Frank Price's very practical framework for thinking about quality in manufacturing:
      Can we make it OK?
      Are we making it OK?
      Have we made it OK?
      Could we make it better?
      which he presented in his excellent book, *Right First Time*.

1.4　Some basic tools

In statistical process control numbers and information will form the basis for decisions and actions, and a thorough data recording system is essential. In addition to the basic elements of a quality system, which will provide a framework for recording data, there exists a set of 'tools' which may be applied to interpret fully and derive maximum use of the data. The simple methods listed below will offer any organization means of collecting, presenting, and analysing most of its data:

- Process flowcharting – what is done?
- Check sheets/tally charts – how often is it done?
- Histograms – what do the numbers look like?
- Graphs – can the numbers be represented pictorially?
- Pareto analysis – which are the big problems?
- Cause and effect analysis and brainstorming – what causes the problems?
- Scatter diagrams – what are the relationships between factors?
- Control charts – which variations to control and how?

A pictorial example of each of these methods is given in Figure 1.3. A full description of the techniques, with many examples, will be given in subsequent chapters. These are written assuming that the reader is neither a mathematician nor a statistician, and the techniques will be introduced through practical examples, where possible, rather than from a theoretical perspective.

Chapter highlights

- Organizations compete on quality, delivery and price. Quality is defined as meeting the requirements of the customer. The supplier/customer interface is both internal and external to organizations.
- Product inspection is not the route to good quality management. Start by asking 'Can we do the job correctly?' and not by asking 'Have we done the job correctly?' – not detection but prevention and control. Detection is costly and neither efficient nor effective. Prevention is the route to successful quality management.
- We need a process to ensure that we can and will continue to do it correctly – this is a model for control. Everything we do is a process – the transformation of any set of inputs into a different set of outputs. Start by defining the process and then investigate its capability and the methods to be used to monitor or control it.

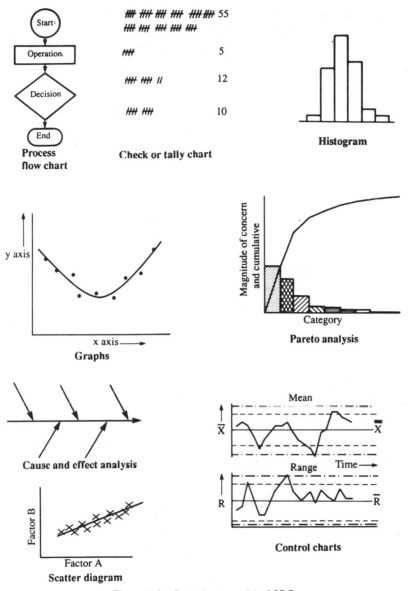

Figure 1.3   *Some basic 'tools' of SPC*

• Control (Are we doing the job correctly?) is only possible when data is collected and analysed, so the outputs are controlled by the control of the inputs and the process. The latter can only occur at the point of the transformation – then the quality is assured.

- There are two distinct aspects of quality – design and conformance to design. Design is how well the product or service measures against its stated purpose or the specification. Conformance is the extent to which the product or service achieves the specified design. Start quality management by defining the requirements of the customer, keep the requirements up to date.
- The costs of quality need to be managed so that their effect on the business is desirable. The measurement of quality-related costs provides a powerful tool to highlight problem areas and monitor management performance.
- Quality-related costs are made up of failure (both external and internal), appraisal and prevention. Prevention costs include the things which are built into the management system to assure quality and include the precise determination of the requirements, planning, a proper management system for quality, and training. This is the area into which more resources should be injected to achieve higher design and conformance achievements. Appraisal costs will always be incurred to allow proper verification, measurement, vendor ratings, etc. Failure is usually the major source of cost and includes scrap, rework, reinspection, waste, as well as repair, warranty, complaints, returns and the associated loss of goodwill, among actual and potential customers. Quality-related costs, when measured from perfection, are seldom less than 10 per cent of sales value.
- The route to improved design, increased conformance and reduced costs is the use of statistically-based methods in decision making within quality management.
- SPC is a set of tools for managing processes and, hence, determining and monitoring the quality of the outputs of an organization. It is also a strategy for reducing variation in products, deliveries, processes, materials, attitudes and equipment. The question which needs to be asked continually is 'Could we do the job better?'
- Continuous improvement implies a continuing willingness to re-examine the existing and accept, where appropriate, the need for change. Accepting the continuing need for reduced variation and planned changes swings an organization away from the short-term to longer-term management objectives, reduces the firefighting, and leaves time and energy for *management* instead of coping.
- SPC exists because there is, and will always be, variation in the characteristics of materials, articles, services, people – there will also be a degree of consistency – variation has to be understood and assessed in order to be managed.
- There are some basic SPC tools. These are: process flowcharting (what is done); check sheets/tally charts (how often it is done);

histograms (pictures of numeric data); graphs (more pictures); Pareto analysis (prioritizing); cause and effect analysis (what causes the problems); scatter diagrams (exploring relationships); control charts (monitoring variation and stability over time). An understanding of all these tools and how to use them requires no prior knowledge of statistics.

2 Understanding the process

2.1 Information about the process

One of the initial steps to understand or improve a process is to gather information about the important activities so that a flowchart may be constructed. A flowchart is a picture of the activities that take place in a process. One of the greatest difficulties here, however, is deciding how many tasks and how much detail should be included. When initially flowcharting a process, people often include too much detail or too many tasks. It is important to consider the sources of information about processes and the following aspects should help to identify the key issues.

- Defining supplier/customer relationships.
- Describing the process and making it tangible.
- Standardizing procedures.
- Designing a new process or modifying an existing process.
- Identifying complexity or opportunities for improvement.

Defining supplier/customer relationships

Since quality is defined by the customer, changes to a process are usually made to increase satisfaction of internal and external customers. At many stages in a process, it is necessary for 'customers' to determine their needs or give their reaction to proposed changes in the process. For this it is often useful to describe the edges or boundaries of the process. This is accomplished by formally considering the inputs and outputs of the process, as well as the suppliers of the inputs and the customers of the outputs. Figure 2.1 is a form that can be used to provide focus on the boundary of any process and to list the inputs and suppliers to the process, as well as the outputs and customers. These lists do not have to be exhaustive, but should capture the important aspects of the process. Knowledge, tools and equipment, maintained by the people inside the process, should not be listed as inputs. The form asks for some

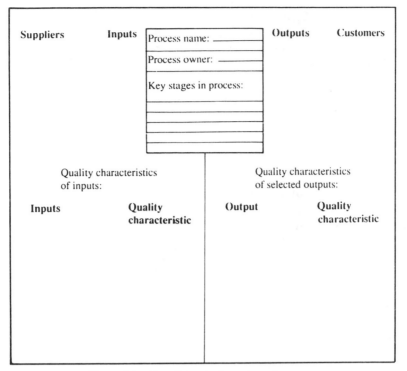

Figure 2.1   *Describing the boundary of a process*

fundamental information about the process itself, such as the name and the 'owner'. The owner of a process is the person at the lowest level in the organization that has the authority to change the process. The owner has the responsibility of organizing and perhaps leading a team to make improvements.

Documentation of the process, through the use of a flowchart, aids the identification of the customers and suppliers at each stage. It is sometimes surprisingly difficult to define these relationships, especially for internal suppliers and customers. Some customers of an output may have also supplied some of the inputs, and there are usually a number of customers for the same output. For example, information on location and amount of stock or inventory may be used by production planners, material handlers, purchasing staff and accountants.

Describing the process and making it tangible

Many processes in need of improvement are now well defined. A production engineering department may define and document in great detail a manufacturing process, but have little or no documentation on the process of design itself. If the process of design is to be improved, then knowledge of that process will be needed to make it tangible.

The first time any process is examined, the main focus should be to put everyone's current knowledge of the process down on paper. A common mistake is to have a technical process 'expert', usually a technologist, engineer, or supervisor, describe the process and then show it to others for their comment. The first information about the process should instead come from a brainstorming session of the people who actually operate or use the process, day-in and day-out. The technical experts, managers, and supervisors should refrain from interjecting their 'ideas' until towards the end of the session. The resulting description will be a reflection of how the process *actually* works. During this initial stage, the concept of what the process could or should be (such as standard operating procedures) is damaging to the main purpose of the exercise. These ideas and concepts should be discussed at a later time.

Flowcharts are important to study manufacturing processes, but they are particularly important for non-manufacturing processes. Because of the lack of documentation of administrative and service processes, it is sometimes difficult to reach agreement on the flowchart for a process. If this is the case, a first draft of a flowchart can be circulated to others who are knowledgeable in the process to seek their suggestions. Often, simply putting a team together to define the process using a flowchart will result in some obvious suggestions for improvement. This is especially true for non-manufacturing processes.

Standardizing procedures

A significant source of variation in many processes is the use of different methods and procedures by those working in the process. This is caused by the lack of documented, standardized procedures, inadequate training, or inadequate supervision. A flowchart is a useful tool to identify parts of the process where varying procedures are being used. The flowchart can also be used to establish a standard process to be followed by all. There have been many cases when standard procedures, developed and followed by operators, with the help of supervisors and technical experts, have resulted in a significant reduction in the variation of the outcomes.

Designing or modifying an existing process

Once a flowchart of a process has been developed, those knowledgeable in the operation of the process should look for obvious areas of improvement or modification. It may be that steps, once considered necessary, are no longer needed. Time should not be wasted improving an activity that is not worth doing in the first place. Before any team proceeds with its efforts to improve a process, it should consider how the process should be designed from the beginning. A flowchart of the new process, compared to the existing process, will assist in identifying areas for improvement. A flowchart can also serve as documentation of a new process. It will help those designing the process to identify weaknesses in the design and prevent problems once the new process is put into use.

Identifying complexity or opportunities for improvement

In any process there are many opportunities for things to go wrong and, when they do, what may have been a relatively simple activity can become quite complex. The failure of an airline computer used to document reservations, assign seats and print tickets can make the usually simple task of assigning a seat to a passenger a very difficult one. Documenting the steps in the process, identifying what can go wrong and indicating the increased complexity when things do go wrong, will identify opportunities for increased quality and productivity.

2.2 Flowcharting

In the systematic planning or examination of any process, whether it is a clerical, manufacturing, or managerial activity, it is necessary to record the series of events and activities, stages and decisions in a form which can be easily understood and communicated to all. If improvements are to be made, the facts relating to the existing method must be recorded first. The statements defining the process should lead to its understanding and will provide the basis of any critical examination necessary for the development of improvements. It is essential, therefore, that the descriptions of processes are accurate, clear and concise.

The flowchart is a very important first step for improving a process. The flowchart 'picture' will assist an individual or team in acquiring a better understanding of the system or process under study than would otherwise be possible. Gathering this knowledge provides a graphic definition of the system and the scope of the improvement effort. The flowchart, then, is a *communication* tool that helps an individual or an improvement team understand a system or process and identify opportunities for improvement.

The usual method of recording and communicating facts is to write them down, but this is not suitable for recording the complicated processes which exist in any organization. This is particularly so when an exact record is required of a long process, and its written description would cover several pages requiring careful study to elicit every detail. To overcome this difficulty certain methods of recording have been developed and the most powerful of these is flowcharting. There are many different types of flowcharts which serve a variety of uses. The classical form of flowcharting, as used in computer programming, can be used to document current knowledge about a process, but there are other kinds of flowcharting techniques which focus efforts to improve a process.

Figure 2.2 is a flowchart of a process by which raw material for a chemical plant was purchased, received, and an invoice for the material was paid. Before an invoice could be paid, there had to be a corresponding receiving report to verify that the material had in fact been received. The accounts department was having trouble matching receiving reports to the invoices because the receiving reports were not available or contained incomplete or incorrect information. A team was formed with members from the accounts, transportation, purchasing and production departments. At the early stages of the project, it was necessary to have a broad overview of the process, including some of the important outputs and some of the problems that could occur at each stage. The flowchart in Figure 2.2 served this purpose.

Figure 2.3 is an example of a flowchart which incorporates another dimension by including the person or group responsible for performing the task in the column headings. This type of flowchart is helpful in determining customer/supplier relationships and is also useful to see where departmental boundaries are crossed and to identify areas where interdepartmental communications are inadequate. The flowchart in Figure 2.3 was drawn by a team working on improving the administrative aspects of the 'sales' process. The team had orginally drawn a flowchart of

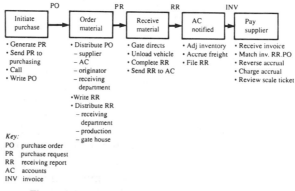
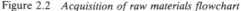

Figure 2.2 *Acquisition of raw materials flowchart*

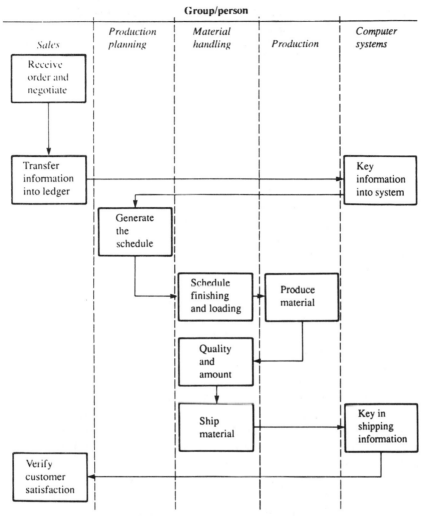

Figure 2.3   *Paperwork for sale of product flowchart*

the entire sales operation using a form similar to the one in Figure 2.2. After collecting and analysing some data, the team focused on the problem of not being able to locate specific paperwork. The flowchart in Figure 2.3 was then prepared to focus the movement of paperwork from area to area.

Classic flowcharts

Certain standard symbols are used on the 'classic' flowchart and these are shown in Figure 2.4. The starting point of the process is indicated by a

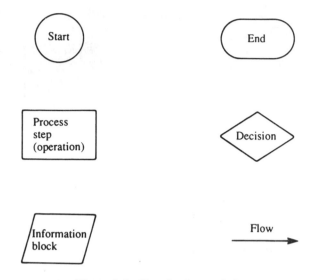

Figure 2.4   *Flowcharting symbols*

circle. Each processing step, indicated by a rectangle, contains a description of the relevant operation, and where the process ends is indicated by an oval. A point where the process branches because of a decision, is shown by a diamond. A parallelogram contains useful information but it is not a processing step. The arrowed lines are used to connect symbols and to indicate direction of flow. For a complete description of the process all operation steps (rectangles) and decisions (diamonds) should be connected by pathways from the start circle to the end oval. If the flow chart cannot be drawn in this way, the process is not fully understood.

Flowcharts are frequently used to communicate the components of a system or process to others whose skills and knowledge are needed in the improvement effort. Therefore, the use of standard symbols is necessary to remove any barrier to understanding or communication.

The purpose of the flowchart analysis is to learn why the current system/process operates in the manner it does, and to prepare a method for objective analysis. The team using the flowchart should analyse and document their findings to identify:

1   The problems and weaknesses in the current process system.
2   Unnecessary steps or duplication of effort.
3   The objectives of the improvement effort.

The flowchart techniques can also be used to study a simple system and how it would look if there were no problems. This method has been called 'imagineering' and is a useful aid to visualizing the improvements required.

It is a salutory experience for most people to sit down and try to draw the flowchart for a process in which they are involved every working day. It is often found that:

1 The process flow is not fully understood.
2 A single person is unable to complete the flowchart without help from others.

The very act of flowcharting will improve knowledge of the process, and will begin to develop the teamwork necessary to find improvements. In many cases the convoluted flow and octopus-like appearance of the chart will highlight unnecessary movement of people and materials and lead to common sense suggestions for waste elimination.

Flowchart construction features

The boundaries of the process must be clearly defined before the flowcharting begins. this will be relatively easy if the outputs and customers, inputs and suppliers are clearly identified. All work connected with the process to be studied must be included. It is most important to include not only the formal, but also the informal activities. Having said that, it is important to keep the flowchart as simple as possible.

Every route through a flowchart must lead to an end point and each process step must have one output line. Each decision diamond should have only two outputs which are labelled Yes and No, which means that the questions must be phrased so that they may be answered in this way.

An example of a 'classic' flowchart for part of a contact lens conversion process is given in Figure 2.5.

2.3   Process examination

A flowchart is a picture of the steps used in performing a function. This function can be anything from a chemical process step to accounting procedures, even preparing a meal. Flowcharts provide excellent documentation and are useful trouble shooting tools to determine how each step is related to the others. By reviewing the flowchart it is often possible to discover inconsistencies and determine potential sources of variation and problems. For this reason, flowcharts are very useful in process improvement when examining an existing process to highlight the problem area. A group of people, with knowledge about the process, should follow the simple steps:

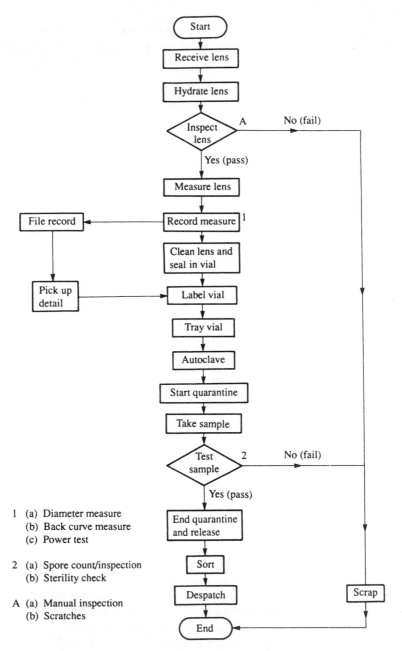

Figure 2.5   *'Classic' flowchart for part of a contact lens conversion process*

1 Draw a flowchart of existing process.
2 Draw a second chart of the flow the process could or should follow.
3 Compare the two charts to highlight the sources of the problems or waste, improvements required, and changes necessary.

A critical examination of the first flowchart is often required, using a questioning technique, which follows a well-established sequence to examine:

the *purpose* for which ⎞
the *place* at which   ⎟
the *sequence* in which ⎬ the activities are undertaken
the *people* by which   ⎟
the *method* by which   ⎠

with a view to ⎧ eliminating ⎫
               ⎪ combining   ⎪
               ⎨ rearranging ⎬ those activities
               ⎪ or   ⎪
               ⎩ simplifying ⎭

The questions which need to be answered in full are:

Purpose: What is actually done?   ⎞
            (or What is actually achieved?)   ⎟ *Eliminate*
                                                      ⎬ unnecessary
            Why is the activity necessary at all?   ⎟ parts of
                                                      ⎟ the job
            What else might be or should be done?   ⎠

Place: Where is it being done?   ⎞

            Why is it done at that particular place? ⎟

                                                      ⎟ *Combine*
            Where else might it or should it be done? ⎟ wherever
                                                      ⎟ possible
Sequence: When is it done?   ⎟ and/or
                                                      ⎟ *rearrange*
            Why is it done at that particular time?   ⎬ operations
                                                      ⎟ for more
            When might or should it be done?   ⎟ effective
                                                      ⎟ results or
People: Who does it?   ⎟ reduction
                                                      ⎟ in waste
            Why is it done by that particular person? ⎟

            Who else might or should do it?   ⎠

Method: How is it done?

Why is it done in that particular way?   } *Simplify* the operations

How else might or should it be done?

Questions such as these, when applied to any process will raise many points demanding explanation.

There is always room for improvement and one does not have to look far to find many real-life examples of what happens when a series of activities is started without being properly planned. Examples of much waste of time and effort can be found in factories and offices all over the world.

2.4   Development of the process

Statistical Process Control (SPC) has played a major part in the efforts of many companies and industries to improve the competitiveness of their products, services, prices and deliveries. But what does SPC mean? A statistician may tell you that SPC is the application of appropriate statistical tools to processes for continuous improvement in quality of products and services, and productivity in the workforce. This is certainly accurate, but at the outset, in many organizations, SPC would be better defined as a simple, effective approach to problem solving, and process improvement.

Every process has problems that need to be solved, and the SPC tools are universally applicable to everyone's job – manager, operator, secretary, chemist, engineer, whatever. Training in the use of these tools should be available to everyone within an organization, so that each 'worker' can contribute to the improvement of quality in his or her work. Usually, the technical people are the major focus of training in SPC, with concentration on the more technical tools, such as control charts. The other simpler basic tools, such as flowcharts, cause and effect diagrams, check sheets, and Pareto charts, however, are well within the capacity of all employees.

Simply teaching individual SPC tools to employees is not enough. Making a successful transition from classroom examples to on-the-job application is the key to successful SPC implementation and problem solving. With the many tools available, the employee often wonders which one to use when confronted with a quality problem. What is often lacking in SPC training is a simple step-by-step approach to developing or improving a process.

Such an approach is represented in the flowchart of Figure 2.6. This

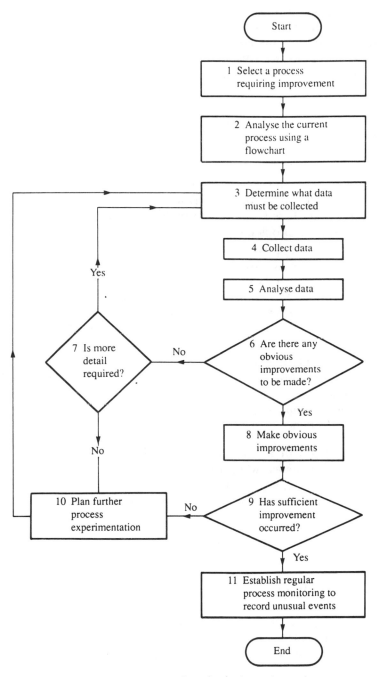

Figure 2.6   *Step-by-step approach to developing or improving a process*

'road map' for problem solving intuitively makes sense to most people, but its underlying feature is that each step has certain SPC techniques that are appropriate to use in that step, and people should not have to wonder which techniques to use. This should reduce the barriers to acceptance of SPC and greatly increase the number of people capable of using it.

The various steps in Figure 2.6 require the use of the basic SPC 'tool kit' introduced in Chapter 1 and which will be described in full in the remaining chapters of this book. This is essential if a systematic approach is to be maintained and satisfactory results are to be achieved. There are several benefits which this approach brings and these include:

- There are no restrictions as to the type of problem selected, but the process originally tackled will be improved.
- Decisions are based on facts not opinions – a lot of the 'emotion' is removed from problems by this approach.
- The quality 'awareness' of the workforce increases because they are directly involved in the improvement process.
- The knowledge and experience potential of the people who operate the process is released in a systematic way through the investigative approach. They better understand that their role in problem solving is collecting and communicating the facts with which decisions are made.
- Managers and supervisors solve problems methodically, instead of by using a 'seat of-the-pants' style. The approach becomes unified, not individual or haphazard.
- Communications across and between all functions are enhanced, due to the excellence of the SPC tools as modes of communication.

The combination of a systematic approach, SPC tools, and outside handholding assistance when required, helps organizations make the difficult transition from learning SPC in the classroom to applying it in the real world. This concentration on *applying* the techniques rather than simply learning them will lead to successful problem solving and process improvement.

Chapter highlights

- One of the first steps in understanding or improving a process is to draw a flowchart. A flowchart uses certain symbols to provide a 'picture' of the sequential activities and decisions in the process.
- Sources of information about the process will include customer requirements, people's knowledge, procedures and documentation,

and details of previous problems. The boundaries of the process must be defined.

- There are various types of flowcharts, including the 'classic' one used in computer programming.
- The correct use of flowcharts to examine processes, using a systematic questioning technique, will lead to improvements.
- SPC is above all a simple, effective approach to problem solving and process improvement. Training in the use of the basic tools, should be available for everyone in the organization. However, training must be followed up to provide a simple stepwise approach to improvement.
- The SPC approach, correctly introduced, will lead to decisions based on facts, an increase in quality awareness at all levels, a systematic approach to problem solving, 'release' of valuable experience, and all-round improvements, especially in communications.

Acknowledgement

The authors would like to thank Exxon Chemical for permission to use some of their flowchart training material in the preparation of this chapter.

3 Data collection and presentation

3.1 The systematic approach

In adopting the definition of quality as 'meeting the requirements', we have already seen the need to consider the quality of design and the quality of conformance to design. To achieve quality requires:

- An appropriate design.
- Suitable resources and facilities (equipment, people, cash, etc.)
- The correct materials.
- An appropriate process.
- Sets of detailed instructions concerning both the operation of the process and the methods used to determine that it is under control.

This has already broken down quality management into a series of component parts. Basically this is quite simply narrowing down each task until it is of a manageable size. If one attempts to tackle the whole subject of quality and its control one will inevitably find that the subject is too complex. Only by seeking the component parts and addressing them can one hope to achieve quality in the complete ensemble that makes up products and services. The first step in any systematic approach is to narrow down the issues and define the parts in such a way that it becomes possible to address them. Too often in discussions of quality the possible agenda is so vast that discussion is unlimited, a talking shop develops and no meaningful progress results.

In applying a systematic approach to process control there are two basic rules:

- Record all data
- Use appropriate techniques

1 *Record all data* – especially at the point of manufacture or operation. If data is not carefully and systematically recorded, it cannot be analysed and put to use. The definition of *statistics* is quite simply the collection,

collation and use of data. Information recorded in a suitable way enables the magnitude of variations, stability and trends to be observed. This allows conclusions to be drawn concerning errors, process capability, vendor ratings, risks etc. Regrettably, the actual observation made is often not recorded – a simple tick or initials merely shows, at most, that an observation was made. The analysis of a set of ticks is almost meaningless. The requirement to record the actual observation (the reading on a measured scale, or a count of things observed) can have a marked effect on the reliability of the data. For example, if a result is only just outside a specified tolerance and experience suggests that, if left alone, the process will probably 'drift back' to meet the requirements, it is easy to write down the 'OK' tick, but the actual recording of a *false* figure is much less likely. The value of this increase in the *reliability of the data*, when recorded properly, cannot be overstated. The practice of recording the result only when it is outside specification is also not recommended, since it ignores the variation going on within the tolerance limits which, hopefully, makes up the largest part of the variation and, therefore, contains the largest amount of information.

2   *Use appropriate techniques* – the essential tools of the 'narrowing down' approach. A wide range of simple, yet powerful, problem-solving and data-handling techniques are available and should form a part of the 'tool kit' used at all points within a process. These include:

- Process flowcharting
- Check sheets/tally charts
- Histograms
- Graphs
- Pareto analysis
- Cause and effect analysis
- Scatter diagrams
- Control charts

3.2   Data collection

Data should form the basis for analysis, decision and action, and its form and presentation will obviously differ from process to process. Information is collected to discover what is going on. It may be used as a part of a product or process control system and it is important to know at the outset what the data are to be used for. For example, if a problem occurs in the amount of impurity present in a product which is manufactured continuously, it is not sufficient to take only one sample per day to discover the variations between the methods of operation used by each shift. Similarly in comparing errors produced by two invoicing procedures, it is essential to have separate data from the outputs of both processes. These statements are no more than common sense, but it is not unusual to find that decisions and action are based on mixed or biased data. In other words, full consideration must be given to the reasons for collecting data, the correct sampling techniques and stratification. There should not be a disproportionate amount of a certain kind of data simply because it can be collected easily. The methods of collecting data and the amount collected must take account of the need for information and not the ease of collection.

Types of quality data

Numeric information on quality will arise from both counting and measurement.

Data arising from counting can only occur in discrete steps. There can only be 0, 1, 2, etc. defectives in a sample of 10 items. The number of faults in a length of fabric, the number of typing errors on a page, the acceptability or unacceptability of the lining on a metal drum are all called *attributes*. One is assessing things as 'OK', or 'not OK' – overweight/ underweight, correct/incorrect, clean/dirty, telephone answered or not answered, order taken or not taken, working or not working, etc. Such a two-way or *binary* classification gives rise to discrete steps in the collected and recorded data.

Data which arises from measurement usually occurs on a continuous scale and is called *variable* data. Variables include temperature, exchange rates, weight, volume, turnover, physical dimensions, age, share prices, efficiency, assay, time, cash flow, sales values, etc. Not all of these variables are measured on completely continuous scales but, in principle, they can be – within limited ranges all values would be possible. Owing to various limitations, including measurement, variable data will often be available on scales which include a large number of small steps rather than being truly continuous.

The statistical principles involved in the analysis of whole numbers are not the same as those involved in continuous measurements. The theoretical background necessary for the analysis of these different types of data will be presented in later chapters.

Recording data

The object of data collection should be its analysis and the extraction, through the use of statistical and other methods, of information on which action can be taken. It follows that data should be obtained in a form which will simplify the subsequent analysis. The first basic rule is to plan and construct the proformas or paperwork for data collection. This can avoid the problems of tables of numbers, the origin and relevance of which has long been forgotten. It is necessary to record not only the purpose of the observation and its characteristics, but also the date, the observer, the sampling plan, the instruments used for measurement, the method, and so on. Computer programs can play an important role in both establishing and maintaining the format for data collection.

Data should be recorded in such a way that it is easy to use. Calculations of totals, averages and ranges are often necessary and the format used for recording the data can make these easier. For example, the format and data recorded in Figure 3.1 has clearly been designed for a situation in which the daily, weekly and grand averages of a percentage impurity are required. Columns and rows have been included for the totals from which the averages are calculated. Fluctuations in the average for a day can be seen by looking down the columns, while variations in the percentage impurity at the various sample times can be reviewed by examining the rows.

Careful design of a data sheet will facilitate easier and more meaningful analysis. A few simple steps in the design are listed below:

- Agree on the exact event to be observed – ensure that everyone is monitoring the same thing(s).
- Decide both how often the events will be observed (the frequency) and over what total period (the duration).

Date	Percentage impurity					Week total	Week average
	15th	16th	17th	18th	19th		
Time							
8 a.m.	0.26	0.24	0.28	0.30	0.26	1.34	0.27
10 a.m.	0.31	0.33	0.33	0.30	0.31	1.58	0.32
12 noon	0.33	0.33	0.34	0.31	0.31	1.62	0.32
2 p.m.	0.32	0.34	0.36	0.32	0.32	1.66	0.33
4 p.m.	0.28	0.24	0.26	0.28	0.27	1.33	0.27
6 p.m.	0.27	0.25	0.24	0.28	0.26	1.30	0.26
Day total	1.77	1.73	1.81	1.79	1.73		
Day average	0.30	0.29	0.30	0.30	0.29	8.83	0.29
Operator	*A. Ridgeworth*						

Week commencing 15 February

Figure 3.1 *Data collection sheet for impurity in a chemical product*

- Design a draft format – keep it simple and leave adequate space for the entry of the observations.
- Tell the observers how to use the format and put it into trial use – be careful to note their initial observations, let them know that it will be reviewed after a period of use and make sure that they accept that there is adequate time for them to record the information required.
- Make sure that the observers record the actual observations and not a tick to show that they made an observation.
- Review the format with the observers to discuss how easy or difficult it has proved to be in use, and also how the data have been of value after analysis.

Again, all that is required is some common sense. Who cannot quote examples of forms which are almost incomprehensible – might they include typical forms from government departments and some service organizations? The authors recall a whole quality circle programme devoted to the redesign of forms used in a bank – a programme which led to large savings.

3.3   Bar charts and histograms

Every day, throughout the world, in offices, factories, on public transport, shops, schools, and so on, data is being collected and accumulated in various forms: data on prices, quantities, exchange rates, numbers of defective items, lengths of pins, temperatures during treatment, weight, number of absentees, etc. Much of the potential information contained in this data may lie dormant or not be used to the full, and often because it makes little sense in the form presented. A vast table or computer printout of figures is not an immediately intelligible document. Our minds cannot read information in such forms unless we have received special training, but the conversion of the information into a picture brings about an immediate advance in the degree of comprehension. The media inundate us with charts, graphs and a wide variety of pictorial presentation of data, and they do this because they know that in this form we can all interpret it. ('A picture paints ten thousand words' – old Chinese proverb.)

Consider, as an example, the data in Table 3.1 which refers to the diameter of pistons. Is it possible to visualize the data as a whole? The eye will tend to concentrate on individual measurements and, in consequence, a large amount of study will be required to give the general 'picture'. A means of vizualizing such a set of data is required.

Look again at the data in Table 3.1. Is the average diameter obvious? Can you tell at a glance the highest or the lowest diameter? Can you

Table 3.1   *Diameters of pistons (mm) – raw data*

56.1	56.0	55.7	55.4	55.5	55.9	55.7	55.4
55.1	55.8	55.3	55.4	55.5	55.5	55.2	55.8
55.6	55.7	55.1	56.2	55.6	55.7	55.3	55.5
55.0	55.6	55.4	55.9	55.2	56.0	55.7	55.6
55.9	55.8	55.6	55.4	56.1	55.7	55.8	55.3
55.6	56.0	55.8	55.7	55.5	56.0	55.3	55.7
55.9	55.4	55.9	55.5	55.8	55.5	55.6	55.2

Table 3.2   *Diameters of pistons ranked in order of size (mm)*

55.0	55.1	55.1	55.2	55.2	55.2	55.3	55.3
55.3	55.3	55.4	55.4	55.4	55.4	55.4	55.4
55.5	55.5	55.5	55.5	55.5	55.5	55.5	55.6
55.6	55.6	55.6	55.6	55.6	55.6	55.7	55.7
55.7	55.7	55.7	55.7	55.7	55.7	55.8	55.8
55.8	55.8	55.8	55.8	55.8	55.9	55.9	55.9
55.9	56.0	56.0	56.0	56.0	56.1	56.1	56.2

estimate the range between the highest and lowest values? Given a specification of 55.0 ± 1.0 mm, can you tell whether the process is capable of meeting the specification, and if it is doing so? Few people can answer these question quickly, but given sufficient time to study the data all the questions can be answered.

If the observations are placed in sequence or ordered from the highest to the lowest diameters, the problems of estimating the average, the highest and lowest readings, and the the range (a measure of the spread of the results) would be simplified. To reorganize the data will take time. The reordered observations are shown in Table 3.2. After only a brief examination of this table it is apparent that the lowest value is 55.0 mm, that the highest value is 56.2 mm and hence that the range is 1.2 mm (i.e. 55.0 mm to 56.2 mm). The average is not immediately obvious although it looks as though it will be about 55.6 or 55.7 mm. Three of the observations are greater than 56.0 mm, the upper tolerance, so the process is not capable of meeting the specification.

Tally charts and frequency distributions

The tally chart and frequency distribution are alternative ordered ways of presenting the data. To construct a *tally chart*, the data may be extracted from the original form given in Table 3.1 or taken from the ordered form of Table 3.2. Let us assume that we are working from the data as initially recorded in Table 3.1. Both a casual glance at the data and the specification suggests that we shall be looking for values lying between something below 55.0 mm and something above 56.0 mm and that the measurements are made in steps of 0.1 mm. So take a piece of paper large enough to allow all the possible values of the diameters between 55.0 mm and 56.0 mm to be listed vertically, and with further room at the top and the bottom of the list for any out of specification results. Now work through the data, one observation at a time, and add a tally mark on the same line as the corresponding value in the original vertical list of possible diameters. If you meet a value below 55.0 mm or above 56.0 mm, add the extra possible values to accommodate the result in the total tally chart (see Table 3.3 for the general layout). Every fifth tally mark is used to form a 'five-bar gate' which makes adding the tallies easier and quicker. The totals from such additions form the *frequency distribution*. Table 3.3 is simply a different representation of the original data in Table 3.1

The tally chart and the frequency distribution both indicate the highest and lowest values, hence the range, and also that the diameters are grouped around a central value with a peak at the most frequently occurring result, known as the mode, at 55.7 mm. Table 3.3 also includes the additional information that there are fifty-six data points.

The tally chart and frequency distribution provide a pictorial presentation of the 'central tendency' or the average, and the 'dispersion' or spead or the range of the results.

The bar chart as a picture

A bar chart is closely related to a tally chart. It is usually constructed with

Table 3.3   *Tally sheet and frequency distribution of diameters of pistons (mm)*

Diameter	Tally		Frequency
55.0	1		1
55.1	11		2
55.2	111		3
55.3	1111		4
55.4	ЦНŦ	1	6
55.5	ЦНŦ	11	7
55.6	ЦНŦ	11	7
55.7	ЦНŦ	111	8
55.8	ЦНŦ	1	6
55.9	ЦНŦ		5
56.0	1111		4
56.1	11		2
56.2	1		1
		Total	56

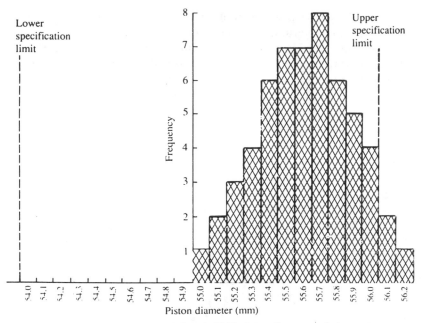

Figure 3.2   *Bar chart of data in Table 2.1 – diameters of pistons*

the measured values on the horizontal axis and the frequency or number of observations on the vertical axis. Above each observed value a bar is drawn with a height corresponding to the frequency. So the bar chart of the data from Table 3.2 will look very much like the tally chart laid on its side – see Figure 3.2.

Like the tally chart, the bar chart shows the lowest and highest values, the range, the centring and the fact that the process is not meeting the specification. It is also fairly apparent that the process is 'capable' of achieving the tolerances since the range available is 2 mm while the spread of the results is only 1.2 mm. Perhaps the idea of capability will be more apparent if you imagine the bar chart of Figure 3.2 being moved to the left so that it is centred around the mid-specification of 55.0 mm. If a process adjustment could be made to achieve this shift, while retaining the same spread of values, all observations would lie within the specification limits and with room to spare.

The basic principle of bar charts can be used for other applications. They can be drawn horizontally and can be lines or dots rather than bars. Figure 3.3 shows a dot plot being used to illustrate the difference in a process before and after an operator was trained to use a milling machine. In Figure 3.3(a) the incorrect method of operation has given rise to a 'bimodal' distribution – one with two peaks. After training, the pattern changes to the single peak or 'unimodal' distribution of Figure 3.3(b). This is illustrating the capability of the operator before and after training. Note how the graphic presentation immediately makes evident the disappearance of the bimodal behaviour as well as the overall reduction in the scatter of the results. A further example of this type of data presentation, which can find ready application at the point of manufacture or operation, is the 'chumbo' chart. This is a device consisting of a set of plastic transparent tubes, appropriately labelled for particular faults or events, down which coloured beads or balls are dropped in order to record the frequency of the events. Such a chart gives an immediate presentation to the operators of the distribution, as it evolves.

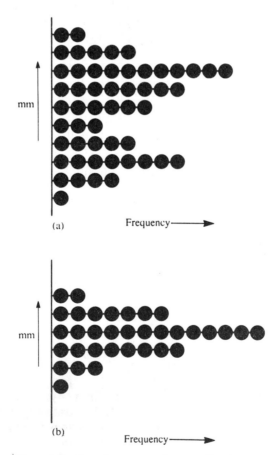

Figure 3.3   *Dot plot – output from a milling machine*

Group frequency distributions and histograms

In the examples of bar charts given above, the number of possible values being observed has been limited. In the case of many variables, and some attributes, the number of possible values which can be observed is very high. For example, most rulers will give several hundred possible values along their whole length and even within a limited range of, say, 2 cm there can be more than 20 possible values to observe. Similarly, when counting the number of absentees in a large population it may vary from, say, 0 to 100. In this case we can improve the picture of the data by studying the frequency at which the observations lie within a limited number of intervals. It is often more useful to present the data in the condensed form of a grouped frequency distribution.

The data shown in Table 3.4 are the thickness measurements of pieces of silicon in mm × 0.001, as sampled from one delivered batch. Table 3.5 was prepared by selecting cell boundaries to form equal intervals called groups or cells, and placing a tally mark in the appropriate group for each observation.

Table 3.4   *Thickness measurements on pieces of silicon (mm × 0.001)*

790	1170	970	940	1050	1020	1070	790
1340	710	1010	770	1020	1260	870	1400
1530	1180	1440	1190	1250	940	1380	1320
1190	750	1280	1140	850	600	1020	1230
1010	1040	1050	1240	1040	840	1120	1320
1160	1100	1190	820	1050	1060	880	1100
1260	1450	930	1040	1260	1210	1190	1350
1240	1490	1490	1310	1100	1080	1200	880
820	980	1620	1260	760	1050	1370	950
1220	1300	1330	1590	1310	830	1270	1290
1000	1100	1160	1180	1010	1410	1070	1250
1040	1290	1010	1440	1240	1150	1360	1120
980	1490	1080	1090	1350	1360	1100	1470
1290	990	790	720	1010	1150	1160	850
1360	1560	980	970	1270	510	960	1390
1070	840	870	1380	1320	1510	1550	1030
1170	920	1290	1120	1050	1250	960	1550
1050	1060	970	1520	940	800	1000	1110
1430	1390	1310	1000	1030	1530	1380	1130
1110	950	1220	1160	970	940	880	1270
750	1010	1070	1210	1150	1230	1380	1620
1760	1400	1400	1200	1190	970	1320	1200
1460	1060	1140	1080	1210	1290	1130	1050
1230	1450	1150	1490	980	1160	1520	1160
1160	1700	1520	1220	1680	900	1030	850

Table 3.5   *Grouped frequency distribution – measurements of silicon pieces*

Cell boundary	Tally						Frequency	Per cent frequency
500–649	11						2	1.0
650–799	ЦНt	1111					9	4.5
800–949	ЦНt	ЦНt	ЦНt	ЦНt	1		21	10.5
950–1099	ЦНt ЦНt	ЦНt ЦНt	ЦНt ЦНt	ЦНt ЦНt	ЦНt	ЦНt	50	25.0
1100–1249	ЦНt ЦНt	ЦНt ЦНt	ЦНt ЦНt	ЦНt ЦНt	ЦНt	ЦНt	50	25.0
1250–1399	ЦНt ЦНt	ЦНt ЦНt	ЦНt ЦНt	ЦНt 111	ЦНt	ЦНt	38	19.0
1400–1549	ЦНt	ЦНt	ЦНt	ЦНt	1		21	10.5
1550–1699	ЦНt	11					7	3.5
1700–1849	11						2	1.0

In the preparation of a grouped frequency distribution and the corresponding histogram, it is advisable to:

1 Make the cell intervals of equal width.
2 If a central target is known in advance, place it in the middle of a cell interval.
3 Preferably, choose the cell boundaries so that they lie between possible observations.
4 Determine the appropriate number of cell intervals from Sturgess rule which can be represented as the mathematical equation:

$$K = 1 + 3.3 \log_{10} N$$

where K = number of intervals
N = number of observations

but which is much simpler if use is made of Table 3.6.

Table 3.6   *Sturgess rule*

Number of observations	Number of intervals
0–9	4
10–24	5
25–49	6
50–89	7
90–189	8
190–399	9
400–799	10
800–1599	11
1600–3200	12

The minimum value of the data in Table 3.4 is 510, the maximum is 1760, there are 200 observations (8 columns and 25 rows), no central target is known, and all observations are to the nearest 10 mm × 0.001. So from Sturgess rule we require about 9 cell intervals between about 500 and 1800 – a convenient interval width is then 150 and a convenient starting point is 500. The application of the above rules then enables the data to be presented as in Table 3.5.

The histogram derived from the data in Table 3.5 is shown in Figure 3.4. The somewhat confusing data, as originally presented in Table 3.4, is now in the form of a picture which shows the central tendency, the spread and the form of the distribution.

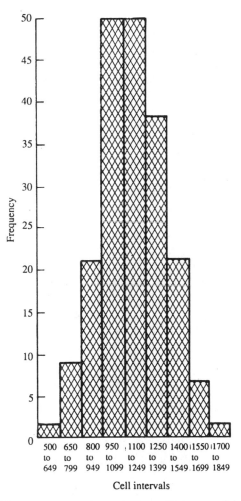

Figure 3.4   *Measurements on pieces of silicon. Histogram of data in Table 3.4*

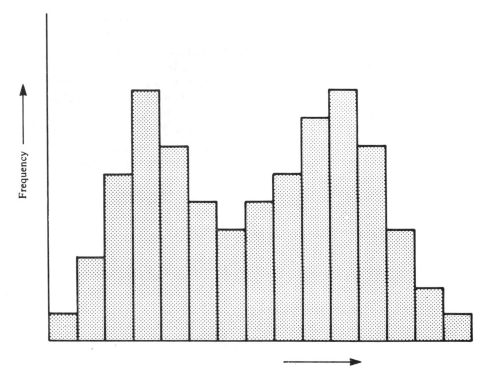

Figure 3.5   *Two peaked histogram – mixed data*

Having seen how to convert data into a picture, and thereby read what is going on more readily, Figures 3.5 and 3.6 show histograms of various types and offer the opportunity to 'read the picture'. Figure 3.5 is an example of mixed data which could have come from two operators, two salesmen, two customers, two machine settings, two different raw materials, two batches being mixed together. Figure 3.6 suggests that some of the observations are being 'adjusted' to bring them back within the tolerance. A method which is frequently used to achieve this, is to take a repeat sample when the first one falls outside the specification and, if the second result is within specification, to record that one only.

There may be other explanations for these examples, and many other forms of histogram are possible. The point is that histograms provide the user with a great deal of information, are easily understood, and are an excellent form of communication of data in a simple display.

The examples shown above include histograms from both continuous data and discrete values and in all cases the picture looks approximately symmetrical about a central value. There is no requirement that distributions should be symmetrical. For example, absenteeism in a small

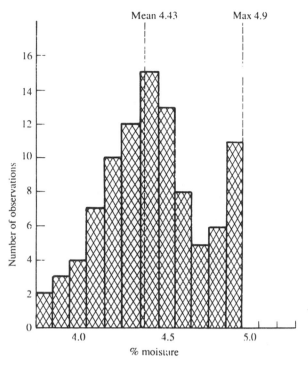

Figure 3.6 *Moisture content of biscuits*

office could most often be zero, so the histogram of absenteeism would peak at zero and show only positive values. Similarly a histogram of the salaries within a normal company will peak towards the lowest level reflecting the fact that most employees are paid salaries towards the bottom of the range with a few below the peak and rather more above it. A histogram of the throwing of dice or the spin of a roulette wheel should show that all values are equally likely and, therefore, the histogram has no peak. Histograms of data recorded on efficiencies, examination marks, purity, yields, etc. are often skewed reflecting the fact that the higher the value the more difficult it becomes to achieve it or improve on it.

The important thing to remember is that a histogram is a simple pictorial presentation of data and that its shape demonstrates the frequency with which events occur and, of course, the chance or probability of any value being found. Referring back to Figure 3.4 one can see that, for the pieces of silicon, there is a high probability that any one piece chosen at random will have a thickness between say 800 and 1600 mm × 0.001 and that there is only a very low probability of finding one, at random, below 500 or above 1800 mm × 0.001.

Other examples of histograms will be discussed along with Pareto analysis and process capability in later chapters.

3.4   Graphs and other pictures

Like histograms, we have all come across graphs. Television presenters use them to illustrate the economic situation; newspapers use them to show trends in anything from average rainfall to the sale of eggs; if you are unfortunate enough to be in hospital, there will be a graph at the bottom of the bed! Graphs can be drawn in many very different ways. The histogram is a type of graph – a bar graph. Graphs also include pie charts, line graphs and pictorial graphs. In all cases they are extremely valuable in quality management in that they convert tabulated data into a picture, thus revealing what is going on within a process, batches of product, customer returns, scrap, rework, and many other aspects of life in manufacturing and service industries.

Line graphs

In line graphs the observations of one parameter are plotted against another parameter and the consecutive points joined by lines. For

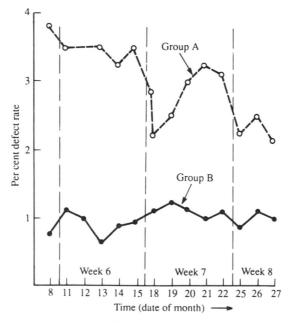

Figure 3.7   *Line graph showing difference in defect rates produced by two groups of operatives.*

example, the various defective rates over a period of time of two groups of workers are shown in Figure 3.7. One parameter, defective rate, is being plotted against another parameter, time, and this is being recorded for the two groups on the same graph, using separate lines and different plot symbols. We can read this picture as showing that Group B perform better than Group A, and also that Group A seem to be making progress at a rate which will enable them to reach Group B's performance level in about another two weeks – this might be part of a learning curve for new operators or new equipment.

Line graphs enable us to recognize patterns which suggest differences, stability, change and rate of change. Where data is available on a continuous basis the line graphs may be curves rather than a series of joined up straight lines.

Pictorial graphs

Often, when presenting results, it is necessary to catch the eye of the

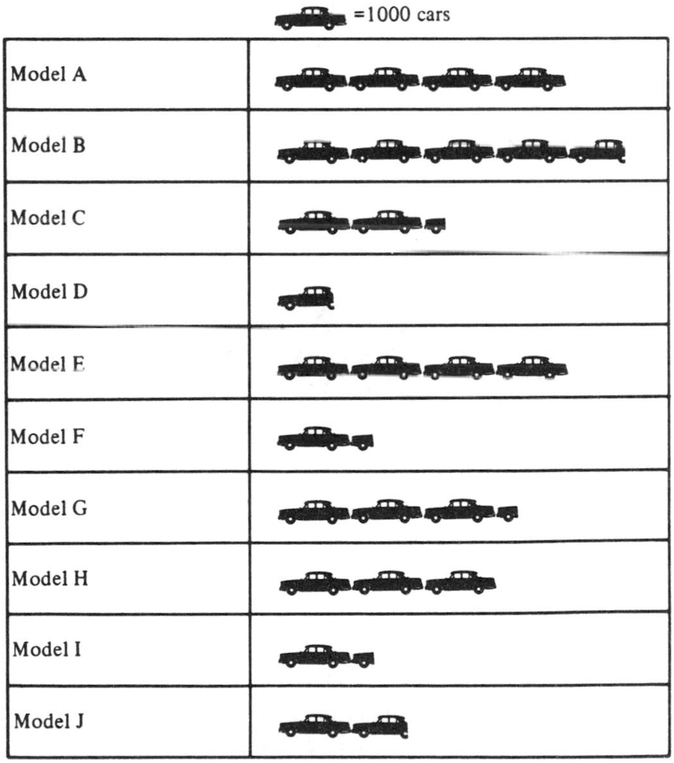

Figure 3.8 *Pictorial graph showing the numbers of each model of car which have been repaired under warranty*

reader. Pictures have a high impact, and the more eye-catching the picture, the higher the impact. Pictorial graphs simply build on the pictorial impact as shown, for example, in Figure 3.8. As soon as the reader sees this, it is clear that the subject is cars and, assuming that cars are of interest to the reader, it is seen to be about different models. One's eye immediately notes that model B is the highest and, only after closer inspection will it be seen that this is bad news, since the graph is of the proportion of cars repaired under warranty, and by this measure Model D is the lowest and best.

Pie charts

Another type of graph is the pie chart in which much information can be illustrated in a relatively small area. Figure 3.9 illustrates an application of a pie chart in which the types and relative importance of defects in furniture are shown. From this it appears that defect D is the largest contributor. Applications of pie charts are limited to the presentation of proportions since the whole 'pie' is normally filled.

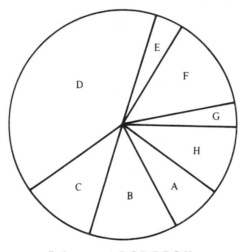

Defect types A,B,C,D,E,F,G,H

Figure 3.9   *Pie chart of defects on furniture*

The use of graphs

All graphs, except the pie chart, are composed of a horizontal and a vertical axis. The scale for both of these must be chosen with some care if the resultant picture is not to mislead the reader. Large and rapid variations can be made to look almost like a straight line by the choice of scale. The nature of the parameters being plotted is also important. Plotting the value of an investment over a number of years will accentuate the growth and may conceal the fact that in some years the growth was either very small or indeed negative. In the pie chart of Figure 3.9 the total elimination of the defect D will make all the others look more important and it may not be immediately obvious that the 'pie' will then be smaller. As illustrated above, the use of pictorial graphs can induce the reader to leap to the wrong conclusion.

Whatever the graph, look at it with care and make sure that the presentation has not been chosen to 'prove' a point which is not totally supported by the data. The abuse of statistics is relatively easy and leads to its reputation for being the next thing to 'damned lies'. Certain groups are very adept at abusing statistics but a closer examination of their data and its presentation is all that is required to identify the abuse.

3.5   Conclusions

This chapter has been concerned with the collection of data and its presentation. In practice, quality improvement programmes can be considerably advanced by the presentation of data. In numerous cases the authors have found that recording performance, and then finding a suitable way of presenting it, is often the first step towards an increased awareness of process behaviour by process operators and managers. The public display of the performance of both the good and the not-so-good can result in renewed efforts being made by the operators of processes. In general, we all seek to do a good job. We also tend to abuse a system which does not seem to care about our performance and which allows us to be less than our best. Equally we all tend to be encouraged by a system which sees our performance as a matter of concern and which takes note of progress as it is made.

In a recent example within the textile industry the defective rate in knitting was reduced from 40 per cent to 4 per cent over a period of about five months. First, data concerning the quality performance of each operator was collected over a few weeks. This gave typical performance figures and not one-off results. The results were then presented in a simple and very public form – on the notice board. The best operator had a defective rate of a steady 2 per cent, others had defective rates much greater than the average of 40 per cent. After only ten weeks the average of all the knitters had declined to 4 per cent, and the only effort made by management, apart from the continued collection and presentation of data, was to assist the very worst operators by some individual tuition.

Collect data, select a good method of presentation, and then present it.

Chapter highlights

- Quality management and process control require a *systematic* approach which includes an appropriate design, resources, materials, process and operating instructions.
- *Narrow the task* of process control to a series of tasks of a manageable size.
- The two basic rules in a systematic approach are *record all data* and *use appropriate techniques*.
- Without records analysis is not possible. Ticks and initials cannot be analysed. Record what is observed and not the fact that there was an observation; this makes analysis possible and also improves the reliability of the data recorded.
- There are two types of numeric data: *variables* which result from measurement, and *attributes* which result from counting.
- The methods of data collection and the presentation format should be designed to reflect the proposed use of data and the requirements of those charged with its recording. Ease of access is also required.
- Tables of figures are not comprehensible – a picture paints ten thousand words. Sequencing data reveals the maximum and the minimum values. Tally charts and counts of frequency also reveal the *distribution* of the data – its *central tendency* and *spread*.
- *Bar charts* are in common use and appear in various forms such as vertical and horizontal bars, line charts, dot charts and chumbo charts. Grouped frequency distributions or *histograms* are another type of bar chart of particular value for continuous variables. The choice of cell intervals can be aided by the use of the Sturgess rule. Reading a histogram provides information about the distribution of the parameters plotted.
- *Line graphs* are yet another way of presenting data as a picture to be read. Graphs include pictorial graphs and pie charts. When reading graphs be aware of the scale chosen, examine them with care, and seek the real meaning – like statistics in general, graphs can be designed to mislead.
- Recording process performance and presenting the results reduce debate and act as a spur to action.
- Collect data, select a good method of presentation, and then present it.

4 Process problem solving

4.1 Introduction

Our personal and professional lives are full of problems, hopefully not all difficulties, but a series of incidents which require us to do something. Given the vast array of problems it is not surprising that we face them with an equally vast array of solutions. When we recognize a problem we tend to seek an immediate solution, apply it and, if that makes the problem go away, move on to the next one. One of the net results of moving directly from the problem to the solution, without any significant data collection and analysis, is that we frequently do not actually solve problems, we simply ignore them. Typical examples of ignoring problems in manufacturing include noting that a product fails to meet specification and deciding to remake it, adjust it, retest it, blend it off, find a customer who will accept it, and a wide variety of other ingenious 'solutions' which do not include finding out the *causes* of the failure and what corrective action could be taken to avoid its recurrence.

Under the pressure of day-to-day events ignoring problems easily becomes the accepted approach. There is often the need for a spur to action to *solve* problems and to recognize that for each problem there is a finite series of possible causes. Finding them may not be easy. Indeed, in the short term, it may be necessary to accept relatively low success rates in locating the real causes of problems and an even lower success rate in finding solutions. The authors' experience is simply that, if no attempt is made to find the actual causes of a problem, it will remain unsolved and continue to perturb the output of the process.

In Chapter 3 methods of collecting and presenting data were discussed and the need to address problems of manageable proportions emphasized. This chapter explores some simple techniques for analysing problems within processes.

4.2   Pareto analysis

In many of the things we do in life, old 'chestnut' problems recur and we frequently find, after analysis, that most of them stem from a relatively limited number of sources. The Italian economist Vilfredo Pareto recognized this concept when he studied the distribution of wealth in his country during the last century. He observed that 80 to 90 per cent of Italy's wealth lay in the hands of 10 to 20 per cent of the population, and that, among the wealthy, some 80 to 90 per cent of their wealth lay in the hands of 10 to 20 per cent of them. This entirely empirical relationship has been found to be true in many other fields. Pareto's observation of this concentrating effect gave rise to the term *Pareto analysis*, sometimes referred to as Pareto's Law or the 80/20 rule.

Pareto analysis procedures

Life in business is not usually short of problems – the overriding problem is often 'where to start'. If it could be established, for example, that 80 per cent of the defects arose from 20 per cent of the causes, or that 80 per cent of debts arose from 20 per cent of the customers, the starting points would be more evident. Pareto analysis is a technique of arranging data according to priority or importance. For example, Table 4.1 gives some data on the reasons for batches of a dyestuff product, Scriptagreen-A, being scrapped or reworked. A definite procedure is needed to transform this data into a spur for action.

It is obvious from the presentation of the data in Table 4.1 that two types of Pareto analysis are possible to identify the areas which should receive attention. One is based on the frequency of each cause of scrap/rework and the other is based on cost. It is reasonable to assume that both types of analysis will be required. The identification of the most frequently occurring reason should lead to some action enabling the total number of batches scrapped or requiring rework to be reduced. This could be necessary to improve plant operator morale which is likely to be adversely influenced by the most frequently occurring and, hence, most evident causes. Analysis using costs will be necessary to seek solutions to those causes which give rise to the greatest costs and, hence, the greatest potential savings. We shall use a generalized stepwise procedure to carry out both of these analyses.

Step 1   List all the elements
All the elements must be listed to avoid the inadvertent drawing of erroneous conclusions. In this case the elements or reasons may be listed as they occur in Table 4.1. They are, in the order in which they occur:

Table 4.1

Scriptagreen-A *Plant B*		*Batches scrapped/reworked*		
		Period 05–07 inclusive		
Batch No.	*Reason for scrap/rework*	*Labour cost (£)*	*Material cost (£)*	*Plant cost (£)*
05–005	Moisture content high	500	50	100
05–011	Excess insoluble matter	500	nil	125
05–018	Dyestuff contamination	4,000	22,000	14,000
05–022	Excess insoluble matter	500	nil	125
05–029	Low melting point	1,000	500	3,500
05–035	Moisture content high	500	50	100
05–047	Conversion process failure	4,000	22,000	14,000
05–058	Excess insoluble matter	500	nil	125
05–064	Excess insoluble matter	500	nil	125
05–066	Excess insoluble matter	500	nil	125
05–076	Low melting point	1,000	500	3,500
05–081	Moisture content high	500	50	100
05–086	Moisture content high	500	50	100
05–104	High iron content	500	nil	2,000
05–107	Excess insoluble matter	500	nil	125
05–111	Excess insoluble matter	500	nil	125
05–132	Moisture content high	500	50	100
05–140	Low melting point	1,000	500	3,500
05–150	Dyestuff contamination	4,000	22,000	14,000
05–168	Excess insoluble matter	500	nil	125
05–170	Excess insoluble matter	500	nil	125
05–178	Moisture content high	500	50	100
05–179	Excess insoluble matter	500	nil	125
05–179	Excess insoluble matter	500	nil	125
05–189	Low melting point	1,000	500	3,500
05–192	Moisture content high	500	50	100
05–208	Moisture content high	500	50	100
06–001	Conversion process failure	4,000	22,000	14,000
06–003	Excess insoluble matter	500	nil	125
06–015	Phenol content > 1%	1,500	1,300	2,000
06–024	Moisture content high	500	50	100
06–032	Unacceptable application	2,000	4,000	4,000
06–041	Excess insoluble matter	500	nil	125
06–057	Moisture content high	500	50	100
06–061	Excess insoluble matter	500	nil	125
06–064	Low melting point	1,000	500	3,500
06–069	Moisture content high	500	50	100
06–071	Moisture content high	500	50	100
06–078	Excess insoluble matter	500	nil	125
06–082	Excess insoluble matter	500	nil	125
06–094	Low melting point	1,000	500	3,500
06–103	Low melting point	1,000	500	3,500

Table 4.1 – cont

Scriptagreen-A Plant B		Batches scrapped/reworked		
		Period 05–07 inclusive		
Batch No.	Reason for scrap/rework	Labour cost (£)	Material cost (£)	Plant cost (£)
06–112	Excess insoluble matter	500	nil	125
06–126	Excess insoluble matter	500	nil	125
06–131	Moisture content high	500	50	100
06–147	Unacceptable absorption spectrum	500	50	400
06–150	Excess insoluble matter	500	nil	125
06–151	Moisture content high	500	50	100
06–161	Excess insoluble matter	500	nil	125
06–165	Moisture content high	500	50	100
06–172	Moisture content high	500	50	100
06–186	Excess insoluble matter	500	nil	125
06–198	Low melting point	1,000	500	3,500
06–202	Dyestuff contamination	4,000	22,000	14,000
06–214	Excess insoluble matter	500	nil	125
07–010	Excess insoluble matter	500	nil	125
07–021	Conversion process failure	4,000	22,000	14,000
07–033	Excess insoluble matter	500	nil	125
07–051	Excess insoluble matter	500	nil	125
07–057	Phenol content > 1%	1,500	1,300	2,000
07–068	Moisture content high	500	50	100
07–072	Dyestuff contamination	4,000	22,000	14,000
07–077	Excess insoluble matter	500	nil	125
07–082	Moisture content high	500	50	100
07–087	Low melting point	1,000	500	3,500
07–097	Moisture content high	500	50	100
07–116	Excess insoluble matter	500	nil	125
07–117	Excess insoluble matter	500	nil	125
07–118	Excess insoluble matter	500	nil	125
07–121	Low melting point	1,000	500	3,500
07–131	High iron content	500	nil	2,000
07–138	Excess insoluble matter	500	nil	125
07–153	Moisture content high	500	50	100
07–159	Low melting point	1,000	500	3,500
07–162	Excess insoluble matter	500	nil	125
07–168	Moisture content high	500	50	100
07–174	Excess insoluble matter	500	nil	125
07–178	Moisture content high	500	50	100
07–185	Unacceptable chromatogram	500	1,750	2,250
07–195	Excess insoluble matter	500	nil	125
07–197	Moisture content high	500	50	100

moisture content high, excess insoluble matter, dyestuff contamination, low melting point, conversion process failure, high iron content, phenol content >1%, unacceptable application, unacceptable absorption spectrum and unacceptable chromatogram.

Step 2 Measure the elements
It is essential to use the same unit of measure for each element. It may be in cash value, time, frequency, number or amount, depending on the element. In the Scriptagreen-A case the elements – reasons for scrap and rework – may be measured in terms of frequency of occurrence, labour cost, material cost, plant cost and total cost. We shall investigate only the first and the last – frequency and total cost. The listing of the elements, the frequency tally charts and the total cost calculations are shown in Table 4.2.

Step 3 Rank the elements
This ordering takes place according to the measures and not to the classification. This is the crucial difference between a Pareto analysis and the more usual frequency plots and is particularly important for numerically classified elements. For example, Figure 4.1 shows, for comparison, the distributions of pin lengths when plotted by length and by ranked frequency. The two distributions are ordered in different ways. In both cases the frequency of occurrence is shown on the vertical or y-axis, while in the first case the horizontal or x-axis shows the pins ranked by length (the classification of the pins) and in the second case the

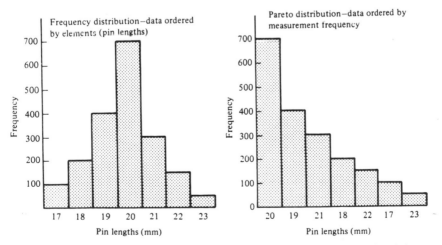

Figure 4.1 *Comparison between frequency and Pareto distribution (pin lengths)*

Table 4.2   *Frequency distribution and total cost of dyestuff batches scrapped/reworked*

Reason for scrap/rework	Tally	Frequency	Cost per batch (£)	Total cost (£)
Moisture content high	LHT LHT LHT LHT LHT LHT 111	23	650	14,950
Excess insoluble matter	LHT LHT LHT LHT LHT LHT 11	32	625	20,000
Dyestuff contamination	1111	4	40,000	160,000
Low melting point	LHT LHT 1	11	5,000	55,000
Conversion process				
failure	111	3	40,000	120,000
High iron content	11	2	2,500	5,000
Phenol content > 1%	11	2	4,800	9,600
Unacceptable application	1	1	10,000	10,000
Unacceptable absorption				
spectrum	1	1	950	950
Unacceptable chromatogram	1	1	4,500	4,500

horizontal axis shows the pins ranked by the frequency with which each length occurs (the frequency of the elements).

To return to Scriptagreen-A, Table 4.3 shows the reasons ranked according to frequency of occurrence while Table 4.4 has them ranked in decreasing order of total cost.

Step 4   Create cumulative distributions
Once the elements have been ranked, the most important ones, on which we may need to concentrate, and the least important, which should attract the least attention, can be identified. Before reaching any conclusions, however, the measures should be cumulated from the highest ranked element to the lowest, and each cumulative figure (frequency or total cost in the case of Scriptagreen-A) shown as a

Table 4.3   *Scrap rework – Pareto analysis of frequency of reasons*

Reason for scrap/rework	Frequency	Cum. freq.	% of total
Excess insoluble matter	32	32	40.00
Moisture content high	23	55	68.75
Low melting point	11	66	82.50
Dyestuff contamination	4	70	87.50
Conversion process failure	3	73	91.25
High iron content	2	75	93.75
Phenol content > 1%	2	77	96.25
Unacceptable:			
Absorption spectrum	1	78	97.50
Application	1	79	98.75
Chromatogram	1	80	100.00

Table 4.4   *Scrap/rework – Pareto analysis of total costs*

Reasons for scrap/rework	Total cost	Cum. cost	Cum. % of grand total
Dyestuff contamination	160,000	160,000	40.0
Conversion process failure	120,000	280,000	70.0
Low melting point	55,000	335,000	83.75
Excess insoluble matter	20,000	355,000	88.75
Moisture content high	14,950	369,950	92.5
Unacceptable application	10,000	379,950	95.0
Phenol content > 1%	9,600	389,550	97.4
High iron content	5,000	394,550	98.65
Unacceptable chromatogram	4,500	399,050	99.75
Unacceptable absorption spectrum	950	400,000	100.00

percentage of the grand total. Tables 4.3 and 4.4 show these calculations for the scrap and rework of Scriptagreen-A. The final columns show, respectively, the cumulative distribution of the frequency of reasons for scrap and rework and the cumulative distribution of the total costs of each cause of scrap and rework.

Step 5   Draw the Pareto curve
The cumulative percentage distributions are plotted on linear graph paper. The cumulative percentage measure is plotted on the vertical axis and either the cumulative percentage of the number of elements or the number of elements itself is plotted along the horizontal axis. Figures 4.2 and 4.3 are the Pareto curves for the frequency of occurrence and the

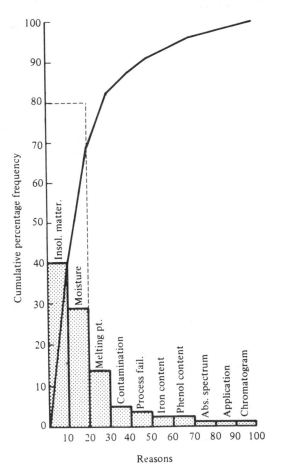

Figure 4.2   *Pareto analysis by frequency – reasons for scrap/rework*

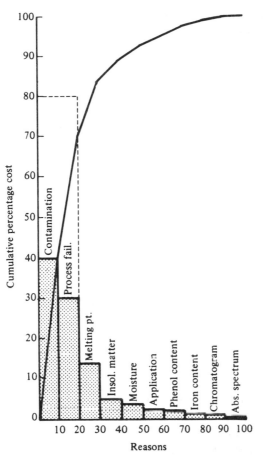

Figure 4.3 *Pareto analysis by costs of scrap/rework*

total costs of the various reasons for the scrapped or reworked batches of Scriptagreen-A. Since in this case the number of causes or elements is ten, the individual elements and the percentage cumulative number of elements can be conveniently shown on the same graph. Figures 4.2 and 4.3 also show the bar chart of the individual frequencies and costs – the difference between these two presentations of the same results will be discussed later.

Step 6   Interpret the Pareto curves

The aim of Pareto analysis in problem solving is to highlight those elements which demand most attention and should be examined first. It is useful to add to the Pareto plot a vertical line at 20 per cent of the cumulative percentage number of elements and a horizontal line at 80 per cent of the cumulative percentage measures. This has been done for the curves in both Figure 4.2 and 4.3 and shows that:

1 20 per cent of the reasons are responsible for about 70 per cent of all the batches being scrapped or reworked. The reasons are, excess insoluble matter (40 per cent) and moisture content high (28.76 per cent), see Figure 4.2.

2 80 per cent of all batches scrapped or reworked result from only three causes – the two noted in 1 above, and low melting point (13.75 per cent), see Figure 4.2.

3 A different 20 per cent of the reasons are responsible for about 70 per cent of the total costs of scrap and rework. The reasons are, contamination (40 per cent) and process failure (30 per cent), see Figure 4.3.

4 80 per cent of the grand total of the costs of scrap and rework result from only three of the causes – the two noted in 3 above, and low melting point (13.75 per cent), see Figure. 4.3.

Either by the criterion of frequency of occurrence or by the criterion of total costs we have identified the 'vital few' and because of the Pareto plot we are able to specify the 20 per cent of vital causes and the 80 per cent of the cumulative effects. It is now quite clear that if the objective is to reduce costs, then contamination should be tackled with priority. Even though this only occurred four times during the eighty incidents analysed, their combined costs were the highest. Concentrating on the problem of excess insoluble matter will have the largest effect on reducing the total number of incidents.

The bar charts do not allow us to readily quantify the impact of the important few. In the Pareto curve, not only can these be quantified, but the importance of the highest ranking few and the triviality of the lowest ranking many are accentuated. As already discussed in Chapter 3, the way in which data is presented can have an important impact on the recipient. The Pareto curve makes it very clear where effort must be concentrated to give maximum effect. For example, if effort were directed towards solving the incidence of high iron content and it was totally successful, the impact on both the total frequency of occurrences and on the total costs would be trivial. Whereas, even a partial solution to the problem of contamination will have a measureable effect on the total costs, as would a partial solution of the insoluble matter problem have on the total frequency of occurrences.

Pareto analysis is a powerful tool but it should not be forgotten that it is based on an empirical rule, which has no mathematical foundation. When using the analysis remember to look carefully at the results and use a combination of common sense and your own knowledge of the particular problems you have analysed to decide on the action to be taken. While the aim is to identify the potential areas of maximum reward, it is not a requirement of the systematic approach that small problems with a known solution must wait until the larger ones have been tackled.

It should not be assumed that the most frequent, the most expensive, or the most important problems are known intuitively by experience – even intelligent guessing can be remarkably wrong. Collecting the data and analysing it is not particularly time consuming, so it should be done on a regular basis to monitor the progress. As the vital few decline in their importance, the other causes expand into the area requiring maximum attention.

Returning to the Scriptagreen-A problem, should we tackle the most frequently occurring cause of scrap and rework or should we tackle the most costly cause? While the Pareto analysis tells us which are the important few when judged by frequency of occurrence or by cost, it does not help us to decide which of these two is the more important. In practice, the process operators are unlikely to be impressed if management are seen to be tackling contamination, a problem which only occurs four times in three months, even though it has the highest total cost. To whom should the contamination problem be delegated bearing in mind its relative infrequency? This is probably a task for the technical staff who can locate the type of contaminant, its possible origin, methods of detection, etc. And to whom would one delegate the most frequently occurring problem? The process operators are the best group to tackle this one since they are likely to be able to identify the possible contributory factors. So it may be possible to tackle both types of 'important few' simultaneously by making use of two different groups.

The real lesson of Pareto analysis is not simply to identify the important few and the trivial many, but to resolve to tackle only the important few and to ignore for now, yes ignore, the trivial many. In practice, quality problems are frequently the subject of firefighting in which the latest problem to occur assumes an unwarranted importance. To apply the results of Pareto analysis effectively requires a systematic approach which enables the trivial many to be ignored while the important few are researched until solutions are found, which either eliminate them or reduce them to one of the trivial many. This demands a discipline which is never easy to establish or maintain. A division of responsibility – a short-term 'firefighter' and a long-term 'important few researcher' – may be the solution. Firefighting may ensure the short-term viability of a process; identifying and solving the important problems leads to significant progress towards never-ending improvement, which ensures the long-term future viability of the process and its operators.

Pareto analysis has very wide applications in management in general, for example, in the analysis of creditors, debtors, absenteeism, sales values by product or customer, contributions by product or customer, delivery performance, distribution, etc., indeed in any area where there are a number of associated activities or items.

In a variant of Pareto analysis, referred to as ABC analysis, the first 20 per cent of the cumulative percentage of elements corresponds to the A tranche, the next 30 per cent is the B tranche and the final 50 per cent the C tranche. It is often found that the tranches give rise to about 80, 15 and 5 per cent respectively of the resultant effects. ABC analysis is used in stock control where there are sometimes advantages in this further division of the less important part of the whole.

As in the case of Scriptagreen-A, data may exist for several measures for each of the elements. After multiple analyses one has to seek to draw conclusions from more than one Pareto curve. Sometimes it may be reasonable to assume that there will be correlations between some of the different measures. For example, in the Scriptagreen case there is no reason to believe that the most frequently occurring problem will be either the most or the least costly, so no obvious correlation of the measures of frequency of occurrence and cost would be anticipated. On the other hand, an analysis of, say, individual product sales value and the individual product contribution to profits might be expected to show an approximate correlation; the higher the total sales of one particular product the higher its expected total contribution to profits. Under circumstances where correlations of this type may exist, a further technique for analysis may be used with advantage. This is called rank plotting and is described in Chapter 14.

4.3   Cause and effect analysis

In the study of quality related and other problems, the *effect* – such as a particular defect or 'out of control' process parameter – is usually known. Cause and effect analysis (C/E analysis) is another simple technique which may be used to elicit the possible contributory factors, or *causes* which give rise to the effect. The technique consists of the use of the expert knowledge of any group of operators, staff, or managers in combination with a C/E diagram, which adds structure to their thinking about specific problems. C/E diagrams are often used in conjunction with brainstorming sessions and have a history of association with the work of quality circles.

When a meeting of any group of people is invited to speculate about the possible causes of an effect, a number of things can happen. The group may include some people who are reticent to offer suggestions; it is not that they do not have ideas, they are simply reluctant to bring them into the open. Others within the group may have very fixed ideas which they seek to promote by repetition; they ride hobby-horses. Still others spend most of their time evaluating suggestions from other participants and either promoting them or dismissing them; in both cases the arguments that they use may either be totally valid or largely invented. Another mode of behaviour consists of initially responding by making suggestions, but then withdrawing once the suggestion has either been accepted or dismissed. Who has not attended a meeting where these various behaviour modes were not present? There is a very simple explanation of why meetings of this type end up by retaining only a few real causes of the effect and a great deal of biased, but not necessarily ill-informed, opinion. Clearly, two things are taking place at the same time; one is the proposal of possible causes to explain the effect, and the other is the evaluation of the causes. When combined in this way discussions move progressively towards evaluation of the causes and away from the original objective, which was to list the possible causes of a specified effect. The C/E diagram, and its use under controlled conditions, enables the listing of possible effects and their evaluation to be separated.

The cause and effect diagram, also known as the Ishikawa diagram (he introduced them to quality circles) and the fishbone diagram (after its appearance), show the effect at the head of a central 'spine' with the causes at the ends of the 'ribs' which branch from it. The basic form is shown in Figure 4.4. Often the principal areas of possible causes are listed first and then expanded into sub-causes and, if necessary, sub-sub-causes. This process could continue forever, being limited only by the ingenuity and patience of the participants. In practice one reaches a point at which it is highly likely that all the major contributing causes have been identified and any new suggestions will probably be of trivial importance in their contribution to the total effect. At this stage the C/E approach

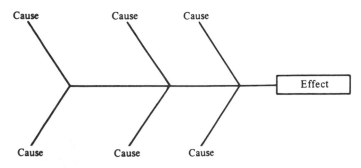

Figure 4.4   *Basic form of cause and effect diagram*

ceases to be useful and one should pass to the collection of data and its analysis, perhaps incorporating the Pareto technique. As in Pareto analysis, a C/E analysis should be repeated from time to time in order to locate new causes which may have occurred, to identify trivial ones not included earlier but which, after improvements, have become significant, and to explore contributory factors previously omitted.

Constructing the cause and effect diagram

The C/E diagram may be a picture resulting from a brainstorming session. During brainstorming, the golden rules are that no one evaluates, and that all suggestions are noted, no matter how trivial, wild, irrelevant or even stupid they may seem. Not surprisingly, these sessions are frequently accompanied by peals of laughter. The atmosphere created is one in which it is 'safe' to suggest anything; there need be no fear of ridicule or criticism. Most people are encouraged by this atmosphere to participate, indeed the success of such a session can best be judged by the level of participation. All suggestions are welcomed because, if for no other reason, any one of them could lead to trains of thought which may go on to the major but more elusive causes of a given defect.

The construction and use of the C/E diagram is best illustrated by an example. The production manager in a tea bag manufacturing firm was extremely concerned about the amount of tea being wasted. Five tonnes of tea was purchased each week but only about 4.5 tonnes was invoiced. Study groups had investigated the problem on more than one occasion but with little or no success. The lack of progress was attributed to a combination of too much talk, arm waving and shouting down – typical symptoms of a non-systematic approach. The problem was handed to a newly-appointed management trainee who used the following stepwise approach:

Step 1   Identify the effect
This sounds simple enough but, in fact, it is often so poorly done that considerable time is wasted in the later steps of the process. It is vital that the effect or problem is stated in clear, concise and agreed terms. This will help to avoid the situation in which the 'causes' are identified and eliminated and yet the problem still exists as a result of it being inadequately defined. Similarly there can be considerable loss of time and effort if the definition of the problem is not agreed by all concerned. The tea bag firm defined the problem as 'Waste tea – the tea lost between its purchase in bulk and its sale in tea bags'. A less explicit definition such as 'waste' would have involved the wastage of perforated paper, cartons, time etc. – the problem would not have been 'narrowed down'.

Step 2   Establish goals

The importance of establishing realistic and meaningful goals may seem to be obvious, but problem solving is not a self-perpetuating endeavour, and most people need to know the yardstick against which their performance will be measured. A goal should, therefore, be stated and in a measurable form which may include a timescale. If there is no measure associated with a goal, monitoring of progress will, at best, be subjective. If goals are not realistic, the enthusiam for achievement may be compromised. Both attempting the impossible and being asked to deal with the trivial are demotivating. In the tea bag case the goal was 'a 50 per cent reduction in the next nine months'. Since the actual tea wastage was averaging 10 per cent, such progress seemed possible. If the current wastage had been 1 per cent, a 50 per cent reduction might have been more difficult to achieve.

Step 3   Construct a diagram framework

A framework in which the causes are to be listed can be very helpful to creative thinking. The authors have found Lockyer's five 'Ps of production management' of use.[1] This states that the five components of any manufacturing task are the:

- *Product* or services, including raw materials and intermediates.
- *Processes* or procedures or methods of transformation.
- *Plant* or equipment or machines used in the transformation process.
- *Programmes* or schedules for ordering, manufacturing and shipping.
- *People*: management, staff and operators both inside and outside the organization.

These are placed at the main ribs of the diagram as shown in Figure 4.5.

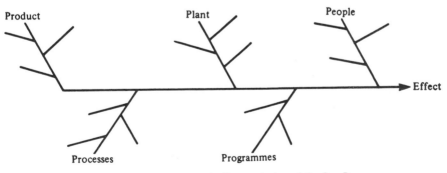

Figure 4.5   *Cause and effect analysis and the five Ps*

---

[1]   See Lockyer, K.G. Muhlemann A.P. and Oakland J.S., *Production and Operations Management*, Pitman, 1988.

The main value of these headings is that when ideas under one heading are exhausted, the thinking process can begin afresh by starting from another heading. The grouping under sub-headings is also of value when subsequent analysis of the diagram is attempted.

Step 4   Record all the causes

A brainstorming session may be used to generate possible causes. The people asked to suggest causes must include those involved in the process from which the effect was identified. In the tea bag example, the management trainee prepared the list of causes by individual discussions with each of the process operators involved.

It may not be easy to know where to start listing causes. Any starting point is valid. A diagram framework, prepared in advance of any questioning or brainstorming process, will immediately suggest a number of possible starting points. The leader of the enquiry into causes must ensure that a record is kept of every possible cause suggested and may choose to nominate a secretary to keep the records. The leader has a dual objective of keeping the suggested causes flowing and ensuring that, at the end of the enquiry session, the record will be complete and capable of editing. The one activity to avoid is any attempt to assess the relative importance of the suggested causes.

Figure 4.6 shows the completed C/E diagram for the waste in tea bag manufacture after editing, which changed the 5P structure originally used.

Step 5   Incubate and analyse the diagram

If it has not been possible to contruct a C/E diagram during the activity of recording the causes, it should now be set up by ordering the suggested causes into the fishbone form. It is vital that every suggestion is included in the final diagram. During editing, repetition should be avoided and all the causes clearly defined.

The draft fishbone diagram should be put on display and left for a few days, during which further additions are invited. After this incubation period, the group responsible for the list of causes should be asked to review them. During the review, which may give rise to further suggestions, criticism and assessment are allowed. After the incubation period the participants are unlikely to recall who suggested what, and it is, therefore, much easier to evaluate the ideas without criticizing those who made them.

Step 6   Decide and take action

The object of the C/E diagram is to assemble the possible causes and present them in a convenient array. There is no point in doing this unless it serves as a spur to action. The typical action required is to collect data

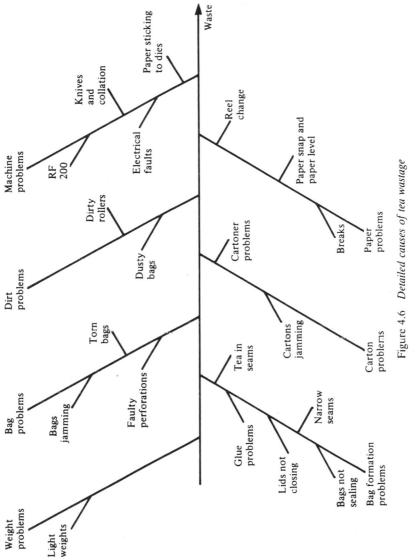

Figure 4.6 *Detailed causes of tea wastage*

with which to begin the evaluation of the causes and their separation into the important and the trivial.

If we return to the teabag example, the trainee went back to the operators, who had assisted her in suggesting causes, and not only reviewed with them the findings but also invited them to collect the required data. It was agreed that this would be done over a two-week period, after which they would analyse the data for presentation and action. To facilitate the data collection, the trainee prepared some simple clipboards and examples of piles of tea weighing 2, 5, 10, 20, 50 and 100 grams. The operators were then asked to record each incident when tea was wasted and an estimate of the quantity of tea lost.

This action plan contained some important components:

- The operators remained in their role as the 'experts' in being invited to suggest the causes, collect the data and to review its analysis;
- Simple aids, such as the clipboard and sample quantities of tea, were provided to assist in the data collection;
- Having engaged the interest and enthusiasm of the operators it could be assumed that they would collect data in a reasonably conscientious manner. This does not mean that there was no risk of double reporting of incidents and no failure to record all the incidents, but both of these risks were small and, since data was being collected over a reasonably long period of time, the relative importance of the contributory causes was unlikely to be affected.

The breakdown of the percentage contribution to the total losses of tea under the framework headings is shown in Table 4.5. What this table does not show is that the total losses of tea accounted for by the operators did not exactly equal the losses shown by the difference between the tea used and the tea sold. This does not invalidate the relative importance of the contributory causes, it merely reflects the fact that not all the events will be accurately recorded.

Table 4.5 *Major categories of causes of tea waste*

Category of cause	Percentage wastage
Weights incorrect	1.92
Bag problems	1.88
Dirt	5.95
Machine problems	18.00
Bag formation	4.92
Carton problems	11.23
Paper problems	56.10

Table 4.5 clearly demonstrates that the major heading is 'paper problems'. A more detailed Pareto analysis showed that among the paper problems the major contributor was 'reel change', which accounted for 24 per cent of all losses. After discussions with the supplier of the reels of perforated paper, and some minor modifications to the machine, the diameter of the reels was doubled and the frequency of reel changes reduced to approximately one-quarter of the original. Prior to this investigation, reel changes were not considered to be a problem – it was accepted as inevitable that a reel would come to an end and that, during the change of reels, tea would be lost. Tackling this particular cause of waste tea did not eliminate the losses at reel change but, by both reducing the frequency of changes, and highlighting the fact that it was better to lose paper rather than tea, a reduction in excess of the expected one-quarter resulted. This is an example of the Hawthorne Effect which was first demonstrated by Mayo on the shopfloor of Western Electric's Hawthorne plant.[1]

Once the reel change problem was on the way to a solution, the next most important cause was tackled. Over the total trial period of nine months wastage was reduced by 75 per cent, which outstripped the targeted 50 per cent.

Applications of cause and effect diagrams

Use of the C/E diagram organizes the free-flowing of ideas into a logical pattern. With a little practice the diagram can be used very effectively when any group seeks to establish the cause of an effect which may be either a problem or a desirable effect. All too often, desirable occurrences are simply welcomed and attributed to chance, when in reality there has to be an explanation of the occurrence. In a game of chance, chance is the explanation of both winning and losing, but in industrial processes, chance plays a smaller part than is commonly supposed. So when something desirable occurs, seek its cause in the above way and, having identified the possible causes and the actual contributors to the effect, use this information to further improve – to decrease the defective rate, to lower the amount of scrap or rework, to speed up the delivery, or generally to continue never-ending improvement.

---

[1] Lighting was improved in a successful attempt to increase productivity. As the light intensity was progressively reduced to its original level, however, the productivity continued to increase. Clearly the attention paid to the employees, not the improved working conditions, was producing the desired effect.

4.4   Scatter diagrams

Scatter diagrams are used to examine two factors or parameters in order to see if there is an association or correlation between them. If there is dependence of one factor on the other, controlling the independent factor will be a method of controlling the dependent factor. For example, if the temperature of a chemical process and the purity of the chemical product are related, one may control the temperature in order to control the purity, and in monitoring the purity one may adjust or correct it by an adjustment to the temperature. If data is available for values of the two factors at consecutive or related times, one may make a plot in which the values for one factor are simply plotted against the values for the other related factor. This will probably not result in a simple line graph, but rather a scatter of points which, in some cases, may lie around a central line or curve. The setting up of such a plot results in a scatter diagram.

For example, Figure 4.7 shows that when the process temperature is set at A, a lower purity results than when the temperature is set at B. Equally if the temperature can only be controlled within the range A to B, the impurity will vary over at least the range C to D. From Figure 4.8, one can see, in spite of the scatter, that to maximize the tensile strength one will have to use a treatment time of about B – both shorter and longer treatment times will result in lower tensile strengths. In both Figures 4.7 and 4.8 there appears to be an association between the 'independent factor' on the horizontal axis and the 'dependent factor' on the vertical axis. A statistical hypothesis test could be applied to the data to determine the statistical significance of the association, which could then be expressed mathematically (see Appendix G). Such a procedure is often quite unnecessary in the industrial situation where all that is required is to establish that a relationship exists and to judge its influence on the operating procedures. The example shown in Figure 4.9 clearly suggests that there is no association between the finished diameter of a plastic pipe and the size of the plastic granules used in its manufacture. Submitting this type of data to a standard statistical package can be dangerous because the mathematics may suggest that a correlation exists and will add details of the 'degree of confidence' and the 'number of degrees of freedom'. This language is only meaningful to those who can speak it. The scatter diagram in Figure 4.9 clearly shows that if there is a correlation between the two factors it has not yet been found, and normally that is all we need to know before returning to a C/E analysis and the collection of alternative and possibly correlating data.

Scatter diagrams can sometimes be misleading in suggesting that there is a causal relationship between the two factors being investigated. For example, Figure 4.10 shows the date plotted against the cumulative

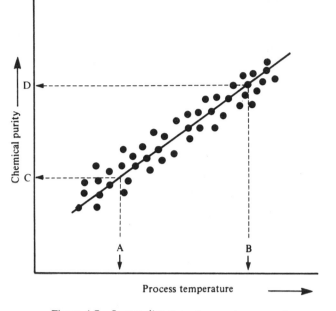

Figure 4.7   *Scatter diagram – temperature v. purity*

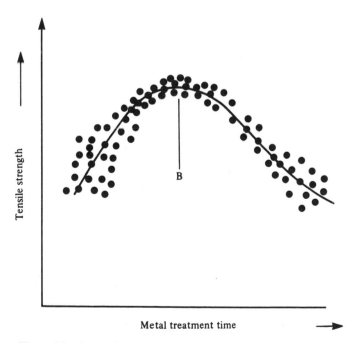

Figure 4.8   *Scatter diagram – metal treatment time v. tensile strength*

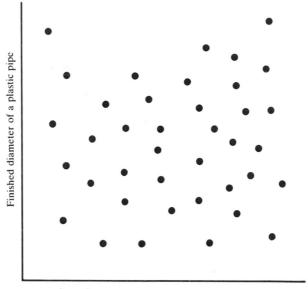

Figure 4.9   *Scatter diagram – no relation between size of granules of plastic and finished diameter of pipe*

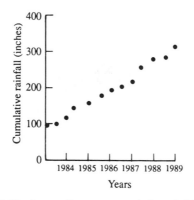

Figure 4.10   *Scatter diagram – cumulative rainfall v. year*

rainfall. At first sight one might be tempted to assume that the rainfall was increasing steadily – in fact all that the plot shows is that from year to year we add both a year and some rainfall. The correlation is not wrong, it is simply not helpful or meaningful. Beware of the media who sometimes make use of this type of presentation to 'prove' points for which there is little evidence. As a further example, Figure 4.11 suggests that there is a correlation between the sales volume and the advertising expenditure.

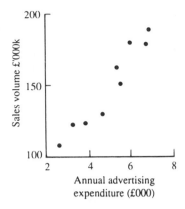

Figure 4.11 *Scatter diagram – sales volume versus advertising expenditure*

These two factors may not be linked by a direct causal effect; they may be separate and independent consequences of some other behaviour, such as an increased awareness of the need to expand sales.

To summarize, remember that when reading scatter diagrams, a relationship may exist but not be directly causal, and when an association does exist it may sometimes not be apparent because other, and greater, causal or random factors are interfering with the ability to detect it.

Some simple steps for setting up a scatter diagram are as follows:

1   Select the dependent and independent factors. The dependent factor could be a 'cause' from a cause and effect diagram, a specification, a measure of quality, or some other result or measure. The independent factor is selected because of its potential relationship with the dependent factor.
2   Either make use of existing data, or set up an appropriate check sheet for recording the data.
3   Decide the time interval between the recording of the two factors and ensure that the methods of measurement are sensitive enough to detect variations. (It may not be a good idea to set off intending to change one of the factors to see what happens to the other one – this type of experiment, conducted outside the laboratory, can be both costly and dangerous.)
4   Once the data record exists, plot the points on the scatter diagram – convention says that the horizontal axis should be used for the independent factor and the vertical axis for the dependent factor, but ignoring this convention has no effect on one's ability to interpret the results.
5   Read the scatter diagram – analyse it and ask if it suggests anything and, if so, what is means.

A very simple correlation test may be applied to a scatter diagram for which there are at least seven points plotted. The steps are:

1   Divide the scatter diagram by a horizontal line with the same number of points on each side of the line and then divide the scatter diagram by a vertical line which also divides the number of points into two equal halves.
2   Count the total number of points in each of the quadrants and add together the numbers falling in diagonally opposite quadrants – this gives two numbers, N the larger and n the smaller.
4   Apply the test:

$$\text{Is } N \text{ greater than } 2 + 0.6n?$$

5   If the answer is 'No', it is safe to assume that a significant correlation has not been found. If the answer is 'yes', it is worth proceeding to a full-scale analysis to establish the significance of the correlation (see both Chapter 9 and Appendix G). In the meantime, examine the scatter diagram and ask what the relationship between the two factors means and how you can make use of this knowledge to improve quality management.

The scatter diagram is yet another of the simple techniques available in the systematic approach to quality management. It is the subject of a more structured approach to variable data in Chapter 9. For the moment we may note its application as another simple problem solving method.

4.5   Control charts

The last of the techniques listed in Chapter 1 as the basic tools of statistical process control is control charts. Plotting graphs, charting, and presenting the data as a picture is common to process control methods used throughout manufacturing and service industries. As already discussed in this and the previous chapter, converting tabulated data into a picture is a vital step towards greater and quicker understanding of processes. So any form of control charting or plotting of time-related data is better than none. In using control charts one seeks to make the best possible use of the data available, to squeeze from it all the information which can be obtained by using simple techniques. This may sound complex and possibly intimidating. In practice, it is not, but to grasp the subject we need first, to understand something about both sampling and variability, then we must see how a simple form of presentation can tell the process operator either 'to carry on', or 'to be careful, look for more information since the process may require adjustment', or 'adjust the process and/or call for help'. Control charts incorporate sampling, variability and time, and present them as a picture which a process operator can easily read.

Sampling

The variability inherent in any process causes the products to differ one from another. Variability is inevitable and no one can produce two exactly *identical* products, although one may lack a technique for distinguishing between them. If the variation is too large it may become difficult, and perhaps impossible, to predict either what is happening or what has occurred by the simple examination of a sample. It is always possible, however, to interpret the results of the examination of a sample, which may have been taken from products, services, processes or from one of the parameters measured during a process.

Anyone who has purchased fruit from a market stall will have lifted two or three of the shining apples on the top in order to examine those underneath. In doing so they will have taken a 'sample' of apples in order to ensure that they are representative of the whole display. In many everyday activities we take samples to estimate the characteristics of the 'population' from which the 'sample' is drawn. This begs the question 'why not look at the whole population, and then the need for an estimate will be removed?' There are a number of answers to this question, and one or more may apply in any particular case.

If challenged to examine this page of text and count, for example, the number of times the letter 'g' occurs, the reader will have no difficulty in accomplishing the task, but the correct answer may well not be obtained. The task is dull and monotonous, as is the task of routine inspection, verification, test, analysis, data inputting, computation, etc. Carrying out monotonous tasks tends to make people respond by ceasing to think about what they are doing. It is for this reason that 100 per cent inspection turns out to be far short of 100 per cent effective. Research has shown that when engaged on routine monotonous work, like that of a routine inspector, about 15 per cent of what is supposed to be detected, is missed. For example, an inspector who is trying to sort good from bad products in which the actual rate of defectives is 10 per cent will, on average, only find 8.5 per cent, that is 85 per cent of the 10 per cent. In the manufacture and verification of critical products, the same limitation exists and is sometimes accommodated by carrying out 100 per cent inspection at least three times. This is standard practice in parts of the aerospace and nuclear industries. Even the examination of X-ray plates is usually carried out by more than one observer, and often with more than one X-ray picture of the same object. Such procedures simply recognize the fallibility of humans, particularly under conditions which are dull and routine. The increasing use of automated inspection, testing, and data capture is a response to human fallibility. Another technique is to make all process operators responsible for their own inspection. In this way inspection can become a vital part of the process and, given special attention, the 85 per cent figure can be pushed closer to 100 per cent effectiveness.

Another frequent objection to the examination of all available product or process parameters is simply one of cost, and yet another occurs when examination or test would destroy the product.

The taking of samples occurs for a variety of reasons, but in all cases, the object is to examine the sample in order to estimate the characteristics of the whole population. Clearly, the extrapolation to the whole population of the observations carried out on the sample will introduce a possible source of error. If there is already variation between the members of the population, there cannot be total agreement between the characteristics of a sample and the characteristics of the whole population. Sampling is itself a source of variation.

There is little hope of finding an acceptable correlation between the sample and the population results unless the sample is selected randomly. This means that each individual item within the whole population must have an equal chance of being selected as a member of the sample. To take only from the top of a pile of fruit, pallets, invoices, orders, or solids in a silo, or liquids in a tank, or the outside of a roll of fabric is not random sampling. A common error in process control is to take samples at very regular intervals, with the accompanying danger of failing to observe cyclical patterns. It is better, for example, to take samples at 9.05, 9.28, 10.02, 10.27 and 10.55 a.m. than exactly at 9.00, 9.30, 10.00, 10.30 and 11.00 a.m.

Statistics, variability and control charting

The remaining chapters of this book are devoted largely to the use of statistics in judging both stability and variability. Statistical techniques can be used to quantify and state the degree of conformance of raw materials, processes, products and services to the previously agreed requirements. They can be used in one-off, short-run/small-batch, high-volume and continuous manufacture. In essence, a representative and random sample is selected from the population, which may be a batch of product, or a continuously-measured process parameter. From a knowledge of the process and an analysis of the sample, it is possible to make decisions, with a known degree of confidence, about either the whole batch or the behaviour of the process parameter. Some people find it easy to make decisions. SPC enables everyone to make decisions and to know the degree of confidence with which the decision may be made, or the risk of making a wrong decision. No technique, not even automated 100 per cent measuring or inspection, can guarantee the validity of the result, there is always some room for doubt.

As will be shown in later chapters, the statistically-based control chart is a device intended to be used at the point of operation, where the process is carried out, and by the operators of that process. They are then asked to assess the current situation by taking a sample and plotting the sample result, or results, on a chart which reflects the measured capability of the process. As shown in Figure 4.12 the control chart has three zones and the action required depends on the zone in which the sample result falls. The possibilities are:

1 'Carry on' or do nothing (stable zone).
2 'Be careful and seek more information, since the process may require adjustment' (warning zone).
3 'Take action', either investigate or, where appropriate, adjust the process (action zone).

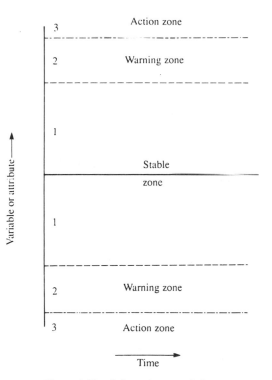

Figure 4.12 *Schematic control chart*

This is rather like a set of traffic lights which signal 'stop', 'caution' or 'go'.

The next chapters deal with the measurement of capability and the use of control charts under a variety of conditions. Armed with the knowledge of these techniques, the later chapters deal with the implementation of SPC, when the need to come back to the use of some of the simpler techniques, discussed in the last three chapters, is again illustrated.

Chapter highlights

- A problem is not solved until its cause is identified and remedial action has been shown to be effective.

- *Pareto analysis* recognizes that a small number of the causes of problems (20 per cent) may result in a large part of the total effect (80 per cent). This principle can be formalized into a procedure for listing the elements or causes, collecting the data, ranking the elements, creating the cumulative distribution, plotting the Pareto curve, analysing the curve and presenting the analysis and conclusions. This leads to a distinction between problems which are among the vital few and the trivial many, a procedure which enables effort to be directed towards the areas of highest potential return. The analysis is simple; the application requires a discipline which only allows effort to be directed to the vital few.

- For each effect there are probably a number of causes. *Cause and effect* (C/E) analysis provides a simple tool to tap the knowledge of experts by separating the generation of possible causes from their evaluation. Brainstorming, to produce C/E diagrams is a part of the work of quality circles and groups. When constructing C/E diagrams it is essential that the evaluation of potential causes of a specified effect are excluded from discussion. Steps in constructing a C/E diagram include identifying the effect, establishing the goals, constructing an initial framework, recording all suggested causes, incubating the ideas prior to a more structured presentation leading to plans for action. Management can assist in this work by providing simple aids, encouragement and interest.

- *Scatter diagrams* are another simple tool used to investigate and show the correlation between two factors or parameters. Random sources of variation will always result in some scatter in the plot of one factor against another, but this need not mask the presence of an association. Correlations may be either causal or associative. The eye is a good first judge of the presence of a relationship, which can be confirmed by the use of a simple test. (A more thorough treatment is given in Chapter 9 and Appendix G.)

- *Control charts* are used to present data in a framework which clearly shows when action is necessary, when further information is required, and when no action is indicated. They are used when judging the operation of a process by the inspection of samples. Variability within a process is inevitable and sampling will contribute to the apparent variation. The use of simple statistics allows an extrapolation from sample results to conclusions about the whole population to be made with known risks of error or misinterpretation. (A thorough treatment of charting for control is the subject of the bulk of this book.)

5   Variables and process variation

5.1   Causes of process variability

At the basis of the theory of process control is a differentiation of the causes of variation of parameters during the operation of a process. Certain variations belong to the category of chance or random variations, about which little may be done, other than to revise the process. This type of variation is the sum of the multitude of effects of a complex interaction of '*random*' or '*common*' causes, many of which are slight. When random variations alone exist, it may not be possible to trace their causes. The set of random causes which produces variation in the quality of products includes random variations in the inputs to the process, draughts, atmospheric pressure or temperature changes, passing traffic or equipment vibrations, electrical or humidity fluctuations and changes in operator physical and emotional conditions. This is analogous to the set of forces which causes a coin to land heads or tails when tossed. When only random variations are present in a process, the process is considered to be 'in statistical control' or 'in control'.

There is also variation in test equipment, and inspection and checking procedures, whether used to measure a physical dimension, an electronic or a chemical characteristic or a property of an information system. The inherent variations in checking and testing contribute to the overall process variability. In a similar way, processes whose output is not an artefact will be subject to random causes of variation: traffic problems, electricity supply, operator performance, the weather, all affect the time likely to complete an insurance estimate, the efficiency with which a claim is handled, etc.

Causes of variation which are relatively large in magnitude, and readily identified are classified as '*assignable*' or '*special*'' causes. When an assignable cause of variation is present, process variability will be excessive and the process is classified as 'out of statistical control' or beyond the expected random variations. For brevity this is usually written 'out of control'.

In Chapter 1 it was suggested that the first question which must be asked of any process is:

'*Can we do* the job correctly?'

Following our understanding of random and assignable causes of variation, this must now be divided into two questions:

1   'Is the process stable, or in control?' – in other words, are there present any assignable causes of variation, or is the process variability due to random causes ony?'
2   'What is the extent of the process variability?' or what is the natural capability of the process when only the random causes of variation are present?

This approach may be applied to both variable and attribute data, and provides a systematic methodology for process examination, control and investigation.

It is important to determine the extent of variability when a process is in control, so that control systems may be set up to detect the presence of assignable causes. A systematic study of a process provides knowledge of the variability or capability of the process and the assignable causes which are potential sources of defective outputs. This information can then be fed back quickly to the product or service design and technology functions. Knowledge of the current state of a process also enables a more balanced judgement of the demands made of equipment, both with regard to the tasks within its capability and its rational utilization.

In Chapter 3 a distinction was made between different types of data – *variables* and *attributes*. It was pointed out that they are governed by different statistical laws, which affect the method of controlling the process. In this and the next five chapters we shall deal with the techniques which are appropriate for the control of variables and in Chapter 11 the methods to be applied for attribute control will be presented. The underlying concepts are similar, it is really only the statistical theories which differ.

Accuracy and precision

In the measurement of process variation by variables, confusion often exists between the accuracy and precision of a process. An analogy may help to clarify the meaning of these terms.

Two men with rifles each shoot one bullet at a target, both having aimed at the bull's eye. By a highly improbable coincidence, each marksman hits exactly the same spot on the target, away from the bull's eye (Figure 5.1). What instructions should be given to the men in order to

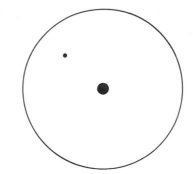

Figure 5.1   *The first coincidental shot from each of two marksmen*

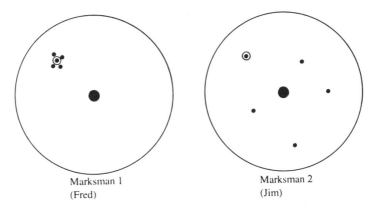

Marksman 1
(Fred)

Marksman 2
(Jim)

Figure 5.2   *The results of five shots each for Fred and Jim – their first identical shots are ringed*

improve their performance? Some may feel that each man should be told to alter his gunsights to adjust the aim: 'Down a little and to the right'. Those who have done some shooting, however, will realize that this is premature, and that a more sensible instruction is to ask the men to fire again – perhaps using four more bullets, *without altering the aim*, to establish the nature of each man's shooting process. If this were to be done, we might observe two different types of pattern (Figure 5.2). Clearly marksman 1 (Fred) is *precise* because all the bullet holes are clustered together – there is little spread, but he is not *accurate* since on average his shots have missed the bull's eye. It should be a simple job to make the adjustment for accuracy – perhaps to the gunsight – and improve his performance to that shown in Figure 5.3.

Marksman 2 (Jim) has a completely different problem. We now see that the reason for the first wayward shot was completely different from the reason for Fred's. If we had adjusted Jim's gunsights after just one

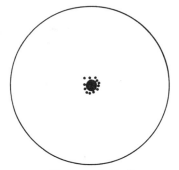

Marksman 1 (Fred)

Figure 5.3 *Shooting process, after adjustment of the gunsight*

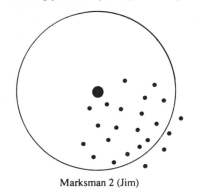

Marksman 2 (Jim)

Figure 5.4 *After incorrect adjustment of gunsight*

shot: 'Down a little and to the right', Jim's whole process would have shifted, and things would have been worse (Figure 5.4). Jim's next shot would then have been even further away from the bull's eye, as the adjustment affects only the accuracy and not the precision.

Jim's problem of spread or lack of precision is likely to be a much greater problem than Fred's lack of accuracy. The latter can usually by amended by a simple adjustment, whereas problems of wide scatter require a deeper investigation into the cause of the variation.

Several points are worth making from this simply analogy:

- There is a difference between accuracy and precision.
- The accuracy of a process relates to its ability to hit the target value.
- The precision of a process relates to the degree of spread of the values.
- The distinction between accuracy and precision may be assessed only by looking at groups of results or values, not by looking at individual ones.

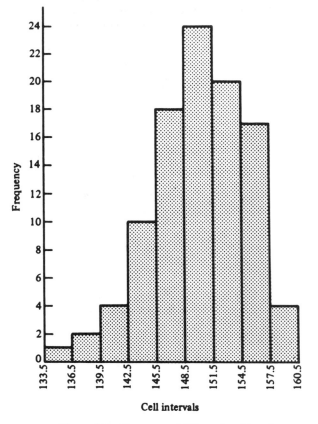

Figure 5.5   *Histogram of 100 steel rod lengths*

- Making decisions about adjustments to be made to a process on the basis of one individual result may give an undesirable outcome, owing to lack of information about process accuracy and precision.
- The adjustment to correct lack of process accuracy is likely to be 'simpler' than the larger investigation usually required to correct problems of spread or dispersion.

The shooting analogy is useful when we look at the performance of a manufacturing process producing goods with a variable property. Consider a steel rod cutting process which has as its target a length of 150 mm. The overall variability of such a process may be determined by measuring a large sample – say 100 rods – from the process (Table 5.1), and shown graphically as a histogram (Figure 5.5). Another method of illustration is a frequency polygon which is obtained by connecting the mid-points of the tops of each column (Figure 5.6).

Table 5.1   *Lengths of 100 steel rods (mm)*

144	146	154	146
151	150	134	153
145	139	143	152
154	146	152	148
157	153	155	157
157	150	145	147
149	144	137	155
141	147	149	155
158	150	149	156
145	148	152	154
151	150	154	153
155	145	152	148
152	146	152	142
144	160	150	149
150	146	148	157
147	144	148	149
155	150	153	148
157	148	149	153
153	155	149	151
155	142	150	150
146	156	148	160
152	147	158	154
143	156	151	151
151	152	157	149
154	140	157	151

When the number of rods measured is very large and the class intervals small, the polygon approximates to a curve, called the frequency curve (Figure 5.7). In many cases, the pattern would take the symmetrical form shown – the bell-shaped curve typical of the '*normal* distribution'. The greatest number of rods would have the average value, but there would be appreciable numbers either larger or smaller than the average length. Rods with dimensions further from the central value would occur progressively less frequently.

It is possible to imagine four different types of process frequency curve, which correspond to the four different performances of the two marksmen – see Figure 5.8. Hence, process 3 is accurate and relatively precise, as the average of the lengths of steel rod produced is on target, and all the lengths are reasonably close to the mean.

If only random causes of variation are present, the output from a process forms a distribution that is stable over time and is, therefore, predictable (Figure 5.9a). Conversely, if assignable causes of variation are present, the process output is not stable over time and is not predictable (Figure 5.9b). For a detailed interpretation of the data, and

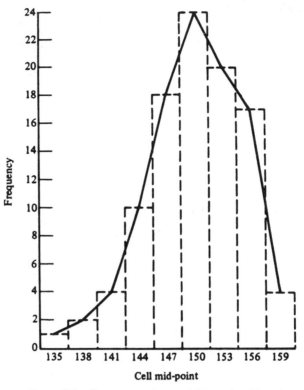

Figure 5.6   *Frequency polygon of 100 steel rod lengths*

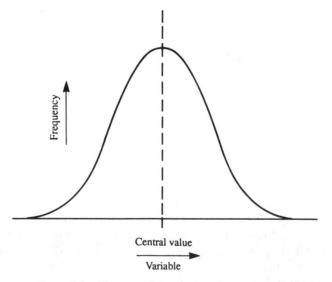

Figure 5.7   *The normal distribution of a continuous variable*

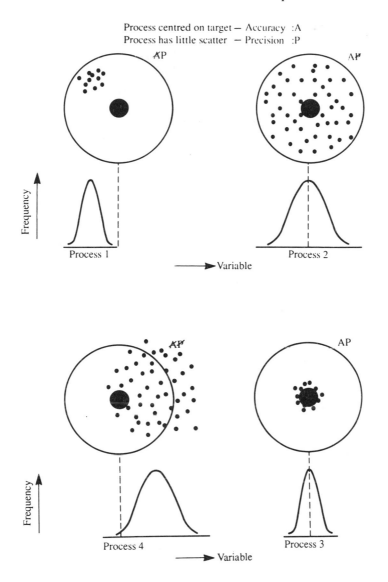

Figure 5.8 *Process variability*

before the detailed design of a process control system can take place, this intuitive analysis must be replaced by more objective and quantitative methods of summarizing the histogram or frequency curve. In particular, some measure of both the location of the central value and of the spread must be found.

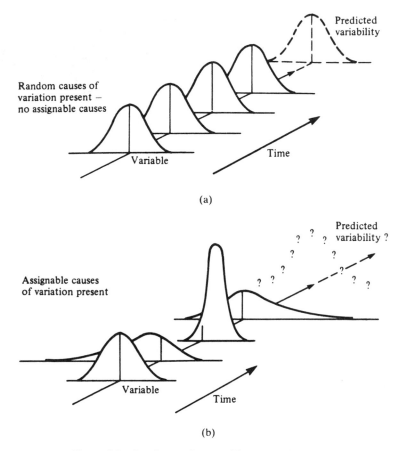

Figure 5.9   *Random and assignable causes of variation*

5.2   Measures of location of central values (accuracy)

Mean (or arithmetic average)

This is very simply the average of the observations, the sum of all the measurements divided by the number of the observations. For example, the mean of the first row of four measurements of rod lengths in Table 5.1: 144 mm, 146 mm, 154 mm and 146 mm is obtained:

$$144 \text{ mm}$$

$$146 \text{ mm}$$

$$154 \text{ mm}$$

$$\underline{146 \text{ mm}}$$

Sum   590 mm

$$\text{Sample mean} = \frac{590 \text{ mm}}{4} = 147.5\text{mm}$$

When the individual measurements are denoted by x_i, the mean of the four observations is denoted by $\overline{X}$.

Hence

$$\overline{X} = \frac{x_1 + x_2 + x_3 + \ldots x_n}{n} = \sum_{i=1}^{n} x_i/n$$

where $\sum_{i=1}^{n} x_i$ = sum of all the measurements in the sample of size n.

The 100 results in Table 5.1 are twenty-five different groups or samples of four rods and we may calculate a sample mean $\overline{X}$ for each group. The twenty-five sample means are shown in Table 5.2.

The mean of a whole population, i.e. the total output from a process, rather than a sample, is represented by the Greek letter μ. We can never know μ, the true mean, but the 'grand' or 'process mean' $\overline{\overline{X}}$, the average of all the sample means, is a good estimate of the population mean. The formula for $\overline{\overline{X}}$ is:

$$\overline{\overline{X}} = \frac{\overline{X}_1 + \overline{X}_2 + \overline{X}_3 + \ldots \overline{X}_k}{k} = \sum_{j=1}^{k} \overline{X}_j/k$$

where k = number of samples taken of size n, and $\overline{X}_j$ is the mean of the jth sample.

Hence, the value of $\overline{\overline{X}}$ for the steel rods is:

$$\overline{\overline{X}} = \frac{147.5 + 147.0 + 144.75 + 150.0 + \ldots + 150.5}{25}$$

$$= 150.1 \text{ mm}$$

Table 5.2   *100 steel rod lengths as twenty-five samples of size 4*

Sample Number	(i)	(ii)	(iii)	(iv)	Sample mean (mm)	Sample Range (mm)
		Rod lengths (mm)				
1	144	146	154	146	147.50	10
2	151	150	134	153	147.00	19
3	145	139	143	152	144.75	13
4	154	146	152	148	150.00	8
5	157	153	155	157	155.50	4
6	157	150	145	147	149.75	12
7	149	144	137	155	146.25	18
8	141	147	149	155	148.00	14
9	158	150	149	156	153.25	9
10	145	148	152	154	149.75	9
11	151	150	154	153	152.00	4
12	155	145	152	148	150.00	10
13	152	146	152	142	148.00	10
14	144	160	150	149	150.75	16
15	150	146	148	157	150.25	11
16	147	144	148	149	147.00	5
17	155	150	153	148	151.50	7
18	157	148	149	153	151.75	9
19	153	155	149	151	152.00	6
20	155	142	150 .	150	149.25	13
21	146	156	148	160	152.50	14
22	152	147	158	154	152.75	11
23	143	156	151	151	150.25	13
24	151	152	157	149	152.25	8
25	154	140	157	151	150.50	17

Median

If the measurements are arranged in order of magnitude, the median is simply the value of the middle item. This applies directly if the number in the series is odd. When the number in the series is even, as in our example of the first four rod lengths in Table 5.1, the median lies between the two middle numbers. Thus, the four measurements arranged in order of magnitude are:

144, 146, 146, 154

The median is the 'middle item'; in this case 146. In general, half the values will not be less than that of the median value, and half will be no more than it. An advantage of using the median is the simplicity with which it may be determined, particularly when the number of items is odd.

Mode

A third method of obtaining a measure of central tendency is the most commonly occurring value, or mode. In our example of four, the value 146 occurs twice and is the modal value. It is possible for the mode to be non-existent in a series of numbers or to have more than one value. When data are grouped into a frequency distribution, the mid-point of the cell with the highest frequency is the modal value. During many operations of recording data, the mode is often not easily recognized or assessed.

Relationship between mean, median and mode

Some distributions, as we have seen, are symmetrical about their central value. In these cases, the values for the mean, median and mode are identical. Other distributions have marked asymmetry and are said to be skewed. Skewed distributions are divided into two types. If the 'tail' of the distribution stretches to the right – the higher values, the distribution is said to be positively skewed; conversely in negatively skewed distributions the tail extends towards the left – the smaller values.

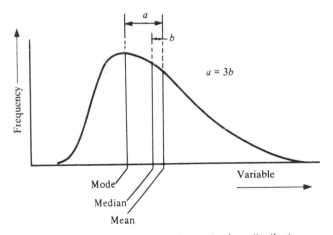

Figure 5.10 *Mode, median and mean in skew distributions*

Figure 5.10 illustrates the relationship between the mean, median and mode of moderately skew distributions. An approximate relationship is:

Mean − Mode = 3 (Mean − Median)

Thus, knowing two of the parameters enables the third to be estimated.

5.3   Measures of spread of values (precision)

Range

The range is the difference between the highest and the lowest observations and is the simplest possible measure of scatter. For example, the range of the first four rod lengths is the difference between the longest (154 mm) and the shortest (144 mm), that is 10 mm. The range is usually given the symbol R_i. The ranges of the twenty-five samples of four rods are given in Table 5.2. The mean range $\bar{R}$, the average of all the sample means, has also been calculated:

$$\bar{R} = \frac{R_1 + R_2 + R_3 + \ldots R_k}{k} = \sum_{i=1}^{k} R_i/k$$

Where $\sum_{i=1}^{k} R_i$ = sum of all the ranges of the samples

k = number of samples of size n

The range offers a measure of scatter which can be used widely, owing to its simplicity. There are, however, two major problems in its use:

1   The value of the range depends on the number of observations in the sample. The range will tend to increase as the sample size increases. This can be shown by considering again the data of steel rod lengths in Table 5.1:

   The range of the first two observations is 2 mm
   The range of the first four observations is 10 mm
   The range of the first six observations is also 10 mm
   The range of the first eight observations is 20 mm

2   Calculation of the range uses only a portion of the data obtained. The range remains the same despite changes in the values lying between the lowest and the highest values. It would seem desirable to obtain a measure of spread which is free from these two disadvantages.

Standard deviation

The standard deviation takes all the data into account and is a measure of the 'deviation' of the values from the mean. It is best illustrated by an example. Consider the deviations of the first four steel rod lengths from the mean:

Value x_i (mm)		Deviation $(x_i - \overline{X})$
144		−3.5 mm
146		−1.5 mm
154	+ 6.5 mm	
146		−1.5 mm

Mean $(\overline{X})$ = 147.5 mm Total = 0

Measurements above the mean have a positive deviation and measurements below the mean have a negative deviation. Hence, the total deviation from the mean is zero, which is obviously a useless measure of spread. If, however, each deviation is multiplied by itself, or squared, since a negative number multiplied by a negative number is positive, the squared deviations will always be positive:

Value x_i (mm)	Deviation $(x_i - \overline{X})$	$(x_i - \overline{X})^2$
144	−3.5	12.25
146	−1.5	2.25
154	+6.5 mm	42.25
146	−1.5	2.25

Sample
Mean $(\overline{X})$ = 147.5 mm   Total = 59.00

The average of the squared deviations may now be calculated and this value is known as the *variance* of the sample. In the above example, the variance or mean squared variation is:

$$\frac{\Sigma(x_i - \overline{X})^2}{n} = \frac{59.0}{4} = 14.75$$

The *standard deviation*, normally denoted by the Greek letter sigma (σ), is the square root of the variance, which then measures the spread in the same units as the variable, i.e. in the case of the steel rods, in millimetres.

$$\sigma = \sqrt{14.75} = 3.84 \text{ mm}$$

$$\text{Generally } \sigma = \sqrt{\sigma^2} = \sqrt{\frac{\Sigma(x_i - \overline{X})^2}{n}}$$

The true standard deviation σ, like μ, can never be known, but for simplicity, the conventional symbol σ will be used throughout this book to represent the process standard deviation. If a sample is being used to estimate the spread of the process, then the sample standard deviation will tend to underestimate the standard deviation of the whole process. This bias is particularly marked in small samples. To correct for the bias, the sum of the squared deviations must be divided by the sample size minus one. In the above example, the *estimated process standard deviation*:

$$s = \sqrt{\frac{59.00}{3}} \qquad = \sqrt{19.67} = 4.43 \text{ mm}$$

The general formula is:

$$s = \sqrt{\frac{\sum\limits_{i-1}^{n}(x_i - \bar{X})^2}{(n-1)}}$$

While the standard deviation gives an accurate measure of spread, it is laborious to calculate. Hand-held calculators capable of statistical calculations may be purchased for a moderate price. A much greater problem is that unlike range, standard deviation is not easily understood.

5.4   Understanding variation – the normal distribution

The meaning of the standard deviation is perhaps most easily explained in terms of the *normal* distribution. If a continuous variable is monitored, such as the lengths of rod from the cutting process, the volume of paint in tins from a filling process, or the weights of tablets from a pelletizing process, that variable will usually be distributed normally about a mean μ. The spread of values may be measured in terms of the population standard deviation, σ, which defines the width of the bell-shaped curve. Figure 5.11 shows the proportion of the output expected to be found between the values of $\mu \pm \sigma$, $\mu \pm 2\sigma$ and $\mu \pm 3\sigma$.

Suppose the process mean of the steel rod cutting process is 150 mm and that the standard deviation is 5 mm, then from a knowledge of the shape of the curve and the properties of the normal distribution, the following facts would emerge:

- 68.3 per cent of the steel rods produced will lie within ± 5 mm of the mean, i.e. $\mu \pm \sigma$.
- 95.4 per cent of the rods will lie within ± 10 mm ($\mu \pm 2\sigma$).
- 99.7 per cent of the rods will lie within ± 15 mm ($\mu \pm 3\sigma$).

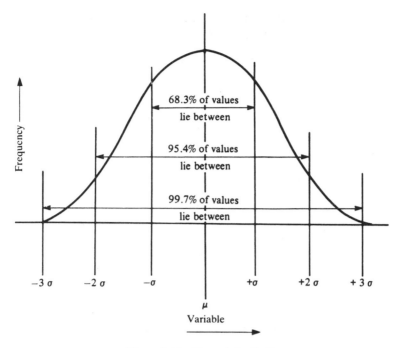

Figure 5.11 *Normal distribution*

We may be confident then that almost all the steel rods produced will have lengths between 135 mm and 165 mm. The approximate distance between the two extremes of the distribution, therefore, is 30 mm, which is equivalent to 6 standard deviations or 6σ.

The mathematical equation and further theories behind the normal distribution are given in Appendix A. This Appendix includes a table on page 340 which gives the probability that any item chosen at random from a normal distribution will fall outside a given number of standard deviations from the mean. The table shows that, at the value μ + 1.96σ, only 0.025 or 2.5 per cent of the population will exceed this length. A similar proportion will be less than μ − 1.96σ. Hence 95 per cent of the population will lie within μ ± 1.96σ.

In the case of the steel rods with mean length 150 mm and standard deviation 5 mm, 95 per cent of the rods will have lengths between:

150 ± (1.96 × 5) mm

i.e. between 140.2 mm and 159.8 mm

Similarly, 99.8 per cent of the rod lengths should be inside the range:

$$\mu \pm 3.09\sigma$$

$$\text{i.e. } 150 \pm (3.09 \times 5)$$

$$\text{or } 135.55 \text{ mm to } 165.45 \text{ mm}$$

Hence, on average only 0.1 per cent, or only one in 1000 rods will be longer than 165.45 mm, and only one in 1000 will be shorter than 134.55 mm. It is clear from this that any process variable, which is normally distributed, is characterized by two parameters; the mean, μ, and the standard deviation, σ.

Using the normal distribution

Estimating proportion defective produced
In manufacturing it is frequently necessary to estimate the proportion of product produced outside the tolerance limits, when a process is not capable of meeting the requirements. The method to be used is illustrated in the following example. 100 units were taken from a margarine packaging unit which was 'in statistical control' or stable. The packets of margarine were weighed and the mean weight, $\overline{X} = 255$ g, and the standard deviation, $\sigma = 4.73$ g. If the product specification demanded a weight of 250 ± 10 g, how much of the output of the packaging process would lie outside the tolerance zone? The situation is represented in Figure 5.12. Since the characteristics of the normal distribution are measured in units of standard deviations, we must first convert the distance between the process mean and the upper specification limit (USL) into σ units. This is done as follows:

$$Z = (\text{USL} - \overline{X}/\sigma)$$

where USL = upper specification limit

$\overline{X}$ = estimated process mean

σ = estimated process standard deviation

Z = number of standard deviations between USL and $\overline{X}$ (termed the standard normal variate)

Hence, $Z = (260 - 255)/4.73 = 1.057$. Using the table of proportions under the normal curve in Appendix A, it is possible to determine that the proportion of packages lying outside the USL was 0.145 or 14.5 per cent. There are two contributory causes for this high level of rejects.

1   The setting of the process, which should be centred at 250 g and not 255 g.
2   The spread of the process.

If the process was centred at 250 g, and with the same spread, one may

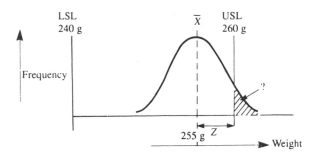

Figure 5.12 *Determination of proportion defective produced*

calculate, using the above method, the proportion of product which
would then lie outside the tolerance band. With a properly centred
process, the distance between both the specification limits and the
process mean would be 10 g. So:

$$Z = (USL - \overline{X})/\sigma = (\overline{X} - LSL)/\sigma = 10/4.73 = 2.11$$

Using this value of Z and the table in Appendix A the proportion lying
outside each specification limit would be 0.0175. Therefore, a total of 3.5
per cent of product would be outside the tolerance band, even if the
process mean was adjusted to the correct target weight.

Setting targets
It is a fairly common procedure in some industries to specify an
acceptance quality level (AQL) – this is the proportion or percentage of
product that the customer is prepared to accept outside the tolerance
band. The characteristics of the normal distribution may be used to
determine the target maximum standard deviation, when the target mean
and AQL are specified. For example, if the tolerance band for a filling
process is 5 ml and an AQL of 2.5 per cent is specified, then for a centred
process:

$$Z = (USL - \overline{X})/\sigma = (\overline{X} - LSL)/\sigma$$
$$\text{and } (USL - \overline{X}) = (\overline{X} - LSL) = 5/2 = 2.5 \text{ ml}$$

We now need to know at what value of Z we will find (2.5 per cent/2)
under the tail – this is a proportion of 0.0125, and from Appendix A this is
the proportion when $Z = 2.24$. So rewriting the above equation we have:

$$\sigma_{max} = (USL - \overline{X})/Z = 2.5/2.24 = 1.12 \text{ ml}$$

In order to meet the specified tolerance band of 5 ml and an AQL of 2.5
per cent, we need a standard deviation, measured on the products, of at
most 1.12 ml.

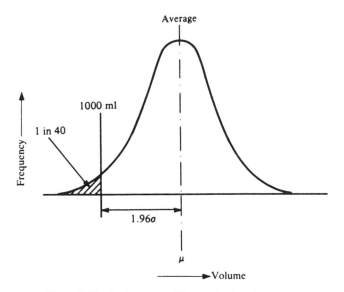

Figure 5.13   *Setting target fill quantity in paint process*

As a final example, let us consider a paint manufacturer who is filling nominal one-litre cans with paint. The quantity of paint in the cans varies according to the normal distribution with a standard deviation of 2 ml. If the stated minimum quantity in any can is 1000 ml, what quantity must be put into the cans on average in order to ensure that the risk of underfill is 1 in 40?

1 in 40 in this case is the same as an AQL of 2.5 per cent or a probability of non-conforming output of 0.025 – the specification is one- sided. The 1 in 40 line must be set at 1000 ml. From Appendix A this probability occurs at a value for Z of 1.96σ. So 1000 ml must be 1.96σ below the average quantity. The process mean must be set at:

$$(1000 + 1.96\sigma) \text{ ml}$$
$$= 1000 + (1.96 \times 2) \text{ ml}$$
$$= 1004 \text{ ml}$$

This is illustrated in Figure 5.13.

A special type of graph paper, normal probability paper, which is also described in Appendix A, can be of great assistance to the specialist in handling normally distributed data.

5.5   Sampling and averages

For successful process control, it is essential that *everyone* understands variation, and how or why it arises. The absence of such knowledge will lead to action being taken to adjust or interfere with processes which, if left alone, would be quite capable of achieving the requirements. Many processes are found to be 'out of statistical control' or unstable, when first examined using SPC techniques. It is frequently observed that this is due to an excessive number of adjustments being made to the process based on individual tests or measurements. This behaviour, commonly known as hunting, causes an overall increase in variability of results from the process, as shown in Figure 5.14. The process is initially set at the target value: $\mu = T$, but a single measurement at A results in the process being adjusted downwards to a new mean μ_A. Subsequently, another single measurement at B results in an upwards adjustment of the process to a new mean μ_B. Clearly if this 'fiddling' behaviour continues throughout the operation of the process, its variability will be greatly and unnecessarily increased, with a detrimental effect on the ability of the process to meet the specified requirements. Indeed it is not uncommon for such behaviour to lead to a call for even tighter tolerances and for the process to be 'controlled' very carefully. This in turn leads to even more

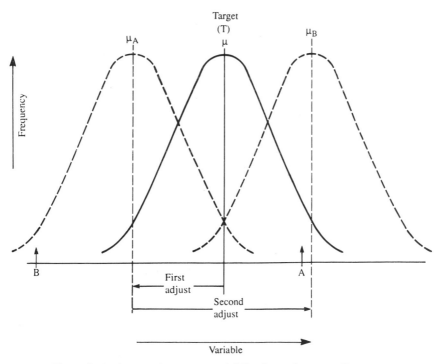

Figure 5.14 *Increase in process variability due to frequent adjustment*

frequent adjustment, further increases in variability, and more failure to meet the requirements.

To improve this situation and to understand the logic behind process control methods for variables, it is necessary to give some thought to the behaviour of sampling and of averages. If the length of a single steel rod is measured, it is clear that occasionally a length will be found which is towards one end of the tails of the process normal distribution. This occurrence, if taken on its own, may lead to the wrong conclusion that the cutting process requires adjustment. If on the other hand, a sample of four or five is taken, it is extremely unlikely that all four or five lengths will lie towards one extreme end of the distribution. If, therefore, we take the average or mean length of four or five rods, we shall have a much more reliable indicator of the state of the process. Sample means will vary with each sample taken, but the variation will not be as great as that for single pieces. Comparison of the two frequency diagrams of Figure 5.15 shows that the scatter of the sample averages is much less than the scatter of the individual rod lengths.

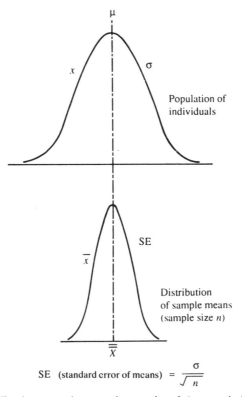

$$SE \ \text{(standard error of means)} \ = \ \frac{\sigma}{\sqrt{n}}$$

Figure 5.15 *What happens when we take samples of size n and plot the means*

In the distribution of mean lengths from samples of four steel rods, the standard deviation of the means, called the standard error of the means, and denotated by the symbol SE, is half the standard deviation of the individual rod lengths taken from the process: In general:

Standard error of means, $SE = \sigma/\sqrt{n}$

and when $n = 4$, $SE = \sigma/2$, i.e. half the spread of the parent distribution of individual items. SE has the same characteristics as any standard deviation, and normal tables may be used to evaluate probabilities related to the distribution of sample averages. We call it by a different name simply to avoid confusion with the population standard deviation.

The smaller spread of the distribution of sample averages provides the basis for a useful means of detecting changes in processes. Any change in the process mean, unless it is extremely large, will be difficult to detect from individual results alone. The reason can be seen in Figure 5.16(a),

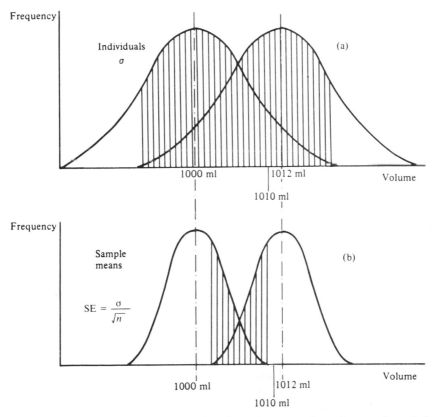

Figure 5.16 *Effect of a shift in average fill level on individuals and sample means. Spread of sample means is much less than spread of individuals*

which shows the parent distributions for two periods in a paint filling process between which the average has risen from 1000 ml to 1012 ml. The shaded portion is common to both process distributions and, if a volume estimate occurs in the shaded portion, say at 1010 ml, it could suggest either a volume above the average from the distribution centred at 1000 ml, or one slightly below the average from the distribution centred at 1012 ml. A large number of individual readings would, therefore, be necessary before such a change was confirmed.

The distribution of sample means reveals the change much more quickly, the overlap of the distributions for such a change being much smaller (Figure 5.16(b)). A sample mean of 1010 ml would almost certainly not come from the distribution centred at 1000 ml. Therefore, on a chart for sample means, plotted against time, the change in level would be revealed almost immediately. For this reason sample means rather than individual values should be used to control the centring of processes.

The central limit theorem

What happens when the measurements of the individual items are not distributed normally? A very important piece of theory in statistical process control is the *central limit theorem*. This states that if we draw samples of size n, from a population with a mean μ and a standard deviation σ, then, as n increases in size, the distribution of sample means approaches a normal distribution with a mean μ and a standard error of the means of: $\sigma/\sqrt{n}$. This tells us that, even if the individual values are not normally distributed, the distribution of that means will tend to have a normal distribution, and the larger the sample size the greater will be this tendency. It also tells us that the grand or process mean, $\overline{X}$, will be a very good estimate of the true mean of the population, μ.

Even if n is as small as 4 and the population is not normally distributed, the distribution of sample means will be very close to normal. This may be illustrated by sketching the distributions of averages of 1000 samples of size four taken from each of two boxes of strips of paper, one box containing a rectangular distribution of lengths, and the other a triangular distribution (Figure 5.17). The mathematical proof of the central limit theorem is beyond the scope of this book. The reader may perform the appropriate experimental work if (s)he requires further evidence. The main point is that, when samples of size $n = 4$ or more are taken from a process which is stable, we can assume that the distribution of the sample means, $\overline{X}$, will be very nearly normal, even if the parent population is not normally distributed. This provides a sound basis for the mean control

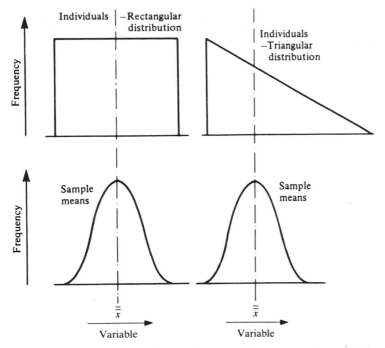

Figure 5.17   *The distribution of sample means from rectangular and triangular distributions*

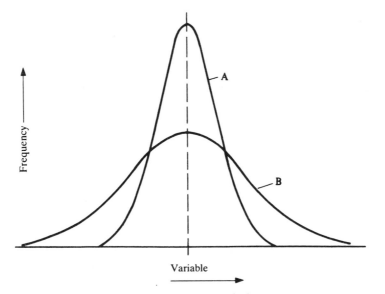

Figure 5.18   *Increase in spread of a process*

chart which, as mentioned in Chapter 4, has decision 'zones' based on predetermined *control limits*. The setting of these will be explained in Chapter 6.

The range chart is very similar to the mean chart, the range of each sample being plotted over time and compared to predetermined limits. The development of a more serious fault than incorrect or changed centring can lead to the situation illustrated in Figure 5.18, where the process collapses from form A to form B, e.g. due to a change in the variation of the material. The ranges of the samples from B will have higher values than ranges of samples taken from A. A range chart should always be plotted in conjunction with the mean chart.

Rational subgrouping of data

We have seen that a subgroup or sample is a small set of observations on a process parameter or its output, taken together in time. The two major problems with regard to choosing a subgroup relate to its size and the frequency of sampling. The smaller the subgroup, the less opportunity there is for variation within it, but the larger the sample size the narrower the distribution of the means, and the more sensitive they become to detecting change.

A rational subgroup is a sample of items or measurements selected in a way that minimizes variation among the items or results in the sample, and maximizes the opportunity for detecting variation between the samples. With a rational subgroup, assignable causes of variation are not likely to be present, but all of the effects of the random causes are likely to be shown. Generally, subgroups should be selected to keep the chance for differences within the group to a minimum, and yet maximize the chance for the subgroups to differ from one another.

The most common basis for subgrouping is the order of output or production. When control charts are to be used, great care must be taken in the selection of the subgroups, their frequency and size. It would not make sense, for example, to take as a subgroup the chronologically ordered output from an arbitarily selected period of time, especially if this overlapped two or more shifts, or a change over from one grade of product to another, or four different machines. A difference in shifts, grades, or machines may be an assignable cause that may not be detected by the variation between samples, if irrational subgrouping is used.

An important consideration in the selection of subgroups is the type of process – one-off, short run, batch or continuous flow, and the type of data available. This will be considered further in Chapter 8, but at this stage it is clear that, in any type of process control charting system, nothing is more important than the careful selection of subgroups.

Chapter highlights

- Variation results from either random (common) causes or assignable (special) causes. When random causes exist, they are likely to be elusive. When only random causes are present the process is 'in statistical control'. When assignable causes are present the process is 'out of control'.

- The capability of a process may be judged by checking that it is in control and assessing the degree of random variation. While this applies to both variables and attributes, variables are considered first.

- There must be a distinction between process accuracy (centring) and precision (spread). Accuracy is always easier to change than precision. Process distributions reflecting stability show evidence of only random causes of variability, and their output is predictable.

- There are several measures of the central value of a distribution (accuracy). These are the mean (the average value), the median (the middle value), and the mode (the most common value). We must know how they are defined and calculated.

- There are several measures of the spread of a distribution of values (precision). These are the range (the highest minus the lowest), the standard deviation and the variance. We must know how they are defined and calculated. The range is limited but it is easy to understand.

- Continuous variables often form a normal or symmetrical distribution. The normal distribution is explained by using the scale of the standard deviation around the mean. Using the normal distribution, the proportion falling in the 'tail' may be used to assess the amount of 'out of specification' product or service, or to set targets.

- A failure to understand and manage variation often leads to unjustified changes to the centring of processes, which results in an unnecessary increase in the amount of variation.

- Variation of the mean values of samples will show less scatter than individual results. The central limit theorem gives the relationship between standard deviation (σ), sample size (n), and standard error of the means (SE) as $SE = \sigma/\sqrt{n}$.

- The grouping of data results in an increased sensitivity to the detection of change, which is the basis of the mean chart.

- The range chart is used to check and control variation.

- The choice of sample size is vital to the control chart system.

6   Process control using variables

6.1   Means, ranges and charts

To control a process using variable data, it is necessary to keep a check on the current state of the accuracy (central tendency) and precision (spread) of the distribution of the data. This may be achieved with the aid of control charts.

All too often processes are adjusted on the basis of a single result or measurement ($n = 1$), a practice which can increase the apparent variability. As pointed out in Chapter 4, a control chart is like a traffic signal, the operation of which is based on evidence from process samples taken at random intervals. A green light is given when the process should be allowed to run without adjustment, only random or common causes of variation being present. The equivalent of an amber light appears when trouble is possible. The red light shows that there is practically no doubt that assignable or special causes of variation have been introduced, the process has wandered and that it must be corrected to prevent defective output.

Clearly, such a scheme can be introduced only when the process is 'in statistical control', i.e. is not changing its characteristics of average and spread. When interpreting the behaviour of a whole population from a sample, often small and typically less than ten, there is a risk of error. It is important to know the size of such a risk.

The American, Shewhart, was credited with the invention of control charts for variable and attribute data in the 1920s, at the Bell Telephone Laboratories, and the term 'Shewhart charts' is in common use. The most frequently used charts for variables are mean and range charts which are used together. There are, however, other control charts for special applications to variables data. These are dealt with in Chapter 8. Control charts for attributes data are to be found in Chapter 11.

We have seen in Chapter 5 that with variable parameters, to distinguish between and control accuracy and precision, we must group results, and that to avoid the question of whether the distribution was or has remained normal, a sample size $n = 4$ or more must be used. This sample size also provides an increased sensitivity with which we can detect changes of the mean of the process and take suitable corrective action.

Is the process in control?

The operation of control charts for sample mean and range to detect the state of control of a process proceeds as follows. Periodically, samples of a given sample size (e.g. four steel rods, five tins of paint, eight tablets, four delivery times) are taken from the process at reasonable intervals, when it is believed to be stable or in control and adjustments are not being made. The variable (length, volume, weight, time, etc.) is measured for each item of the sample and the sample mean and range recorded on a chart, the layout of which resembles Figure 6.1 The layout of the chart makes sure the following information is presented:

- Process and operator identification
- Product specification
- Statistical data
- Data collected or observed
- Sample means and ranges
- Plot of the sample mean values
- Plot of the sample range values

The grouped data on steel rod lengths from Table 5.2 have been plotted on mean and range charts, without any statistical calculations being performed, in Figure 6.2. Such a chart should be examined for any 'fliers', for which, at this stage, only the data itself and the calculations should be checked. The sample means and ranges are not constant. They vary a little about an average value. Is this amount of variation acceptable or not? Clearly we need an indication of what is acceptable, against which to judge the sample results.

Mean chart

We have seen in Chapter 5 that if the process is stable, we expect most of the individual results to lie within the range $\overline{X} \pm 3\sigma$. Moreover, if we are sampling from a stable process most of the sample means will lie within the range $\overline{\overline{X}} \pm 3SE$. Figure 6.3 shows the principle of the mean control chart. Here we have turned the bell onto its side and extrapolated the + and − 2SE and + and − 3SE lines as well as the grand or process mean line. We can use this to assess the degree of variation on the twenty-five estimates of the mean rod lengths, taken

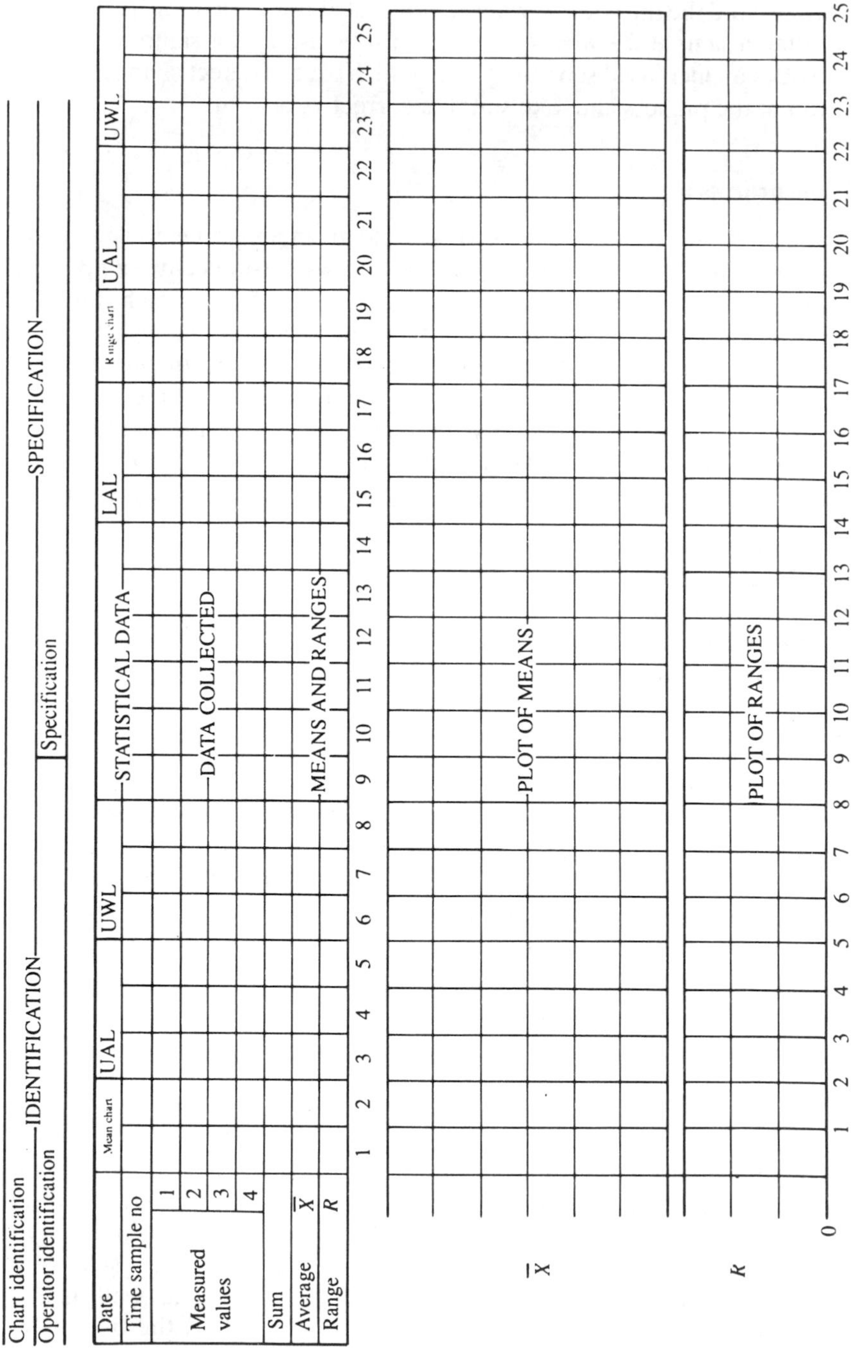

Chart identification ———————IDENTIFICATION———————   ———————SPECIFICATION———————
Operator identification   Specification

Date	Mean chart	UAL			UWL						STATISTICAL DATA					LAL				Range chart UAL				UWL	
Time sample no		1	2	3	4	5	6	7	8	9	10	11	12	13	14	15	16	17	18	19	20	21	22	23	24	25
Measured values	1									DATA COLLECTED													
	2																						
	3																						
	4																						
Sum																							
Average	$\bar{X}$									MEANS AND RANGES													
Range	R																						

$\bar{X}$ PLOT OF MEANS

R PLOT OF RANGES

0 1 2 3 4 5 6 7 8 9 10 11 12 13 14 15 16 17 18 19 20 21 22 23 24 25

Figure 6.1   *Layout of mean and range charts*

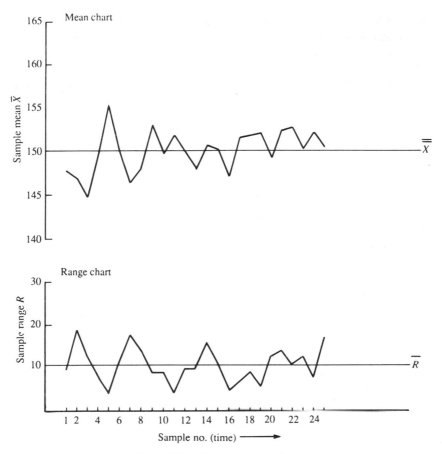

Figure 6.2 *Mean and range chart*

over a period of supposed stability. This can be used as the 'template' to decide whether the means are varying by an expected or unexpected amount, judged against the known degree of random variation. We can also plan to use this in a control sense to estimate whether the means have moved by an amount sufficient to require us to make a change to the process.

If the process is running satisfactorily, we expect from our knowledge of the normal distribution that 99.7 per cent of the means of successive samples will lie between the lines marked upper action and lower action. These are set at a distance equal to 3SE on either side of the mean. The chance of a point falling outside either of these lines is approximately 1 in 1000, unless the process has altered during the sampling period.

Figure 6.3 also shows warning limits which have been set 2SE each side of the process mean. The chance of a sample mean plotting outside either of these limits is about 1 in 40, i.e. it is expected to happen but only once in approximately forty samples.

So, as indicated in Chapter 4, there are three zones on the mean chart (Figure 6.4). If the mean value based on four results lies in zone 1 – and remember it is only an estimate of the actual mean position of the whole family – this is a very likely place to find the estimate, if the true mean of the population has not moved.

If the mean is plotted in zone 2 – there is, at most, a 1 in 40 chance that this arises from a process which is still set at the calculated process mean value, $\overline{\overline{X}}$. If the results of the mean of four lies in zone 3 there is only about a 1 in 1000 chance that this can occur without the population having moved, which means that the process must be unstable, or 'out of control'. The chance of two consecutive sample means plotting in zone 2 is approximately $1/40 \times 1/40 = 1/1600$, which is even lower than the chance of a point in zone 3. Hence, two consecutive warning signals also suggest that the process is out of control.

The presence of unusual patterns, such as runs or trends, even when all sample means and ranges are within zone 1, may also be evidence of changes in process average or spread. This may be the first warning of unfavourable conditions which should be corrected even before points occur outside the warning or action lines. Conversely, certain patterns or trends could be favourable and should be studied for possible permanent improvement of the process.

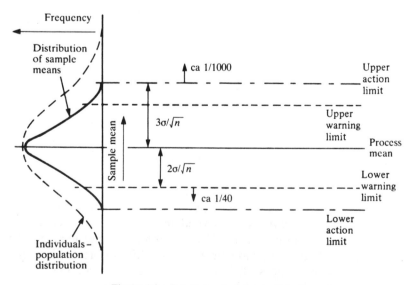

Figure 6.3   *Principle of mean control chart*

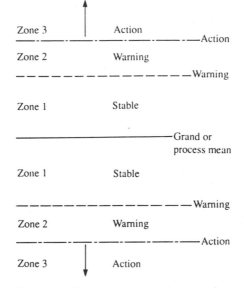

Figure 6.4 *The three zones on the mean chart*

Runs are often signs that a process shift has taken place or has begun. A run is defined as a succession of points which are above or below the average. A trend is a succession of points on the chart which are rising or falling, and may indicate gradual changes, such as tool wear. The rules concerning the detection of runs and trends are based on finding a series of *seven* points in a rising or falling trend (Figure 6.5), or in a run above or below the mean value (Figure 6.6). These are treated as out of control signals.

The reason for choosing seven is associated with the risk of finding one point above the average, but below the warning line being 0.475. The probability of finding seven points in such a series will be $(0.475)^7 = $ ca 0.005. This indicates how a run or trend of seven has approximately the same probability of occurring as a point outside an action line (zone 3). Similarly, a warning signal is given by *five* consecutive points rising or falling, or in a run above or below the mean value.

The formulae for setting the action and warning lines on mean charts are:

Upper action line @	$\overline{\overline{X}} + 3\sigma/\sqrt{n}$
Upper warning line @	$\overline{\overline{X}} + 2\sigma/\sqrt{n}$
Process or grand mean @	$\overline{\overline{X}}$
Lower warning line @	$\overline{\overline{X}} - 2\sigma/\sqrt{n}$
Lower action line @	$\overline{\overline{X}} - 3\sigma/\sqrt{n}$

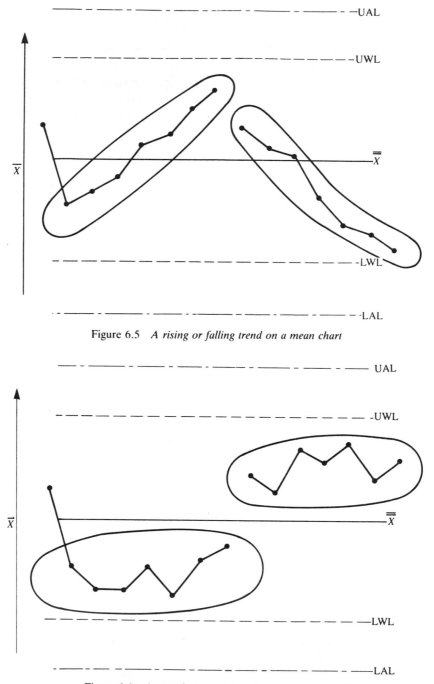

Figure 6.5   *A rising or falling trend on a mean chart*

Figure 6.6   *A run above or below the process mean value*

It is, however, possible to simplify the calculation of these control limits for the mean chart. In statistical process control for variables, the sample size is usually less than ten, and it becomes possible to use the alternative measure of spread of the process – the mean range of samples, $\bar{R}$. Use may then be made of Hartley's conversion constant (d_n or d_2) for estimating the process standard deviation. The individual range of each sample, R_i is calculated and the average range ($\bar{R}$) is obtained from the individual sample ranges:

$$\bar{R} = \sum_{i=1}^{n} \frac{R_i}{k} \qquad \text{where } k = \text{the number of samples of size } n.$$

Then,

$$\sigma = \frac{\bar{R}}{d_n} \text{ or } \frac{\bar{R}}{d_2} \qquad \text{where } d_n \text{ or } d_2 = \text{Hartley's constant}$$

Substituting $\sigma = \bar{R}/d_n$, in the formulae for the control chart limits, they become:

Action lines at $\bar{\bar{X}} \pm \left(\dfrac{3}{d_n\sqrt{n}} \right) \bar{R}$

Warning lines at $\bar{\bar{X}} \pm \left(\dfrac{2}{d_n\sqrt{n}} \right) \bar{R}$

As 3, 2, d_n and n are all constants for the same sample size, it is possible to replace the numbers and symbols within the dotted rings with just once constant.

Hence

$$\frac{3}{d_n\sqrt{n}} = A_2$$

And

$$\frac{2}{d_n\sqrt{n}} = 2/3\, A_2$$

The control limits now become:

Action lines at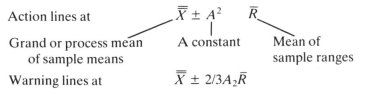

$$\bar{\bar{X}} \pm A^2 \quad \bar{R}$$

| Grand or process mean of sample means | A constant | Mean of sample ranges |

Warning lines at $\bar{\bar{X}} \pm 2/3 A_2 \bar{R}$

The constants d_n, A_2, and 2/3 A_2 for sample sizes $n = 2$ to $n = 12$ have been calculated and appear in Appendix B. For sample sizes up to $n = 12$, the range method of estimating σ is relatively efficient. For values of n greater than 12, the range loses effectiveness rapidly as it ignores all the information in the sample between the highest and lowest values. For the small sample sizes ($n = 4$ or 5) often employed on variables control charts, it is entirely satisfactory.

Using the data on lengths of steel rods in Table 5.2, we may now calculate the action and warning limits for the mean chart for that process:

Process mean, $\bar{\bar{X}} = \dfrac{147.5 + 147.0 + 144.75 + \dots + 150.5}{25}$

$= 150.1$ mm

Mean range $\bar{R} = \dfrac{10 + 19 + 13 + 8 + \dots + 17}{25}$

$= 10.8$ mm

From Appendix B, for a sample size, $n = 4$; d_n or $d_2 = 2.059$

Therefore, $\sigma = \dfrac{\bar{R}}{d_n} = \dfrac{10.8}{2.059} = 5.25$ mm

and Upper action line $= 150.1 + (3 \times 5.25/\sqrt{4})$
$= 157.98$ mm
Upper warning line $= 150.1 + (2 \times 5.25/\sqrt{4})$
$= 155.35$ mm
Lower warning line $= 150.1 - (2 \times 5.25/\sqrt{4})$
$= 144.85$ mm
Lower action line $= 150.1 - (3 \times 5.25/\sqrt{4})$
$= 142.23$ mm

Alternatively, the simplified formulae may be used if A_2 and $2/3A_2$ are known:

$A_2 = \dfrac{3}{d_n\sqrt{4}}$

$= \dfrac{3}{2.059\sqrt{4}} \qquad = 0.73$

and $2/3A_2 = \dfrac{2}{d_n\sqrt{4}}$

$= \dfrac{2}{2.059\sqrt{4}} \qquad = 0.49$

Alternatively, the values of 0.73 and 0.49 may be obtained directly from Appendix B.

Now,

Action lines at		$\overline{\overline{X}} \pm A_2 \overline{R}$
therefore, upper action line	=	150.1 + (0.73 + 10.8) mm
	=	157.98 mm
and lower action line	=	150.1 − (0.73 × 10.8) mm
	=	142.22 mm

Similarly,

Warning lines at		$\overline{\overline{X}} \pm 2/3A_2 \overline{R}$
therefore, upper warning line	=	150.1 + (.049 × 10.8) mm
	=	155.40 mm
and lower warning line	=	150.1 − (0.476 × 10.8) mm
	=	144.81 mm

Range chart

The control limits on the range chart are asymmetrical about the mean range since the distribution of sample ranges is a positively skewed distribution (Figure 6.7). The table in Appendix C provides four constants $D'_{.001}$, $D'_{.025}$, $D'_{.975}$, $D'_{.999}$ which may be used to calculate the control limits for a range chart. Thus:

Upper action line at	$D'_{.001}\overline{R}$
Upper warning line at	$D'_{.025}\overline{R}$
Lower warning line at	$D'_{.975}\overline{R}$
Lower action line at	$D'_{.999}\overline{R}$

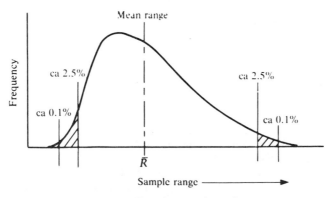

Figure 6.7 *Distribution of sample ranges*

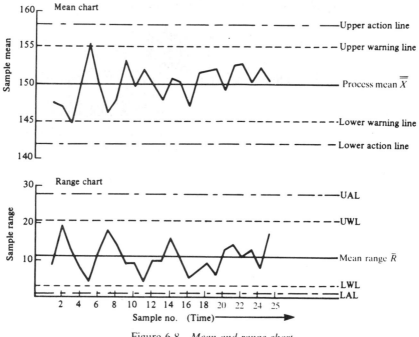

Figure 6.8   *Mean and range chart*

For the steel rods, the sample size is four and the constants are thus:

$$D'_{.001} = 2.57 \qquad D'_{.025} = 1.93$$
$$D'_{.999} = 0.10 \qquad D'_{.975} = 0.29$$

As the mean range, $\overline{R}$ is 10.8 mm the control limits for range are:

Action lines at	$2.57 \times 10.8 =$	27.8 mm
and	$0.10 \times 10.8 =$	1.1 mm
Warning lines at	$1.93 \times 10.8 =$	20.8 mm
and	$0.29 \times 10.8 =$	3.1 mm

The action and warning limits for the mean and range charts for the steel rod cutting process have been added to the data plots in Figure 6.8.

Although the statistical concepts behind control charts for mean and range may seem complex to the non-mathematically inclined, the steps in setting up the charts are remarkably simple:

Steps in assessing process stability
1   Select a series of random samples of size n (greater than 4 but less than 12) to give a total number of individual results between 50 and 100.
2   Measure the variable x for each individual item.

3 Calculate $\bar{X}$, the sample mean and R, the sample range for each sample.
4 Calculate the process mean $\bar{\bar{X}}$ – the average value of $\bar{X}$
      and the mean range $\bar{R}$ – the average value of R.
5 Plot all the values of $\bar{X}$ and R and examine the charts for any possible miscalculations.
6 Look up: d_n, A_2, $2/3A_2$, $D'_{.999}$, $D'_{.975}$, $D'_{.025}$, $D'_{.001}$ (see Appendix B and C).
7 Calculate the values for the action and warning lines for the mean and range charts. A typical $\bar{X}$ and R chart calculation form is shown in Table 6.1.

Table 6.1 $\bar{X}$ *and R chart calculation form*

Process: _____   Date:

Variable measured:
Number of subgroups (K):
Dates of data collection:
Number of samples/measurements per subgroup: (n)

---

1 Calculate grand or process mean $\bar{\bar{X}}$:

$$\bar{\bar{X}} = \frac{\Sigma \bar{X}}{K} = \text{———} =$$

---

2 Calculate mean range:

$$\bar{R} = \frac{\Sigma R}{K} = \text{———} =$$

---

3 Calculate limits for $\bar{X}$ chart:

UAL/LAL	$= \bar{\bar{X}} \pm (A_2 \times \bar{R})$
UAL/LAL	$= \pm ($ $)$
UAL/LAL	$= \pm$
UAL	$=$ LAL $=$

UWL/LWL	$= \bar{\bar{X}} \pm (2/3A_2 \times \bar{R})$
UWL/LWL	$= \pm ($ $)$
UWL/LWL	$= \pm$
UWL	$=$ LWL $=$

---

4 Calculate limits for R chart:

UAL	$= D'_{.101} \times \bar{R}$	LAL	$= D'_{.999} \times \bar{R}$
UAL	$= \times$	LAL	$= \times$
UAL	$=$	LAL	$=$
UWL	$= D'_{.025} \times \bar{R}$	LWL	$= D'_{.975} \times \bar{R}$
UWL	$= \times$	LWL	$= \times$
UWL	$=$	LWL	$=$

---

8 Draw the limits on the mean and range charts.
9 Examine charts again – is the process in statistical control?

6.2   Are we in control?

At the beginning of the section on mean charts it was stated that samples should be taken to set up control charts, when it is believed that the process is in statistical control. Before the control charts are put into use or the process capability is assessed, it is important to confirm that, when the samples were taken, the process was indeed 'in statistical control' i.e. the distribution of individual items was reasonable stable. Most industrial processes are not in control when first examined using control chart methods and the assignable causes of the out of control periods must be found and corrected.

Assessing the state of control

A process is in statistical control when all the variations have been shown to arise from random or common causes and none of the variations are attributable to assignable, non-random or special causes. The randomness of the variations can best be illustrated by collecting at least fifty observations of data and grouping these into samples or sets of at least four observations; presenting the results in the form of both mean and range control charts – the limits of which are worked out from the data. If the process from which the data was collected is 'in statistical control' there will be:

- No mean or range values which lie outside the action limits (zone 3, Figure 6.4).
- No more than about 1 in 40 values between the warning and action limits (zone 2).
- No incidence of two consecutive mean or range values which lie outside the same warning limit on either the mean or the range chart (zone 2).
- No run or trend of five or more which also infringes a warning or action limit (zone 2 or 3).
- No runs of more than six sample means which lie either above or below the grand mean (zone 1).
- No trends of more than six values of the sample means which are either rising or falling (zone 1).

If a process is 'out of control', the assignable causes will be located in time and must now be identified and eliminated. The process can then be re-examined to see if it is in statistical control. If the process is shown to be in statistical control the next task is to compare the limits of this control with the tolerance sought.

The means and ranges of the twenty-five samples of four lengths of steel rods, which were plotted in Figure 6.2 may be compared with the calculated control limited in this way, using Figure 6.8.

We start by examining the range chart in all cases, because it is the range which determines the position of the range chart limits *and* the 'separation' of the limits on the mean chart. The range is in control – all the points lie inside the warning limits, which means that the spread of the distribution remained constant – the process is in control with respect to range and spread.

For the mean chart there are two points which fall in the warning zone – they are not consecutive and of the total number of points plotted on the charts we are expecting 1 in 40 to be in each warning zone when the process is stable. There are not forty results available and we have to make a decision. It is reasonable to assume that the two plots in the warning zone have arisen from the random variation of the process and do not indicate an out of control situation.

There are no runs or trends of seven or more points on the charts and, from Figure 6.8, the process is judged to be in statistical control, and the mean and range charts may now be used to control the process.

During this check on process stability, should any sample points plot outside the action lines, or several points appear between the warning and action lines, or any of the trend and run rules be contravened, then the control charts must not be used, and the assignable causes of variation must be investigated. When these non-random or special causes of variation have been identified and eliminated, either another set of samples from the process is taken and the control chart limits recalculated, or approximate control chart limits are recalculated by simply excluding the out of control results for which assignable causes have been found and corrected. The exclusion of samples representing unstable conditions is not just throwing away bad data. By excluding the points affected by known causes, we have a better estimate of variation due to random causes only; the probability of the decision to discard the data being wrong is 1 in 1000.

A clear distinction must be made between the tolerance limits set down in the product specification and the limits on the control charts. The former are based on the functional requirements of the products, the latter are based on the stability and actual capability of the process. The process may be unable to meet the specification requirements but still be in a state of statistical control (Figure 6.9). A comparison of process capability and tolerance can only take place, with confidence, when it has been established that the process is in control statistically.

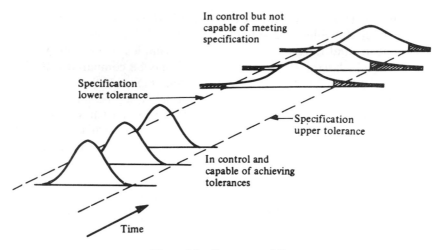

In control but not
capable of meeting
specification

Specification
lower tolerance

Specification
upper tolerance

In control and
capable of achieving
tolerances

Time

Figure 6.9   *Process capability*

Capability of the process

So with both the mean and the range charts in statistical control, we have
shown that the process was stable for the period during which samples
were taken. We now know that the variations were due to random causes
only, but how much scatter is present, and is the process capable of
meeting the requirements? We know that, during this period of stable
running, the results were scattered around a process mean of
$\overline{\overline{X}} = 150.1$ mm, and that, during this period, the mean range
$\overline{R} = 10.8$ mm. From this we have calculated that the standard deviation
was 5.25 mm, and it is possible to say that about 99.7 per cent of the
output from the process will lie within three standard deviations on either
side of the mean, i.e. between $150.1 \pm 3 \times 5.25$ mm or 134.35 to
165.85 mm.

   If a specification for the rod cutting process had been set, it would be
possible at this stage to compare the capability of the process with the
requirements. It is important to recognize that the information about
capability and the requirements come from different sources – they are
totally independent. The specification does not determine the capability
of the process and the process capability does not determine the
requirements, but they do need to be known, compared and found to be
compatible. The quantitative assessment of capability with respect to the
specified requirements is the subject of Chapter 7.

6.3   Do we continue to be in control?

When the process has been shown to be in control, the mean and range charts may be used to make decisions about the state of the process during its operation. Just as for testing whether a process was in control, we can use the three zones on the charts for controlling or managing the process:

Zone 1   If the points plot in this zone it indicates that the process has remained stable and actions/adjustments are unnecessary, indeed they may increase the amount of variability.

Zone 3   Any points plotted on these zones indicate that the process should be investigated and that, if action is taken, the latest estimate of the mean and its difference from the original process mean or target value should be used to assess the size of the correction.

Zone 2   A point plotted in this zone suggests there may have been an assignable change and that another sample must be taken in order to check.

Such a second sample can only lie in one of the three zones as shown in Figure 6.10.

- If it lies in zone 1 then the previous result was a statistical event which has approximately a 1 in 40 chance of occurring every time we estimate the position of the mean.
- If it lies in zone 3 there is only approximately a 1 in 1000 chance that it can get there without the process mean having moved, so the latest estimate of the value of the mean may be used to correct it.
- If it again lies in zone 2 then there is approximately a $1/40 \times 1/40 = 1/1600$ chance that this is a random event arising from

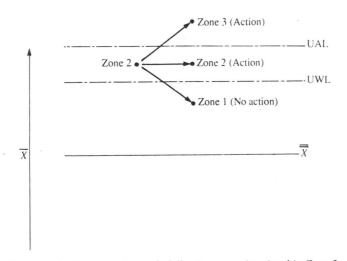

Figure 6.10   *The second sample following a warning signal in Zone 2*

an unchanged mean, so we can again use the latest estimate of the position of the mean to decide on the corrective action to be taken.

This is a simple list of instructions to give to an operator of any process. The first three options corresponding to points in zones 1, 2 and 3 respectively are: 'do nothing', 'take another sample', or 'investigate or adjust the process'. If a second sample is taken following a point in zone 2, it is done in the certain knowledge that this time there will be one of two conclusions: 'do nothing', or 'investigate/adjust'. In addition, when the instruction is to adjust the process, it is accompanied by an estimate of by how much, and this is based on four or five observations not one. The rules given on page 118 for detecting runs and trends should also be used in controlling the process.

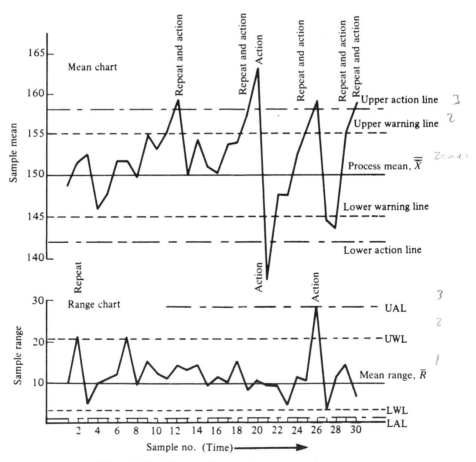

Figure 6.11   *Mean and range chart in process control*

Figure 6.11 provides an example of this scheme in operation. It shows mean and range charts for the next thirty samples taken from the steel rod cutting process. The process is well under control, i.e. within the action lines, until sample 11, when the mean almost reaches the upper warning line. A cautious person may be tempted to take a repeat sample here although, strictly speaking, this is not called for if the rules are applied rigidly. This decision depends on the time and cost of sampling, amongst other factors. Sample 12 shows that the cautious approach was justified for its mean has plotted above the upper action line and corrective action must be taken. This action brings the process back into control again until sample 18 which is the fifth point in a run above the mean – another sample should be taken immediately, rather than wait for the next sampling period. The mean of sample 19 is in the warning zone and these two consecutive 'warning' signals indicate that corrective action should be taken. However, sample 20 gives a mean well above the action line, indicating that the corrective action caused the process to move in the wrong direction. The action following sample 20 results in overcorrection and sample mean 21 is below the lower action line.

The process continues to drift upwards out of control between samples 21 to 26 and from 28 to 30. The process equipment was investigated as a result of this – a worn adjustment screw was slowly and continually vibrating open, allowing an increasing speed of rod through the cutting machine. This situation would not have been identified as quickly in the absence of the process control charts. This simple example illustrates the power of control charts in both process control and in early warning of equipment trouble.

It will be noted that 'action' and 'repeat' samples have been marked on the control charts. In addition, any alterations in materials, the process, operators, or any other technical changes should be recorded on the charts when they take place. This practice is extremely useful in helping to track down causes of shifts in mean or variability. The chart should not, however, become overcluttered; simple marks with cross references to plant or operators' notebooks are all that is required. In some organizations it is common practice to break the pattern on the $\bar{X}$ and R charts, by not joining points which have been plotted either side of action being taken on the process.

It is vital that any process operator should be told how to act for warning zone signals (repeat the sample), for actions signals on the mean (stop, investigate, call for help, adjust, etc.) and action signals on the range (stop, investigate or call for help – there is no possibility of 'adjusting' the process spread – this is where management must become involved in the investigative work).

6.4   International control charts for variables

Instead of calculating upper and lower warning lines at two standard errors, the American automotive and other industries use simplified control charts and set an 'upper control limit' (UCL) and a 'lower control limit' (LCL) at three standard errors either side of the process mean. To allow for the use of only one set of control limits, the UCL and LCL on the corresponding range charts are set in between the 'action' and 'warning' lines. The general formulae are:

$$\text{Upper control limit} \quad \pm \quad D_4\overline{R}$$
$$\text{Lower control limit} \quad \pm \quad D_2\overline{R}$$

Where n is 6 or less, the lower control limit will turn out to be less than 0 but, because the range can not be less than 0, the lower limit is not used. The constants d_2 and D_4 may be found directly in Appendix C for sample sizes of two to twelve. A sample size of five is commonly used in the automotive industry.

Such control charts are used in a very similar fashion to those designed with action and warning lines. Hence, the presence of any points beyond either upper or lower control limit is evidence of an out of control situation, and provides a signal for an immediate investigation of the assignable cause. Because there are no warning limits on these charts, some additional guidance is usually offered to assist the process control operation. This guidance is more complex and may be summarized as:

1   Approximately two-thirds of the data points should lie within the middle third region of each chart – for mean and for range. If substantially more or less than two-thirds of the points lie close to $\overline{\overline{X}}$ or $\overline{R}$, then the process should be checked for possible changes.
2   If random causes of variation only are present, the control charts should not display any evidence of runs or trends in the data. The following are taken to be signs that a process shift or trend has been initiated:

   • Seven points in a row on one side of the average.
   • Seven lines between successive points which are continually increasing or decreasing.

3   There should be no occurrences of two mean points out of three consecutive points on the *same* side of the centre line in the zone corresponding to two standard errors (SE) from the process mean, $\overline{X}$.
4   There should be no occurrences of four mean points out of five consecutive points on the *same* side of the centreline in the zone between one and two standard errors away from the process mean $\overline{\overline{X}}$.

As already pointed out, it is useful practice for those using the control chart system with warning lines to also apply the simple checks described above. The control charts with warning lines offer a less stop or go situation than the UCL/LCL system, so there is less need for these additional checks. The more complex the control chart system rules, the less likely it is that they will be adhered to. The temptation to adjust the process when a point plots near to a UCL or LCL is real. If it falls in a warning zone, there is a clear signal to check, not to panic and, above all, not to adjust. It is the authors' experience that the use of warning limits and zones give prcess operators and managers clearer rules and quicker understanding of variation and its management.

The precise points on the normal distribution at which 1 in 40 and 1 in 1000 probabilities occur is at 1.96 and 3.09 standard deviations from the process mean respectively. Using these refinements, instead of the simpler 2 and 3 standard deviations, makes no significant difference to the control system. The British Standards on control charts currently in use quote the 1.96 and 3.09 values.

There are clearly some differences between the various types of control charts for mean and range. Far more important than any operating discrepancies is the need to understand and adhere to whichever system has been chosen.

6.5   Choice of sample size and frequency of sampling

In the example used to illustrate the design and use of control charts, twenty-five samples of four steel rods were measured to set up the charts. Subsequently, further samples of size 4 were taken at regular intervals to control the process. This is a common sample size, but there may be justification for taking other sample sizes. Some guidelines may be helpful:

1   The sample size must be at least two to give an estimate of residual variability, but a minimum of four is preferred unless the distribution of the population is known to remain normal.
2   As the sample size increases, the mean control chart limits become closer to the process mean. This makes the control chart more sensitive to the detection of small variations in the process average.
3   As the sample size increases, the inspection costs per sample may increase. One should question whether the greater sensitivity justifies any increase in cost.
4   The sample size should not exceed twelve if the range is to be used to measure process variability. With larger samples the resulting mean

range ($\overline{R}$) does not give a good estimate of the standard deviation and sample standard deviation charts should be used.

5   When each item has a high monetary value and destructive testing is being used, a small sample size – say four or five – is desirable and satisfactory for control purposes.

6   A sample size of $n = 5$ is often used because of the ease of calculation of the sample mean (multiply sum of values by two and divide the result by ten or move the decimal point one digit to left). However, with the advent of inexpensive handheld calculators this is no longer necessary.

7   The technology of the manufacturing process may indicate a suitable sample size. For example, in the control of a paint filling process the filling head may be designed to discharge paint through six nozzles into six cans simultaneously. In this case, it is obviously sensible to use a sample size of six – one can from each identified filling nozzle, so that a check on the whole process and the individual nozzles may be maintained.

There are no general rules for the frequency of taking samples. It is very much a function of the product being made and the process used. In general, it is recommended that samples are taken quite often at the beginning of a process capability assessment and process control. When it has been confirmed that the process is in control, the frequency of sampling may be reduced. It is important to ensure that the frequency of sampling is determined in such a way that no bias exists and that, if autocorrelation is a problem, it does not give false indications on the control charts. The problem of how to detect and handle autocorrelated data is dealt with in Appendix I.

In certain types of operation, measurements are made on samples taken at different stages of the process, when means of such samples are expected to vary or follow a predetermined pattern. Examples of this are to be found in chemical manufacturing, where process parameters change as the raw materials are converted into products or intermediates. It may be desirable to plot the sample means against time to observe the process profile or progress of the reaction, and draw warning and action control limits on these graphs, in the usual way. Alternatively, a chart of means of differences from a target value, at a particular point in time, may be plotted with a range chart.

Control chart performance – average run length to detection of change

In the design and use of control charts for mean and range it is usually assumed that, when the process is 'in control', the points plotted on the charts should appear to have been sampled at random from a normal distribution. The Central Limit Theorem tells us that the distribution of the sample means is likely to be 'quite normal', even if the distribution of the population is not. This provides a good reason for basing decisions on means or groups of data rather than individual values. There is, however, another reason: the performance or power of the control chart in detecting changes.

When process parameters are normally distributed, it is possible to derive an 'operating characteristic' or OC curve for the $\overline{X}$ and R charts, which shows the probability of detecting, or not detecting, a specified change in the process, by a single sample. The detailed calculations for control chart OC curves are rather complex and will not be presented here. (See Scheffe, H., 'Operating Characteristics of Average and Range Charts', *Industrial Quality Control*, May 1949, pp 13–18.) It is possible, however, to examine generalized curves and to use them to indicate the so-called 'average run length' or ARL to detection of change. This is a statement of how many sample points, on average, will be plotted on the charts before a specific change is detected. The calculations, OC and ARL curves are given in Appendix H.

6.6   Summary of SPC for variables using $\overline{X}$ and R charts

If data is recorded on a regular basis, SPC for variables proceeds in three main stages:

1   An examination of the 'state of control' of the process. *(Are we in control?)*   A series of measurements are carried out and the results plotted on $\overline{X}$ and R control charts to discover whether the process is changing due to assignable causes. Once any such causes have been found and removed, the process is said to be 'in statistical control' and the variations then result only from the random or common causes.
2   A 'process capability' study. *(Are we capable?)*   It is never possible to remove all random or common causes – some variations will remain. A process capability study shows whether the remaining variations are acceptable and whether the process will generate products or services which match the specified requirements.
3   Process control using charts. *(Do we continue to be in control?)*   The $\overline{X}$ and R charts carry 'control limits' which form traffic light signals or decision rules and give operators information about the process and its state of control.

Chapter highlights

● Control charts are used to monitor, through means ($\overline{X}$) and ranges (R), processes which are in control. Individuals charts do not allow the distinction between accuracy and precision to be made.

- There is a recommended method of collecting data for a process capability study and prescribed layouts for $\overline{X}$ and R control charts, which include warning and action lines (limits). The control limits of the mean and range charts are based on statistical calculations.

- Mean chart limits are derived using the process mean $\overline{\overline{X}}$, the mean range $\overline{R}$, and either A_2 constants or by calculating the standard error (SE) from $\overline{R}$. The range chart limits are derived from $\overline{R}$ and D' constants.

- The interpretation of the plots are based on rules for action, warning, and trend signals. Mean and range charts are used together to control the process.

- A set of detailed rules is required to assess the state of control of a process and to establish a state of statistical control. The capability of the process can be measured in terms of the standard deviation (σ), and its spread compared with the specified tolerances. A process capability study may be spelt out in a series of simple steps.

- Mean and range charts may be used to monitor the performance of a process. There are three zones on the charts which are associated with rules for determining action, if any, to be taken. The operation of a process is thus assisted by the use of mean and range charts. The importance of the process operator being in charge, and knowing what to do and why, must be stressed.

- There are various forms of the charts originally proposed by Shewhart. These include 'American charts' without warning limits, which require slightly more complex guidance in use.

- The sample size must be greater than one to distinguish between accuracy and precision, and greater than four, unless the process parameter is known to be normally distributed. Larger sample sizes give greater sensitivity to change, but may cost more.

- No general rules for the frequency of sample taking exist, but frequency is clearly related to the inherent period of process stability. The concept of average run length (ARL) to detection is useful in the assessment of 'performance' of control charts.

- SPC for variables is in three stages:

  1 Examination of 'state of control' of the process using $\overline{X}$ and R charts.
  2 A process capability study, comparing spread with specifications.
  3 Process control using the charts.

7 Process capability for variables and its measurement

7.1 Will it meet the requirements?

In managing variables the usual aim is not to achieve for every steel rod exactly the same length, every piston the same diameter, every tablet the same weight, a perfectly constant flow or temperature, but to reduce the variation of products and process parameters around a target value, and to operate parameters within specified limits or tolerances. No adjustment of a process is called for as long as there has been no identified change in its accuracy or precision. This means that, in controlling a process, it is necessary first to establish that it is in statistical control, and then to compare its centring, degree of random scatter and the sensitivity of the detection of change with the specified target and tolerance around that value.

As already discussed, if a process is not in statistical control, the time at which the assignable causes occurred may be located by use of control charts. Some assignable causes can be eliminated and others will arise from the inevitable drift of process parameters. The latter changes are a part of the normal process behaviour and include the effects of toolwear, progressive fouling of heat exchangers, ageing, wearing out, and the progressive exhaustion of, for example, a chemical during reaction. Only when all the assignable causes have been brought under control or eliminated can process capability be assessed. The ability to meet the specification may then be considered by comparing the width of the tolerance zone with the width of the distribution which represents the random variations plus the width necessary to allow for either the means or the ranges to change and for such a change to be detected and corrected.

For example, if the specification for the lengths of the steel rods discussed in Chapters 6 and 7 had been set at 150 $\pm$ 10 mm and on three different machines the processes were found to be in statistical control, centred correctly but with different standard deviations of 2 mm, 3 mm

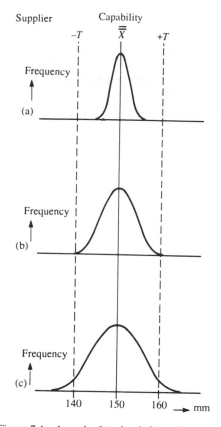

Figure 7.1   *Length of steel rods from three sources*

and 4 mm, we could represent the results as shown in Figure 7.1(a), (b) and (c). Figure 7.1(a) shows that when the standard deviation is 2 mm, the bell width of six standard deviations is 12 mm, and the process sits inside the tolerance band of 20 mm with some degree of 'comfort'. There is room for the process to wander a little and provided that any wandering of mean or change in the range is detected early, the edge of the process bell will not infringe the tolerance limits. With a standard deviation of 3 mm (Figure 7.1(b)) the room for movement before the tolerance limits are infringed is seriously reduced, and with a standard deviation of 4 mm (Figure 7.1(c)) out of tolerance product is inevitable. Measures of process capability and what they mean is the subject of this chapter.

Referring back to Figure 7.1(c), the inevitability of reject material arises because the width of the tolerance zone and the width of the process distribution are incompatible. The process is in statistical control but it is not capable of meeting the requirements. The process could be made capable by either reducing the degree of random variation and the standard deviation, or by increasing the width of the tolerance zone.

Increasing the tolerance band

The argument against increasing the tolerance band is that it is not compatible with the 'never-ending-improvement' concept. The argument for considering an increase in the tolerance band is that the first step to progress can sometimes be the simple act of starting to admit the truth to oneself as well as to one's customers.

If a process capability study reveals that an established process is in statistical control but not capable, there is a fair chance that in making this discovery one will question how it has been possible to successfully deliver product in the past. A part of the answer is revealed by our knowledge of the normal distribution. If the capability is as shown in Figure 7.1(c), the proportion of product outside the tolerance zone will not be very large. In this case the proportion in the tail of the normal distribution is 0.062 or 0.62 per cent. So there will be 0.62 per cent of the total population above the upper tolerance and similarly 0.62 per cent below the lower tolerance. A total of only 1.24 per cent of the output is outside the tolerance zone. As single random samples are taken from such a stable or 'in statistical control' process, the probability of finding a result outside either of the tolerance limits is 0.62 per cent, or 1 in 161. When such an infrequent event occurs it is common practice in a wide range of industries to check it and 'take another sample'. Of course the probability of the second samply lying outside either of the tolerance limits is again 1 in 161, so the probability of the repeat result lying within specification is 159 in 161. The second sample and analysis are almost certain to give an in-specification result and 'justify' passing the product. The probability of both the first and the second samples giving an out-of-tolerance result is about 1 in (161 × 161) or 1 in 2592!

This type of error is sufficiently common to have a name – it is called a Type 1 error. The first 'doubtful' result has exactly the same value as the second 'OK' result, and it is an error to assume otherwise. Does anyone ever question an 'OK' result and take another sample to be sure? In the case quoted above there will be very few instances where the out-of-tolerance product or parameter will actually be detected and, when it is, the 'take a repeat sample' routine will almost always allow the first result to be ignored. This simple consequnce of the normal distribution applies to all distributions which tail off at the edges, although the actual areas under the tail of non-normal distributions will be slightly different.

Where the Type 1 error of accepting out-of-specification material has been the established practice in assessing the characteristics of a product or process, it may make good sense to recognize that the written specification has not been respected, but that there has been a hidden specification which reflects the *actual* capability of the process, and which it would be sensible to put into use. We shall see later in Section 7.4 that a knowledge of the standard deviation, and how it was estimated, can be used to calculate the specification limits which could be achieved with very high levels of confidence. Such a mutual understanding of process variability and capability can be a step towards a more constructive and honest relationship between a supplier and a customer. Without such a relationship, there is no easy path to increasing the capability of meeting the requirements. Of course, modifying the specification to reflect reality is only the first in a series of steps. It will need to be followed by the more difficult steps which will reduce the total variations and increase the capability of the process and the degree of customer satisfaction.

As already mentioned, existing processes are often found to be in statistical control but not capable. Why should this be so? It usually arises as the result of customers pressing the specification limits closer together and continuing to do so until they meet resistance from the supplier chain, or one producer in particular. Such resistance becomes real only when non-conformance can be detected with reasonable ease and certainty. If neither the producer nor the customer understands the nature of variations, this will not occur until the specification limits enter well inside the process distribution. This results in a specification which is grossly incompatible with the process capability.

Assessing the variation

Referring back to Figure 7.1, it can be seen that there is a need to accommodate within the tolerance band both the truly random variations which exist within a process, and the ability of the control techniques to detect and adjust for changes in the nominal settings. The combined effect of these two factors has been defined by Taguchi as 'noise'. Clearly noise consists of two separate components each of which needs to be assessed, monitored and adjusted during process control. While the part played by the control techniques in contributing to the capability of a process is the subject of this chapter, the subject of how to understand and tackle the truly random component of the noise is the subject of Chapter 9.

The discussion of capability may be simplified by formalizing a relationship between the random component of the process variability and the tolerances. There are a number of possible process capability indices which do this and which we must now consider.

7.2   Process capability indices

Cp

In formalizing a relationship between the random component of the variations within a process and the tolerances we make use of the standard deviation, σ, of the process. We may express the approximate width of the random variations as equal to six standard deviations, 6σ, and recall that this will cover 99.7 per cent of all the actual values, if the distribution is normal. In order to manufacture within the specification, the distance between the upper specification limit (USL) and the lower specification limit (LSL), must be less than the width of the base of the bell. So a comparison of 6σ with (USL − LSL), or 2*T*, is an immediately obvious process capability index:

$$Cp = \frac{\text{USL} - \text{LSL}}{6\sigma} \text{ or } \frac{2T}{6\sigma}$$

To use this index it is necessary to know both the width of the tolerance zone, (USL − LSL) or 2*T*, and the standard deviation which, given the mean range, can be determined from the formula:

$$\sigma = \bar{R}/d_n \text{ or } \bar{R}/d_2$$

Clearly any value of *Cp* below 1 means that the width of the random variations is already greater than the specified tolerance band so the processs is incapable. For increasing values of *Cp* the process becomes increasingly capable. The practical application of the *Cp* index lies in expressing the capability of a process assuming that it is correctly centred about the mid-specification.

Cpk

It is possible to envisage a tolerance band width of 12σ and a bell of width 6σ, so a *Cp* of 2, but with all the observed values lying outside the tolerance band. This does not invalidate the use of *Cp* as an index to measure the 'potential capability' of a process when centred, but suggests the need for another index which takes account of both the degree of the random variations and the centring. Such an index is the *Cpk*, which is rapidly becoming industry's means of communicating process capabilities.

The *Cpk* measures the distance between the process mean and both the upper and the lower specification limits and expresses this as a ratio of half of the bell width:

$$Cpk = \text{the lesser of } \frac{\text{USL} - \overline{\overline{X}}}{3\sigma} \text{ or } \frac{\overline{\overline{X}} - \text{LSL}}{3\sigma}$$

A *Cpk* of 1 or less means that the width of the distribution bell and its centring is such that it infringes one of the tolerance limits and the process is incapable. As in the case of the *Cp* index, increasing values of the *Cpk* index correspond to increasing capability. It may be possible to increase the *Cpk* value by centring the process so that its mean value and the mid-specification, or target, coincide. A comparison of the *Cp* and the *Cpk* will show no difference if the process is centred on the target value, and a difference if it is not.

The *Cpk* can be used when there is only a minimum or maximum value specified – a one-sided specification. This, of course, occurs quite frequently and the *Cp* index cannot be used in this situation.

The *Cpk* is rapidly becoming a standard term used throughout industry to describe process capability. As can easily happen with any technical parameter, its meaning is not always understood by those who use it, and cases of serious abuse arise. In an effort to overcome some of the possible misunderstandings of the *Cpk* a number of variants have been introduced in specific industries.

A couple of examples will indicate how *Cp* and *Cpk* indices are calculated.

1   The process parameters from twenty samples of size $n = 4$ are:

$$\text{Mean range } (\overline{R}) = 91 \text{ mg, Process mean } (\overline{\overline{X}}) = 2500 \text{ mg}$$
$$\text{Specified requirements USL} = 2650 \text{ mg, LSL} = 2350 \text{ mg}$$
$$\sigma = \overline{R}/d_n = 91/2.059 = 44.2 \text{ mg}$$

$$Cp = \frac{\text{USL} - \text{LSL}}{6\sigma} \text{ or } \frac{2T}{6\sigma} = \frac{2650 - 2350}{6 \times 44.2} = \frac{300}{265.2} = 1.13$$

$$Cpk = \text{lesser of } \frac{\text{USL} - \overline{\overline{X}}}{3\sigma} \text{ or } \frac{\overline{\overline{X}} - \text{LSL}}{3\sigma}$$

$$= \text{lesser of } \frac{2650 - 2500}{3 \times 44.2} \text{ or } \frac{2500 - 2350}{3 \times 44.2}$$

$$= 1.13$$

Conclusion – the process is centred and of low capability since the indices are only just greater than 1.

2 The process parameters from twenty samples of size $n = 4$ are:

$$\text{Mean range } (\bar{R}) = 91 \text{ mg, Process mean } (\bar{\bar{X}}) = 2650 \text{ mg}$$
$$\text{Specified requirements USL} = 2750 \text{ mg, LSL} = 2250 \text{ mg}$$
$$\sigma = \bar{R}/d_n = 91/2.059 = 44.2 \text{ mg}$$

$$Cp = \frac{\text{USL} - \text{LSL}}{6\sigma} \text{ or } \frac{2T}{6\sigma} = \frac{2750 - 2250}{6 \times 44.2} = \frac{500}{265.2} = 1.89$$

$$Cpk = \text{lesser of } \frac{2750 - 2650}{3 \times 44.2} \text{ or } \frac{2650 - 2250}{3 \times 44.2}$$
$$= \text{lesser of } 0.75 \text{ or } 3.02 = 0.75$$

Conclusion – the *Cp* at 1.89, well over 1, indicates a potential for higher capability than in Example 1, but the *Cpk* shows that this potential is not being realized because the process is not centred.

It is important to emphasize that in all process capability studies, no matter how precise they may appear, the results are only ever approximations – we never actually *know* anything, progress lies in obtaining successively closer approximations to the truth. In the case of the process capability this is true because:

- There is always some variation due to sampling
- No process is ever fully in statistical control, or rather if it were we would not be able to detect it.
- No output exactly follows the normal distribution or indeed any other standard distribution.

Interpreting process capability indices without knowedge of the source of the data on which they are based can, and regrettably does, give rise to serious misinterpretation.

7.3 Interpreting capability indices

Process capability indices can be meaningful only to those who understand the scale on which they are measured.

So far we have derived the standard deviation, σ, from the mean of measures of the range and recognized that this estimates the short-term random variations within a process. This short term is the period over which the process remains relatively stable. But we know that processes do not remain stable for all time and so we need to allow within the specified tolerance limits for:

- Movement of the mean.
- The detection of changes of the mean.

- Possible changes in the scatter (range).
- The detection of changes in the scatter.
- The possible complications of non-normal distributions.

Let us first suppose that there is nothing we can do about the degree of random variations and investigate the ability of the mean chart to detect changes of the process mean. As the process mean moves we shall seek, by using the mean chart, to detect the change. For a process of high capability we should be able to detect all changes of the mean which are likely to give rise to out-of-specification output. The limiting case will be one of a very high probability of detecting a change of the process mean when the distribution bell of the scatter of the individual results just touches either of the specification limits. Clearly we should like to detect such a change of the process mean from the results of the first sample examined after the change took place.

Consider Figure 7.2(a) in which the individuals results are assumed to be scattered normally about a process mean, $\overline{\overline{X}}$, and with a standard deviation of σ. Assuming a sample size of $n = 4$, the means will be distributed normally about the same process mean, but with a distribution whose width will be 1.5σ (the standard error of the means SE = $\sigma/\sqrt{n}$). This will be the position of the action line of the mean chart, as shown in Figure 7.2(b).

Let us suppose that the USL is set at $+4\,\sigma$ above $\overline{\overline{X}}$ (Figure 7.2(c))

$$\text{Then the } Cpk = (\text{USL} - \overline{\overline{X}})/3\sigma = 4\sigma/3\sigma = 11.33$$

Now suppose that the process drifts up to the point where the tail of the distribution of the individuals results just touches the USL. The process mean will now be centred at the $+1\sigma$ postion as shown in Figure 7.2(c), and the probability of observing a mean result above the action limit will be represented by the shaded area. This area starts at one standard deviation of the means (1SE) from the process mean and we know from the properties of the normal distribution that in this shaded tail only 0.159 (15.9 per cent) of the observed means will occur. Hence, the probability of detecting this 1σ change of the process mean, by an action signal on the first sample after the distribution of the individuals just touches the USL, is 0.159. This means that *on average* this change of the process mean will not be detected until after 6.3 sample mean results have been plotted. This is the average run length to detection (ARL) which we met in Chapter 6. So with $n = 4$ and a Cpk of 1.33 our chances of detecting non-conformance when we first hit it are *not good* at 15.9 per cent and, therefore, a Cpk of 1.33 is *not good*. On average the change of the process mean will only be detected 6.3 consecutive mean results after the change. It follows that a manufacturer with a Cpk of 1.33 will only occasionally

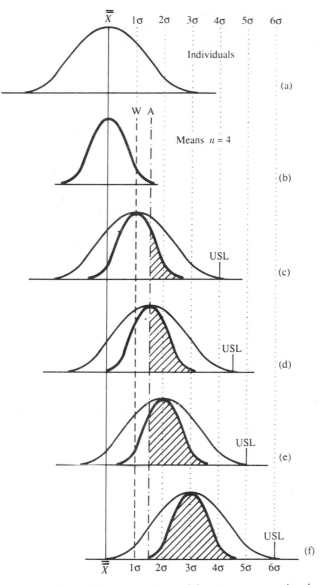

Figure 7.2   *Probabilities of detecting a change of the process mean using the mean chart*

detect non-conforming output when it occurs, even when using mean charts.

Now examine the situation when the USL is set at 4.5σ above $\bar{\bar{X}}$ (Figure 7.2(d)).

The $Cpk = (USL - \bar{\bar{X}})/3\sigma = 4.5\sigma/3\sigma = 1.5$

Now the process mean will be at the $+1.5\sigma$ position when the tail of the individuals results just touches this new USL. The probability of detection of this change by a mean chart action signal on the first sample is again represented by the shaded area and is now 0.5 or 50 per cent. The average run to detection (ARL) is then 2. Clearly this is better than with a *Cpk* of 1.33 and although *still not good* there is some hope that the movement of the process mean will be detected reasonably quickly – on average at the second sample after the change.

Let us move on to the case where the USL is set at 5σ above $\overline{\overline{X}}$ (Figure 7.2(e)).

$$\text{The } Cpk = (\text{USL} - \overline{\overline{X}})/3\sigma = 5\sigma/3\sigma = 1.67$$

In the limiting case the process mean will now be at $+2\sigma$ and the probablity of detection of this change is again represented by the shaded area (see Figure 7.2(e)). The unshaded area is 0.159, so the shaded area is $(1 - 0.159)$ or 0.841. This is the probability of detecting the change from the first sample taken after the change. The ARL is 1.2. This begins to be promising with a good chance that non-conforming output will be detected and perhaps eliminated. So with $n = 4$ and a *Cpk* of 1.67 the probability of detecting non-conformance when we first hit it becomes more acceptable. However, we know that the producer will still not detect, and hence not even have the facility to eliminate non-conforming product on about 16 per cent of the occasions when it occurs.

In the final position shown in Figure 7.2(f), the USL is 6σ from $\overline{\overline{X}}$, the *Cpk* is 2 and the probability of detection is now 0.999 or 99.9 per cent. So when $n = 4$, and with a *Cpk* of 2, the producer will detect a swing of the process mean when only 0.15 per cent of the output is outside the specification limits. Since this event will presumably not occur often, the proportion of defective product will be considerably below the 1 in 1000 level. We may have entire confidence in a producer or a process for which the *Cpk* is 2 when and where mean and range charts are in use for control of the process, and $n = 4$ or more.

We have now established the following points on the *Cpk* scale, *when $n = 4$:*

- *Cpk* < 1 – a *hopeless situation* in which the producer is not capable and at all times there will be non-conforming output from the process.
- *Cpk* = 1 – also a situation in which the producer is *not capable* since any change within the process will result in some undetected non-conforming output.
- *Cpk* = 1.33 – still *far from acceptable* since non-conformance is not likely to be detected.
- *Cpk* = 1.5 – *not yet satisfactory* since non-conforming output will occur and the chances of detecting it are still not good enough.

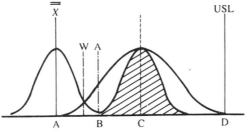

Figure 7.3

- *Cpk* = 1.67 – *promising*, non-conforming output will occur but there is a very good chance that it will be detected.
- *Cpk* = 2 – *total confidence in the producer*, provided only that we also know that mean and range charts are in regular use and that *n* = 4 or more.

Note These calculations of *Cpk*s, and our understanding of their values are based on standard deviations estimated from random variations occurring when the process is in statistical control.

The limiting case of a *Cpk* of 2 when *n* = 4, which is shown in Figure 7.2(f), has been redrawn in Figure 7.3. From this one can see that for a process with a target mean of $\overline{\overline{X}}$, shown at A, the upper action line on the mean chart will be set at $(\overline{\overline{X}} + 3SE)$, shown at B. In the limiting case, when the distribution of the individuals just touches the upper specification limit, the process mean has moved to the position C. The narrower distribution of the sample means is also centred at C, and its lower extremity now touches the upper action limit. So the distance from A to C is made up of AB and BC, both of which are the half widths of the distribution of the means. In other words AC is 6SE and, of course, the width of DC is 3σ.

Hence the limiting *Cpk*, for all values of '*n*', is:

$$\frac{\overline{\overline{X}} - USL}{3\sigma} = \frac{\text{distance A to D}}{3\sigma} = \frac{6SE + 3\sigma}{3\sigma}$$

Since $SE = \sigma/\sqrt{n}$, the limiting value of the *Cpk* is:

$$\frac{6\sigma/\sqrt{n} + 3\sigma}{3\sigma} = \frac{2}{\sqrt{n}} + 1$$

We may present these conclusions in graphical form by plotting *P(d)*, the probability of detecting non-conforming output, which results from a change in the mean value, at the first sample plotted on the mean chart following the change of mean, *P(d)*, against the *Cpk*. This precise definition of *P(d)* is more simply expressed as a degree of confidence that we may have in the ability of a producer to detect, and hence have the possibility of eliminating, non-conforming output from his process. On such a confidence scale, unity represents total confidence, and zero represents no confidence. Figure 7.4 shows the plot of *P(d)* against *Cpk* for various values of the sample size, *n*.

The *n* = 4 operating characteristic curve (OC curve) reproduces the results discussed above and provides the added conclusions:

- For a *Cpk* below 1 the degree of confidence does not change – it is always a state of no confidence.
- That for a *Cpk* of, say, 1.8 or above the degree of confidence is already high, but this does not make any allowance for the possibility of undetected changes of the range, non-normality etc.
- For a *Cpk* of more than 2 there is virtually no room for any further improvement in confidence, but again this does not allow for possible undetected changes in the range or the shape of the distribution.

We can also see from the family of OC curves in Figure 7.4 that, as the sample size increases, the *Cpk* values at which both 'high' confidence and 'almost total' confidence occur, decrease. This conclusion is subject to the qualification that there may be undetected changes of the range, or more exactly the random component of variation. (For the *n* = 25, 49 and 100 curves the sample size is too large to permit adequate control using the

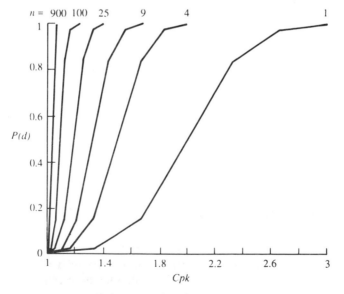

Figure 7.4

range as a measure of the random variation and a standard deviation chart would have to replace the range chart.)

It is clear that the degree of confidence is the same for a *Cpk* of 2 when $n = 4$, as it is for a *Cpk* of 1.67 when $n = 9$ and a *Cpk* of 1.4 when $n = 25$, etc. and it therefore follows that a *Cpk* value has little meaning unless the control chart sample size is also specified. With values of 'n' below 4, Figure 7.4 shows that even larger values of the *Cpk* would be required for 'high' and 'almost total' confidence. For $n = 1$, individual observations not grouped into samples, the limiting *Cpk* is 3 – the specification limits would need to be a minimum of $\pm 9\sigma$ from a correctly centred mean.

The family of OC curves in Figure 7.4 also shows how large sample sizes can be used to maintain control closer to the limits imposed by the random variations. With a sample size of 900 the limiting *Cpk* is close to unity. Where data is continuously recorded, for example temperatures, pressures, flows etc., separate measures may be available every microsecond and such large samples can then be used to give a very high degree of sensitivity to changes in the process mean. In such a semi-continuous control system if adjustments are made after each observation, perhaps through an automatic control loop, the OC curve will be that for $n = 1$. The failure to group data will result in an unnecessary loss in the sensitivity to change inherent within the data.

One final word concerning these OC curves. There are three distinct and different zones:

1   When the *Cpk* is less than 1, progress can only be made by identifying the components of the random variations and reducing them. This is seldom easy (see Chapter 9).
2   When the *Cpk* is between 1 and 2, the opportunity may exist to make the process entirely capable of meeting the requirements by selecting the sample size.
3   When the *Cpk* is above 3, process control is good and there may be no need, nor urgency, to improve it. It is unlikely that there will be major and immediate gains in improving such processes. Even individuals charts may be adequate for reasonable control. There is an opportunity to concentrate effort on processes in zones 1 and 2 above.

7.4   The use of control chart and process capability data

As we have seen, the sample size and the *Cpk* indicate immediately the potential for control. The *Cpk* values so far calculated have been based on estimates obtained over relatively short periods of data collection and should be more properly known as the *Cpk* (potential). Knowledge of the

$Cpk_{\text{(potential)}}$ is available only to those who have direct access to the process and can assess the short-term variations which are typically measured during process capability studies.

One can always calculate an estimate of the standard deviation from any set of data. Put the data into a calculator with a statistical mode and press the standard deviation button (marked variously as s, σ_n and σ_{n+1}). No matter what the origin of the data you will always be given the mathematical answer. For example, a customer can measure the variation within a delivered batch of material, or between batches of material, as supplied over time, and use the data to calculate the corresponding standard deviations. This will provide some knowledge of the process from which the examined product was obtained. The customer can also estimate the process mean values and, coupled with the specification, calculate a Cpk using the usual formula. This practice is recommended, provided that one knows how to interpret the results.

An example will be used to illustrate the various methods of calculating Cpks. A pharmaceutical company carried out a process capability study on the weight of tablets produced and showed that the process was in statistical control with a process mean ($\overline{\overline{X}}$ of 2504 mg and a standard deviation (σ), based on the mean range $\overline{R}$ from samples of size $n = 4$ of 91 mg. The specification was USL = 2800 mg and LSL = 2200 mg. Hence:

$$\sigma = \overline{R}/d_n = 91/2.059 = 44.2 \text{ mg}$$

and

$$Cpk_{\text{(potential)}} = (\text{USL} - \overline{\overline{X}})/3\sigma = 296/3 \times 44.2 = 2.21$$

On a far from typical day, the mean and range chart used to control the process were as shown in Figure 7.5. In a total of twenty-three samples, there were four warning signals and six action signals, from which it is clear that during this day the process was no longer in statistical control. The data from which this chart was plotted are given in Table 7.1.

On the following morning, the data and the mean and range charts were available at the daily product classification meeting. With such an erratic process performance during the previous day, clearly there was going to be some debate about the product and its future. The extremes of opinion expressed varied from:

- The process was not in control so scrap or rework all the output, to
- The table of results shows no observations outside specification, so give the tablets a good mix and despatch them all.

A laboratory assistant present at the meeting used the recorded tablet

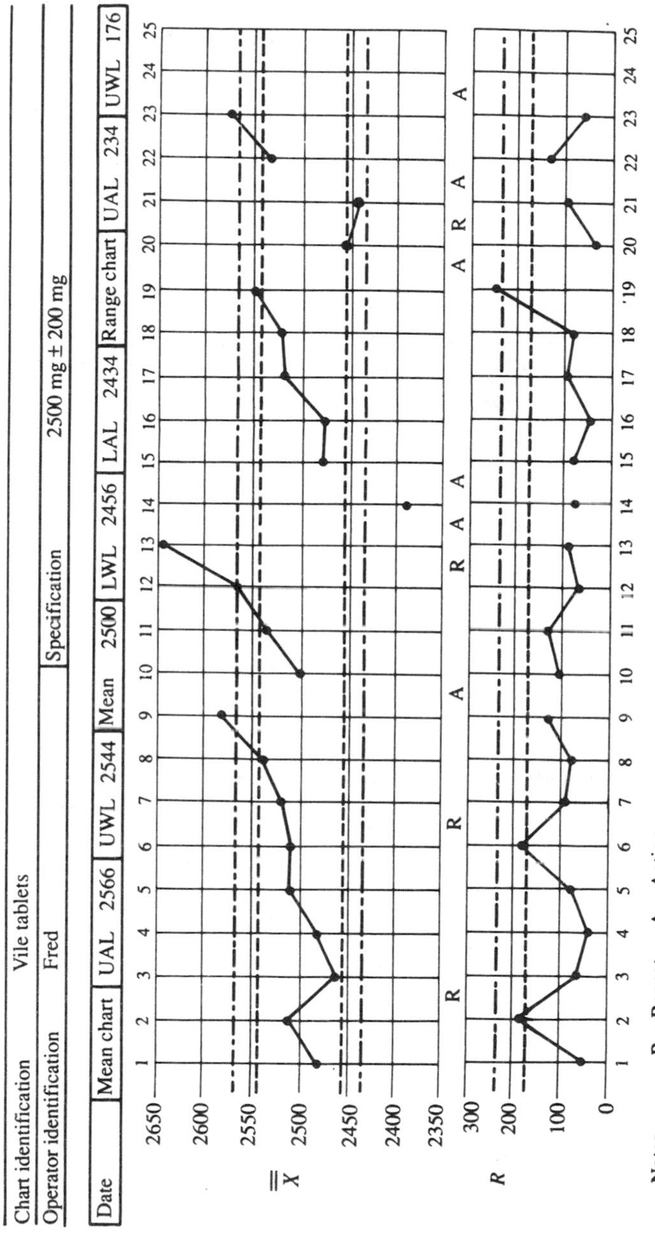
Chart identification Vile tablets
Operator identification Fred
Specification 2500 mg ± 200 mg
Date
Mean chart UAL 2566 UWL 2544 Mean 2500 LWL 2456 LAL 2434
Range chart UAL 234 UWL 176
2650
2600
2550
2500
2450
2400
2350
300
200
100
0
R R R A R A A A R A A
Notes
R = Repeat A = Action

Figure 7.5 *Mean and range control charts – tablet weights*

Table 7.1   *Tablet weights*

Sample number	Weight (mg)			
1	2501	2561	2512	2468
2	2416	2602	2482	2526
3	2487	2494	2428	2443
4	2471	2462	2504	2499
5	2510	2543	2464	2531
6	2558	2412	2595	2482
7	2518	2540	2555	2461
8	2481	2540	2569	2571
9	2504	2599	2634	2590
10	2451	2463	2525	2559
11	2556	2457	2554	2588
12	2544	2598	2531	2586
13	2591	2644	2666	2678
14	2353	2373	2425	2410
15	2460	2509	2433	2511
16	2447	2490	2477	2498
17	2523	2579	2488	2481
18	2558	2472	2510	2540
19	2579	2644	2394	2572
20	2446	2438	2453	2475
21	2402	2411	2470	2499
22	2551	2454	2549	2584
23	2590	2600	2574	2540

weights in Table 7.1 and his calculator to compute their grand mean as 2513 mg and their standard deviation as 68 mg. He then calculated a *Cpk* as:

$$\frac{\text{USL} - \overline{\overline{X}}}{3\sigma} = \frac{2800 - 2513}{3 \times 68} = 1.41$$

The standard deviation which he had calculated reflected various components including the inherent random variations, all the assignable causes apparent from the mean and range chart, and the limitations introduced by using a sample size of four. It clearly reflected more than the inherent random variations and so the *Cpk* which resulted from its use was not a good measure of the $Cpk_{(potential)}$. He had in fact worked out the $Cpk_{(potential)}$ – a capability index of the day's output and a useful way of monitoring, over a period, the actual performance of any process. The symbol *Ppk* is sometimes used to represent the $Cpk_{(production)}$ but this is by no means standard practice.

The $Cpk_{(production)}$ includes the random and the assignable sources of variation and cannot be less than the $Cpk_{(potential)}$ – if it appears to be less than the $Cpk_{(potential)}$, it can only mean that the $Cpk_{(potential)}$ has improved from that previously calculated, as a result of a decrease in the amount of random variation. A record of the $Cpk_{(production)}$ reveals how the production performance varies and takes account of both the process centring and the spread.

If the day's output had been mixed and sent to the customer, and he had taken random samples from the delivery he should have obtained estimates of the grand mean, the standard deviation and hence the $Cpk_{(delivered)}$ similar to the above result. He could not know, however, whether the Cpk of 1.41 reflected a regular and moderated pattern of vaiation throughout the day or limited periods of seriously off-target production. No amount of closer inspection of a *mixed sample* would tell him. Only when the data is kept in chronological order can the patterns in the mean and range chart be detected.

At this stage the laboratory assistant became aware of the continuing, and increasingly noisy, debate on the classification of the previous day's output. With his increased conviction that the control charts were telling him useful things about what happened to the process on the previous day, he suggested that the charts be used to classify the product and only product from 'good' periods be despatched, if necessary, after mixing. He defined 'bad' product as that produced in periods prior to an action signal as well as any periods prior to warning signals which were followed by action signals. Reading from the charts in Figure 7.5 this means eliminating the product from the periods preceding samples 8, 9, 12, 13, 14, 19, 20, 21, 23.

While another debate ensured on this proposal, the laboratory assistant referred back to Table 7.1 and excluded from it the weights corresponding to those periods now proposed for elimination from the product for despatch. Of the ninety-two original weights he was left with 56 tablet weights from which he calculated the process mean at 2503 mg, the standard deviation at 49.4 mg and the Cpk at:

$$Cpk = (USL - \overline{\overline{X}})/3\sigma = (2800 - 2503)/3 \times 49.4 = 2.0$$

This is the $Cpk_{(delivery)}$. If this selected output from the process were despatched, the customer would find a similar process mean, standard deviation and $Cpk_{(delivery)}$ and should be content. It is not surprising that the Cpk should be increased by the elimination of the product known to have been produced using 'out of control' periods. The term Csk is sometimes used to present the $Cpk_{(delivery)}$ but this is not standard practice.

This additional result of analysis of the available data was greeted by the meeting with some enthusiasm since it offered action which could apparently be justified. The debate was relaunched when it was pointed out that this still condemned about 30 per cent of the day's output, so it was suggested that the product corresponding to the periods prior to warning signals be reintroduced for despatch.

Clearly one can now also include in the calculation of mean, standard deviation, and Cpk the results corresponding to samples 8, 12, 20 and 21. This leaves the process mean unchanged at 2503 mg, increases the standard deviation to 52.2 mg and reduces the Cpk to 1.90. The customer will again probably find similar results by sampling, and should be less impressed by a Cpk of 1.9 than he would have been by a Cpk of 2.

If, however, a $Cpk_{(potential)}$ of 2 is required (when $n = 4$) what is the corresponding acceptable limit of a $Cpk_{(delivery)}$? There is no direct answer to this question. The origin of the decrease of the Cpk to 1.9 could be either that there was an acceptable and continuous amount of non-random variation during manufacture, or that, for a short period during the operation of the process, out-of-specification product was produced but not detected and hence not eliminated from the despatches. Only the producer can know the $Cpk_{(potential)}$ and the method of product classification used. Not only the product, but the justification of its classification should be available to the customer. The only way in which the latter may be achieved is by letting the customer have copies of the control charts and also the justification of the $Cpk_{(potential)}$. Both of these requirements are rapidly becoming standard in those industries which understand and have assimilated the concepts of process capability and the use of control charts for variables.

There are two important points for emphasis:

- The use of control charts not only allows the process to be controlled, it also provides all the information required to complete product classification.
- The producer, through the data coming from the process capability study and the control charts, can judge the performance of a process – the process performance cannot be judged equally well from the product alone.

So if a customer knows that a supplier has a $Cpk_{(potential)}$ of at least 2 when using sample sizes of four, or equivalent values of the limiting $Cpk_{(potential)}$ for other sample sizes, and that the supplier uses control charts for both control and classification, then he can have entire confidence in the supplier's process and method of product classification. To be absolutely exact, in the above case the risk of non-conforming product being undetected may, on infrequent occasions rise to as high as 1 in 1000. If this risk is too high, it may be redefined by selecting either a larger sample size and maintaining the $Cpk_{(potential)}$ or retaining the sample size and specifying a higher $Cpk_{(potential)}$.

In parts of the electronic components industry, which is among the most advanced in low defect rates, the probability of delivering non-conforming product has been reduced to parts per million. There is nothing inherently more stable about the manufacture of electronic components. Success in this case followed commercial pressures which resulted in the need to adopt and develop good quality management. Such management is available to all suppliers of either artefacts or services.

7.5 Modified or relaxed control charts

Processes exist for which the $Cpk_{(potential)}$ is very large. The highest Cpk the authors have so far encountered is in the chemical industry, for trace elements – a Cpk of 192 with $n = 4$. In mechanical engineering high Cpks are not uncommon. For example, a capstan lathe used at a slow cutting speed, may well be inherently capable of a Cpk of 2 for $n = 4$, when the tolerance band is, say, $\pm$ 0.01 mm. On less critical dimensions, the cutting speed will probably be increased so that the random spread increases as some precision is sacrificed because of the wider tolerances. When the machine is called upon to machine to a tolerance of, say, $\pm$ 1 mm the Cpk will probably be much greater than 2.

Under such conditions there is often no point in readjusting the mean of the process to keep it at mid-specification. It can be allowed to drift between the upper and lower specification limits until it meets the point where infringement becomes significantly possible. If the standards of the industry call for sample sizes of four and $Cpk_{(potential)}$s of at least 2, the control limits on the mean chart can be relaxed (or modified) to assimilate this condition. Figure 7.6 illustrates how this may be done for the upper specification limit (USL).

The process mean must be adjusted when it reaches a value which is $(3\sigma + 6SE)$ from either of the specification limits. At this value the mean of the sample means will also be at $(USL - (3\sigma + 6SE))$ and the warning and action limits need to be set at 2SE and 3SE above this limiting value of the process mean. Hence:

$$\text{Modified action lines are at} \quad USL - 3\sigma - 3\sigma/\sqrt{n}$$
$$\text{and} \quad LSL + 3\sigma + 3\sigma/\sqrt{n}$$
$$\text{and Modified warning limits are at} \quad USL - 3\sigma - 4\sigma/\sqrt{n}$$
$$\text{and} \quad LSL + 3\sigma + 4\sigma/\sqrt{n}$$

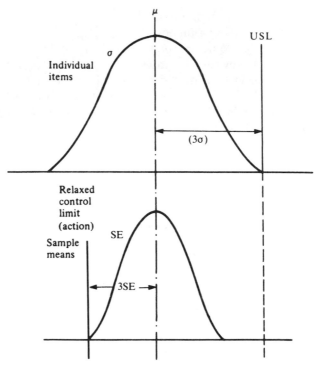

Figure 7.6   *Relaxed control charts*

'Relaxed' range charts are not suggested. Relaxed mean charts should always be used together with conventional range charts. Failure to monitor the process spread could invalidate the above calculations and, hence, the control by the relaxed mean charts. It is not that a 'relaxed spread' is not permitted but that, if this is advantageous, it must be monitored and the mean control charts must take account of the relaxed spread.

Under these circumstances, the use of individuals charts ($n = 1$) may be acceptable provided that a *Cpk* of 3 is used, i.e. that the action limits are set at 9σ inside the specification limits. This procedure cannot be recommended, however, since the random component of the variations cannot be monitored without grouping results together. In addition, with values of 'n' less than 4, it must be shown that the distribution of the random variations is always normal.

Chapter highlights

● Process capability is assessed by comparing the width of the tolerance band with the sum of the spread of the process due to random variations, and the ability to detect and adjust for changes in either the accuracy or precision of a process.

- For processes which are in statistical control, but not capable, the initial move towards establishing capability may be to establish an open dialogue between customers and suppliers which includes the widening of the tolerance band.

- Capability can be assessed by a comparison of the standard deviation and the width of the tolerance band. This gives a process capability index.

- The *Cp* index is the ratio of the tolerance band to six standard deviations. The *Cpk* index is the ratio of the band between the process mean and the closest tolerance limit, to three standard deviations. *Cp* measures the potential capability of the process, if centred; *Cpk* measures the capability of the process including its actual centring. The *Cpk* index can be used for one-sided specifications.

- Values of the standard deviation, and hence the *Cp* and *Cpk*, depend on the origin of the data used as well as the method of calculation. Unless the origin of the data is known the interpretation of the results will be confused.

- The $Cpk_{(potential)}$ measures the capability of the process and indicates the confidence one may have in the ability of the process operator to control the process, and classify the output so that the presence of non-conforming output is at an acceptable level.

- For all sample sizes a $Cpk_{(potential)}$ of 1 or less is totally unacceptable, since the presence of unidentified non-conforming output is inevitable.

- If the $Cpk_{(potential)}$ is between 1 and 3, there is a sample size at which the control of the process and the elimination of non-conforming output will be acceptable. The sample size, however, may be unacceptably large and/or expensive.

- When the $Cpk_{(potential)}$ exceeds 3, control methods other than mean and range charts may be contemplated.

- If the standard deviation is estimated from data collected during normal running of the process it will give rise to a $Cpk_{(production)}$, which will be less than the $Cpk_{(potential)}$. The $Cpk_{(production)}$ is a useful index of the process performance during normal production. Similarly, if the standard deviation is based on data taken from deliveries of an output it will result in a $Cpk_{(delivery)}$ which will be less than the $Cpk_{(potential)}$, may be greater than the $Cpk_{(production)}$ as a result of output selection, and can be a useful index of the delivery performance.

- Since there are no predetermined values of $Cpk_{(production)}$ or $Cpk_{(delivery)}$ which correspond to an acceptable level of non-conforming output customers must seek from their suppliers information concerning the potential of their processes, the methods of control in use and the methods of product classification practised.
- The calculation and interpretation of all these process capability indices are illustrated by an example.
- For processes of very high capability the use of relaxed mean control chart limits is recommended.

8 Other types of control charts for variables

8.1 Life beyond the mean and range chart

Statistical process control is based on a number of basic principles, all of which apply to both batch and continuous processes of the type commonly found in the manufacture of bulk chemicals, pharmaceutical products, speciality chemicals, processed foods, metals and, in general, the products of the process industries. One of these principles is that within manufacturing processes variations are inevitable. Variations are due to one of two types of causes, either 'random' or 'assignable' causes. Random causes cannot easily be identified individually and set the limits of the 'precision' of a process, while assignable causes reflect specific changes which either occur or are introduced as part of 'the control'.

There is a need to separate the two types of variation when interpreting data. If it is known that the difference between an individual observed result and the target value is simply a part of the inherent randomness, there is no readily available means for correcting or adjusting for it. If the observed difference is known to be due to an assignable cause then correction or control of this cause is possible and probably desirable. Adjustments and control by instruments, computers, operators, instructions, etc., can only correct for accuracy – no one has a simple control device or discipline for adjusting the inherent randomness.

These principles and the ensuing use of mean and range control charts are readily understood and accepted in the context of mass production where data is available on a large scale (dimensions of thousands of mechanical components, weight control of tablets, ampoules, bottles etc.). The use of such control procedures is often thought not to apply to situations in which a new item of data is available either in isolation or infrequently. This is the case in batch processes where analysis of the final product may reveal for the first time the characteristics of what has been manufactured. In continuous processes data is also often available on a one result per period basis (e.g. one analysis per hour or per shift).

If it is accepted that variation is an inevitable part of all manufacturing and operations processes, it must equally be accepted that stability should also be part of all such processes. Apart from a very limited range of activities which are explosive in nature, only those processes which are stable and reproducible tend to be used. For batch type processes this means that, each time a new batch is manufactured, the whole recipe of components and the methods of operation are reproduced in the expectation that the products of successive batches will be similar and lie within predetermined specification limits. In continuous processes stability of the output is also sought by either holding conditions constant or by seeking to change them in a way which will hold the output constant (the output may either be the product to be manufactured or a parameter important to the product or process control).

The distinction between batch, continuous batch and continuous processes is not always clear. They are essentially the same in their objective to maintain conditions and output, and differ only in their timescale.

With batch processes the speed with which data is built into the data bank is relatively slow, but there is time to reflect between batches and decide whether corrective action is or is not necessary. With continuous processes, data diarrhoea is a common problem. When things go wrong it may not be possible to shut down the process and then trouble pours out at an embarrasing rate.

Even if batches are made very infrequently, the two types of variation are still present and failure to distinguish between them will result in induced variations by overcontrol. How to handle the single result of an isolated batch, or observation made during a continuous process or a batch process is an important subject and is dealt with in this chapter.

The control charts for variables, first formulated by Shewhart, make use of the mean and the range of samples to determine whether a process is in a state of statistical control. Some control chart techniques exist which make use of other measures.

Charts for individuals (i-chart) or run charts

The simplest variable chart which may be plotted is one for individual measurements. Figure 8.1 shows measurements of lengths from a process making lithographic plates. The specification tolerances in this case are 640 mm + 1.5 mm and these are shown on the chart. If the process is known to be normally distributed, the mean and the ± 2σ and ± 3σ lines may be added in order to provide some idea of the expected spread of results. When using the conventional sample mean chart the tolerances are not included, since the distribution of the means is much narrower

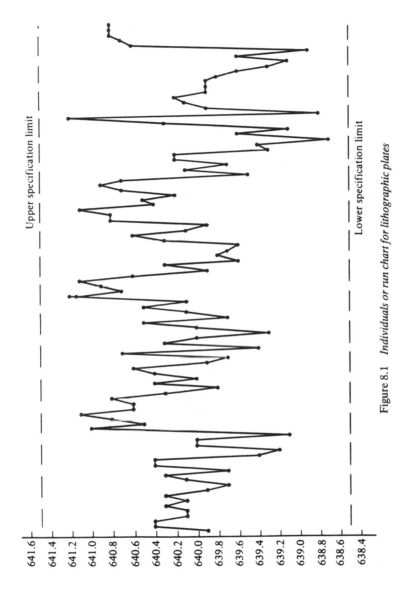

Figure 8.1 *Individuals or run chart for lithographic plates*

than that of the process population, and confusion may be created if the tolerances are shown. The inclusion of the specification tolerances on the individuals chart may, however, lead to overcontrol of the process as points are plotted near to the specification lines and adjustments are made.

Charting with individual item values is always better than nothing. It is, however, much less satisfactory than the charting of means and ranges, both because of its relative insensitivity to substantial changes in process average and the lack of any distinction between changes in accuracy and precision. Compare, for example, the individual chart of Figure 8.1 with the charts for mean and range of sample size 5 in Figure 8.2. The latter uses exactly the same data as the former, but with grouping of the results. The control limits for Figure 8.2 have been calculated using the standard method described in Chapter 6.

What is quite clear from the mean and range control charts, and is not at all apparent from the individuals chart, is that the process is not in statistical control, that it was changing as the plates were being taken from the process. Also there is no clear information from Figure 8.1 about the spread or capability of the process. To detect changes in process spread, the individuals chart must be accompanied by a range chart which, in turn, requires subgrouping of the values (minimum subgroup size, $n = 2$). Alternatively, a moving range chart, described in Section 8.3, may be used in conjunction with the chart for individual values.

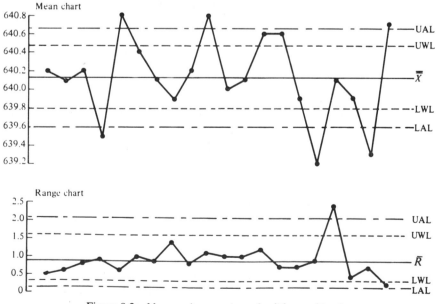

Figure 8.2   *Mean and range charts for lithographic plates*

The rules for the setting up and interpretation of i-charts are similar to those for conventional mean and range charts. Measurements are taken from the process over a period of expected stability. The mean $(\overline{\overline{X}})$ of the measurements is calculated, together with the range or moving range between adjacent observations ($n = 2$), and the mean range, $\overline{R}$. The control chart limits are found in the usual way.

i-chart:
    Action lines   @ $\overline{\overline{X}} \pm 3\sigma$ or $\overline{\overline{X}} \pm 3\overline{R} \Big/ d_n$

    Warning lines @ $\overline{\overline{X}} \pm 2\sigma$ or $\overline{\overline{X}} \pm 2\overline{R} \Big/ d_n$

    Central line $\overline{\overline{X}}$

Moving range chart:
    Action lines   @ $D'_{.001}\overline{R}$
    Warning lines @ $D'_{.025}\overline{R}$

When plotting the individual results on the i-chart, the rules for out of control situations are:

- Any points outside the 3σ limits.
- Two out of three successive points outside the 2σ limits.
- Seven points in a run on one side of the mean.

Owing to the insensitivity of i-charts, it is desirable to draw horizontal lines at $\pm 1\sigma$ either side of the mean and take action if four out of five points plot outside these limits.

In general, the chart for individual measurements is inferior to other types of control chart because it gives neither a clear picture of the type of changes taking place nor adequate evidence of small assignable causes of variation. An improvement is the combined individual-sample mean chart, which may show the specification tolerances and the upper and lower control limits. Such a chart, which would be provided by superimposing Figures 8.1 and 8.2, does however increase complexity and the chances of errors in interpretation.

Mean charts with varying sample size

The mean charts described in Chapter 6 have constant sample sizes and it is desirable that, where possible, this should be so. In some situations changes in sample sizes cannot be avoided. For example, when deliveries are being made it is frequently impossible to control the number of items being delivered. If one item or 'scoop' sample is to be taken from each homogeneous batch, the sample sizes will vary each time. For statistical

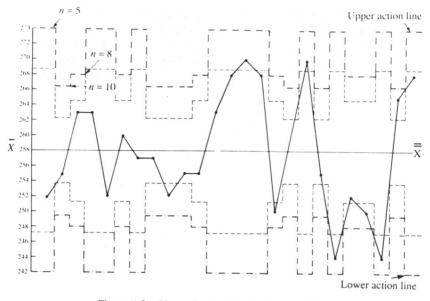

Figure 8.3 *Mean chart with variable action limits*

perfection, this situation requires the calculation of different action and warning lines for each sample size. Figure 8.3 shows a mean chart with variable sample sizes and the corresponding variable control limits, which become closer to the process mean as the sample size increases. Such a chart requires many control chart limit calculations and careful explanation to process operators. For these reasons, this type of chart should be used only where it is not possible to maintain a constant sample size and is then best used incorporating some form of computerization for the limits.

The method of calculating the variable control limits may be simplified by calculating the process standard deviation for each sample, σ_i from the formula

$$\sigma_i = R_i/d_n$$

where: R_i = the sample range,

d_n = Hartley's constant for the sample size n (See Appendix B)

An estimate of the overall process standard deviation may then be obtained by finding the average of the values of σ_i. Hence:

$$\sigma = \sum_{i=1}^{k} \sigma_i/k = \sum_{i=1}^{k} (R_i/d_n)/k$$

where k = the number of samples of variable size.

This method is not precisely correct in statistical theory, but it provides an excellent working approximation. The control chart limits for each sample size may be calculated as before, namely:

Action lines at:

$$\overline{\overline{X}} \pm 3\sigma/\sqrt{n} \text{ and}$$

Warning lines at:

$$\overline{\overline{X}} \pm 2\sigma/\sqrt{n}$$

where n is the average value of k samples.

8.2 Median, mid-range and multi-vari charts

As we saw in Chapter 5, there are several measures of central tendency of variables data. An alternative to sample average is the median, and control charts for this may be used in place of mean charts. The most convenient method for producing the *median chart* is to plot the individual item values for each sample in a vertical line and to ring the median – the middle item value. This has been used to generate the chart shown in Figure 8.4, which is derived from the data plotted in a different way in Figure 8.1. The method is only really convenient for odd number sample sizes. It allows the tolerances to be shown on the chart, provided the process data is normally distributed.

The control chart limits for this type of chart can be calculated from the median of sample ranges, which provides the measure of spread of the process. Grand or process median $(\widetilde{\widetilde{X}})$ – the median of the sample medians, and the median range $(\widetilde{R})$ – the median of the sample ranges, for the lithographic plate data, previously plotted in Figures 8.1 and 8.2, are 640.1 mm and 0.85 mm respectively. The control limits for the median chart are calculated in a similar way to those for the mean chart, using the factors A_4 and 2/3 A_4. Hence median chart action lines appear at:

$$\widetilde{\widetilde{X}} \pm A_4\widetilde{R}$$

and the warning lines at:

$$\widetilde{\widetilde{X}} \pm 2/3 \, A_4\widetilde{R}$$

Use of the factors, which are reproduced in Appendix D, requires that the samples have been taken from a process which has a normal distribution.

A chart for medians should be accompanied by a range chart so that the spread of the process is monitored. It may be convenient, in such a case, to calculate the range chart control limits from the median sample range, $\widetilde{R}$ rather than the mean range, $\overline{R}$. The factors for doing this are given in Appendix D, and used as follows:

$$\text{Action line at } D^m{}_{.001} \, \widetilde{R}$$
$$\text{Warning line at } D^m{}_{.025} \, \widetilde{R}$$

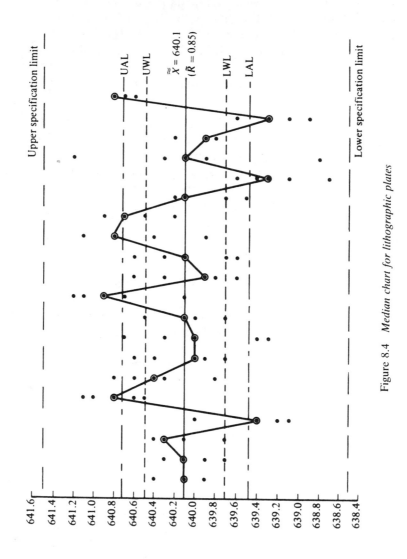

Figure 8.4 *Median chart for lithographic plates*

The advantage of using sample medians over sample means is that the former are very easy to find, particularly for odd sample sizes where the method of circling the individual item values on a chart is used. No arithmetic is involved. The main disadvantage, however, is that the median does not take account of the extent of the extreme values – the highest and lowest. Thus, the medians of the two samples below are identical, even though the spread of results is obviously different. The sample means take account of this difference and provide a better measure of the central tendency.

Sample number	Item values	Median	Mean
1	134, 134, 135, 139, 143	135	137
2	120, 123, 135, 136, 136	135	130

This failure of the median to give weight to the extreme values can be an advantage in situations where 'outliers' – item measurements with unusually high or low values – are to be treated with suspicion.

A technique similar to the median chart is the *chart for mid-range*. The middle of the range of a sample may be determined by calculating the average of the highest and lowest values. The mid-range of the sample of five: 553, 555, 561, 554, 551, is:

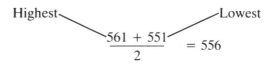

$$\underset{\text{Highest}}{} \quad \frac{561 + 551}{2} \quad \underset{\text{Lowest}}{} = 556$$

The central line on the mid-range control chart is the median of the sample mid-ranges $\widetilde{M}_R$. The estimate of process spread is again given by the median of sample ranges and the control chart limits are calculated in a similar fashion to those for the median chart. Hence:

$$\text{Action lines at:}$$
$$\widetilde{M}_R \pm A_4 \widetilde{R}$$
$$\text{Warning lines at:}$$
$$\widetilde{M}_R \pm 2/3 \, A_4 \widetilde{R}$$

An example of this type of chart was observed in a pin-producing factory. Operators periodically took samples of size $n = 100$ from the manufacturing process and stood the pins on their heads in a jig. The tallest and the smallest were measured, from which were calculated and plotted the sample ranges and mid-ranges.

Certain quality characteristics exhibit variation which derives from more than once source. For example, if cylindrical rods are being formed, their diameters may vary from piece to piece and along the length of each rod, due to taper. Alternatively, the variation in diameters may be due in part to the ovality within each rod. Such multiple variation may be represented on the *multi-vari chart.*

In the multi-vari chart, the specification tolerances are used as control limits. Sample sizes of three to five are commonly used and the results are

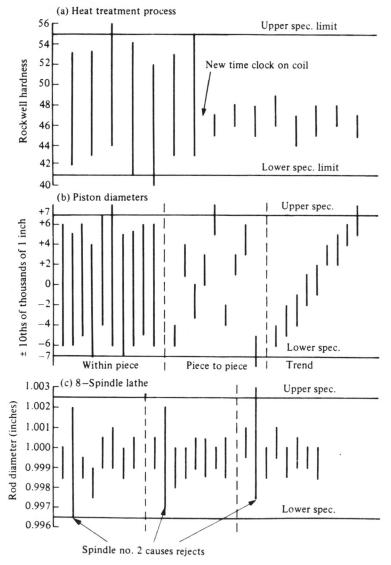

Figure 8.5   *Multi-vari charts*

plotted in the form of vertical lines joining the highest and lowest values in the sample, thereby representing the sample range. An example of such a chart used in the control of a heat treatment process is shown in Figure 8.5(a). The longer the lines, the more variation exists within the sample. The chart shows dramatically the effect of an adjustment, or elimination or reduction of one major cause of variation.

The technique may be used to show within piece or batch, piece to piece, or batch to batch variation. Detection of trends or drift is also possible. Figure 8.5(b) illustrates all these applications in the measurement of piston diameters. The first part of the chart shows that the variation within each piston is very similar and relatively high. The middle section shows piece to piece variation to be high but relatively small variation within each piston. The last section of the chart is clearly showing a trend of increasing diameter, with little variation within each piece.

One application of the multi-vari chart in the mechanical engineering, automotive, and process industries is for trouble shooting of variation caused by the position of equipment or tooling used in the production of similar parts, for example, a multi-spindle automatic lathe, parts fitted to the same mandrel, multi-impression moulds or dies, parts held in string-milling fixtures. Use of multi-vari charts for parts produced from particular, identifiable spindles or positions can lead to the detection of the cause of faulty components and parts. Figure 8.5(c) shows how this can be applied to the control of ovality on an eight-spindle automatic lathe.

8.3 Moving mean and moving range charts

The key to the separation of random and assignable causes lies in grouping results together and using this grouping to assess changes both in the average value and the scatter of grouped results – reflections of the centring of the process and the spread due to random causes. Many alternative attempts have been made to avoid grouping results but there is no known way of segregating random and assignable causes without doing so. This applies to all processes, including batch and continuous.

When only one result is available at the conclusion of a batch process or when an isolated estimate is obtained of an important parameter on an infrequent basis, clearly one cannot simply ignore the result until more data is available with which to form a group. Equally, it is impractical to contemplate taking, say, four samples instead of one and repeating the analysis several times in order to form a group – the costs of doing this would be prohibitive in many cases, and statistically this would be different from the grouping of less frequently available data.

An important technique for handling data which is difficult or time consuming to obtain, and therefore not available in sufficient numbers to enable the use of conventional mean and range charts, is the *moving mean and moving range chart*. In the chemical industry, for example, the nature of certain production processes and/or analytical methods causes long time intervals between consecutive results. We have already seen in this chapter that plotting of individual results offers a poor method of control, which is relatively insensitive to changes in process average and affords little information about the spread of the process. On the other hand, waiting for several results in order to plot a conventional mean chart may allow many tonnes of material to be produced outside specification before one point can be plotted.

In a polymerization process, one of the important process control measures is the unreacted monomer. Individual results are usually obtained once every twenty-four hours, with a delay of four hours for analysis between results. Typical data from such a process appears in Table 8.1.

If the individual chart of this data (Figure 8.6) was being used alone for control during this period, the conclusions would probably include:

16 April   concern and perhaps a repeat sample
18 April   panic – do something
23 April   panic – do something
29 April   concern and perhaps a repeat sample
From about 30 April a gradual decline in the values being observed

When using the individuals chart, each decision tends to be based on the one isolated last result. So the fundamental rule of not reaching a decision

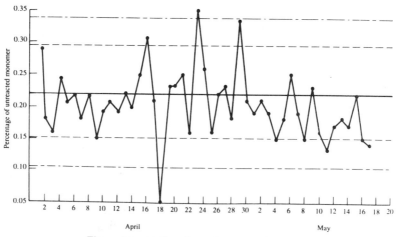

Figure 8.6   *Daily values of unreacted monomer*

Table 8.1   *Data on per cent of unreacted monomer at an*
intermediate stage in a polymerization process

Date		Daily value	Date		Daily value
April	1	0.29		25	0.16
	2	0.18		26	0.22
	3	0.16		27	0.23
				28	0.18
	4	0.24		29	0.33
	5	0.21		30	0.21
	6	0.22	May	1	0.19
	7	0.18			
	8	0.22		2	0.21
	9	0.15		3	0.19
	10	0.19		4	0.15
				5	0.18
	11	0.21		6	0.25
	12	0.19		7	0.19
	13	0.22		8	0.15
	14	0.20			
	15	0.25		9	0.23
	16	0.31		10	0.16
	17	0.21		11	0.13
				12	0.17
	18	0.05		13	0.18
	19	0.23		14	0.17
	20	0.23		15	0.22
	21	0.25			
	22	0.16		16	0.15
	23	0.35		17	0.14
	24	0.26			

and taking action on an isolated result is broken. But it is not realistic to wait for another three days, or to wait for a repeat of the analysis three times and then group data in order to make a valid decision based on the examination of a mean and range chart.

The alternative of moving mean and moving range charts uses the data differently and is generally preferred for the following reasons:

1   Grouping data together, we will not be reacting to individual results and overcontrol is less likely.
2   We will avoid the temptation to adjust the only available 'accuracy' knob for both errors of precision (when touching it will make things worse) and errors of accuracy (when touching it may make the right adjustment).
3   In using the moving mean and moving range technique we shall be making more meaningful use of the latest piece of data – two points,

one on each of two different charts telling us different things will be plotted from each individual result.

4   There will be a calming effect on the process.

Knowledge of the Central Limit Theorem (Chapter 5) tells us that a suitable sample size is four or more. The calculation of the moving means and moving ranges ($n = 4$) for the polymerization data is shown in Table -8.2. For each successive group of four, the earliest result is discarded and replaced by the latest. In this way it is possible to obtain and plot a 'mean' and 'range' every time an individual result is obtained – in this case every twenty-four hours. These have been plotted on charts in Figure 8.7.

The purist statistician would require that these points be plotted at the mid-point, thus the moving mean for the first four results should be placed on the chart between 2 and 3 April. In practice, however, the point is usually plotted at the last result time, in this case 4 April. In this way the moving average and moving range charts indicate the current situation, rather than being behind time.

There may be some doubts about using data collected over a period of, say, four days in order to decide what to do. If we are dealing with stable processes, we are expecting the period between samples to reflect the inherent period of stability. In a process which has been in use for some time, the interval between samples will have been determined by a number of factors – laboratory facilities, costs, etc., but above all the established 'fact' that experience has shown that we need to know the result about once per day, or whatever. If the inherent period of stability is shorter than the sampling period, the scatter previously experienced will already have led to more frequent sampling. In practice, because the best use may not have been made of the data, the actual period between samples will always be shorter than the inherent period of stability. This is usually by a factor of at least two and often up to ten. So, making

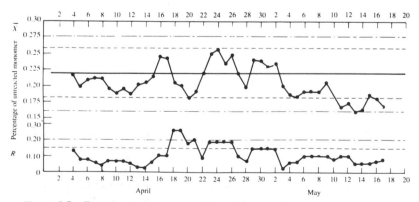

Figure 8.7   *Four-day moving mean and moving range charts (unreacted monomer)*

Table 8.2 *Moving means and moving ranges for data on unreacted monomer (Table 8.1)*

Date		Daily value	Four-day moving total	Four-day moving mean	Four-day moving range	Combination (for conventional mean and range control charts)
April	1	0.29				
	2	0.18				
	3	0.16				
	4	0.24	0.87	0.218	0.13	A
	5	0.21	0.79	0.198	0.08	B
	6	0.22	0.83	0.208	0.08	C
	7	0.18	0.85	0.213	0.06	D
	8	0.22	0.83	0.208	0.04	A
	9	0.15	0.77	0.193	0.07	B
	10	0.19	0.74	0.185	0.07	C
	11	0.21	0.77	0.193	0.07	D
	12	0.19	0.74	0.185	0.06	A
	13	0.22	0.81	0.203	0.03	B
	14	0.20	0.82	0.205	0.03	C
	15	0.25	0.86	0.215	0.06	D
	16	0.31	0.98	0.245	0.11	A
	17	0.21	0.97	0.243	0.11	B
	18	0.05	0.82	0.205	0.26	C
	19	0.23	0.80	0.200	0.26	D
	20	0.23	0.72	0.180	0.18	A
	21	0.25	0.76	0.190	0.20	B
	22	0.16	0.87	0.218	0.09	C
	23	0.35	0.99	0.248	0.19	D
	24	0.26	1.02	0.255	0.19	A
	25	0.16	0.93	0.233	0.19	B
	26	0.22	0.99	0.248	0.19	C
	27	0.23	0.87	0.218	0.10	D
	28	0.18	0.79	0.198	0.07	A
	29	0.33	0.96	0.240	0.15	B
	30	0.21	0.95	0.238	0.15	C
May	1	0.19	0.91	0.228	0.15	D
	2	0.21	0.94	0.235	0.14	A
	3	0.19	0.80	0.200	0.02	B
	4	0.15	0.74	0.185	0.06	C
	5	0.18	0.73	0.183	0.06	D
	6	0.25	0.77	0.193	0.10	A
	7	0.19	0.77	0.193	0.10	B
	8	0.15	0.77	0.193	0.10	C
	9	0.23	0.82	0.205	0.10	D
	10	0.16	0.73	0.183	0.08	A
	11	0.13	0.67	0.168	0.10	B
	12	0.17	0.69	0.173	0.10	C
	13	0.18	0.64	0.160	0.05	D
	14	0.17	0.65	0.163	0.05	A
	15	0.22	0.74	0.185	0.05	B
	16	0.15	0.72	0.180	0.07	C
	17	0.14	0.68	0.170	0.08	D

judgements based on the results which stretch over four periods – in the polymerization case, four days – is seldom likely to exceed the inherent period of stability. An analysis of historic data using the normal mean and range charts technique will tell us if four days is, or is not, a period of likely stability.

An earlier stage in controlling the polymerization process would have been to look at the data available from an earlier period, say during February and March, and to analyse this to find both the process mean and the mean range and to establish the mean and range chart limits for the moving mean and moving range charts. The process was found to be in statistical control during February and March and capable of meeting the requirements of producing product with less than 0.35 per cent monomer impurity. These observations had a process mean of 0.22 per cent and, with groups of $n = 4$, a mean range of 0.079 per cent. So the control chart limits, which are the same for both conventional and moving mean and moving range charts, would have been calculated before starting to plot the moving mean and moving range data onto charts. The calculations are shown below:

Moving mean and mean charts limits

$$\left.\begin{array}{l} n = 4 \\ \overline{\overline{X}} = 0.22 \\ \overline{R} = 0.079 \end{array}\right\} \begin{array}{l} \text{from the} \\ \text{results for} \\ \text{February/March} \end{array} \quad \left.\begin{array}{ll} A_2 = 0.73 \\ \\ 2/3A_2 = 0.49 \end{array}\right\} \begin{array}{l} \text{From tables} \\ \text{(Appendix B)} \end{array}$$

$$\text{UAL} = \overline{\overline{X}} + A_2\overline{R}$$
$$= 0.22 + (0.73 \times 0.079) = 0.2777$$

$$\text{UWL} = \overline{\overline{X}} + 2/3A_2\overline{R}$$
$$= 0.22 + (0.49 \times 0.079) = 0.2587$$

$$\text{LWL} = \overline{\overline{X}} - 2/3A_2\overline{R}$$
$$= 0.22 - (0.49 \times 0.079) = 0.1813$$

$$\text{LAL} = \overline{\overline{X}} - A_2\overline{R}$$
$$= 0.22 - (0.73 \times 0.079) = 0.1623$$

Moving range and range chart limits

$$\left.\begin{array}{l} D'_{.001} = 2.57 \\ \\ D'_{.025} = 1.93 \end{array}\right\} \text{from tables (Appendix C)}$$

$$\text{UAL} = D'_{.001}\,\overline{R}$$
$$= 2.57 \times 0.079 = 0.2030$$

$$\text{UWL} = D'_{.025}\,\overline{R}$$
$$= 1.93 \times 0.079 = 0.1525$$

The moving mean chart has a smoothing effect on the results compared with the individual plot. This enables trends and changes to be observed more readily. The larger the sample size the greater the smoothing effect. So a sample size of six would smooth even more the curves of Figure 8.7. A disadvantage of increasing sample size, however, is the lag in following any trend – the greater the size of the grouping, the greater the lag. This is shown quite clearly in Figure 8.8 in which sales data have been plotted using moving means of three and nine individual results. With such data the technique may be used as an effective forecasting method.

Recall that, in our polymerization example, one new piece of data becomes available each day and if moving mean and moving range charts were being used the result would be reviewed day-by-day. An examination of Figure 8.7 shows that:

- There was absolutely no abnormal behaviour of either the mean or the range on 16 April
- The abnormality on 18 April was not caused by a change in the mean of the process, which remained undisturbed, but of the range which shows an action situation. This implies that there was reason to 'panic' and do something but that 'something' was to investigate the assignable cause which gave rise to the action signal on the range chart. An extremely low result for the unreacted monomer is unlikely because it implies almost total polymerization and/or distillation. When this result was plotted on the range chart, an investigation was initiated to reveal that the plant chemist had picked up the bottle containing the previous day's sample from which the unreacted monomers had already been extracted during analysis – so when he erroneously repeated the analysis the result was unusually low. This type of error is a human one – the process mean had not changed and the charts have showed this.
- The plots for 19 April again show an action on the range chart. This is because the new mean and range plots are not independent of the previous ones. In reality, once we are satisfied that there was an assignable cause, we could eliminate the individual 'flier' result from the series. If this had been done we would not now have another action on the range.
- The warning signals on 20 and 21 April are also due to the same isolated low result which is not removed from the series until 22 April. The fact that the points on a moving mean and range chart are not independent influences the way in which the data is handled and the charts interpreted. We should have excluded the flier or 'outlier' from the series and either replaced it by a new result coming from the repeated analysis or simply discarded it from the series and grouped

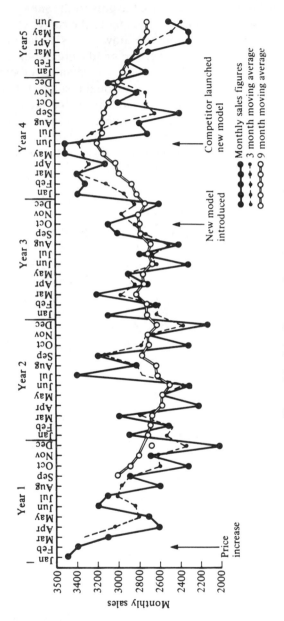

Figure 8.8   *Sales figures and moving average charts*

the next day's results with the three previously available ones. Even when no explanation can be found for this assignable cause, the risk of error in discarding such a result is only 1 in 1000.

If the same data had been grouped for conventional mean and range charts, with a sample size of $n = 4$, the decision as to the date of starting the grouping would have been entirely arbitrary. The first sample group might have been 1, 2, 3 and 4 April; the next 5, 6, 7, 8 April and so on; this is identified in Table 8.2 as combination A. Equally, 2, 3, 4, 5 April might have been combined; this is combination B. Similarly, 3, 4, 5, 6 April leads to combination C; and 4, 5, 6, 7 April will give combination D.

A moving mean chart with $n = 4$ is as if the points from four conventional mean charts were superimposed. This is shown in Figure 8.9. The plotted points on this chart are exactly the same as those on the moving mean and range plot previously examined. Two of the four sets of points have now been joined up in their independent A and B series. Note that in each of the series the problem at 18 April is seen to be on the range and not on the mean chart. As we are looking at four separate mean and range charts superimposed on each other it is not surprising that the limits for the mean and range and the moving mean and moving range charts are identical. When plotting the moving mean and moving range, however, it is necessary to exclude a result coming from an assignable cause or its effect will be carried over to points plotted later. This is a new rule. With mean and range charts, because the points are independent, there is no requirement to shuffle the basic data. With moving mean and moving range charts, however, unless the result giving rise to an assignable cause is eliminated from the series, it will influence later non-independent points.

Returning to Figures 8.6 and 8.7, was there anything abnormal about the result on 23 April? The mean may appear to be somewhat high – in

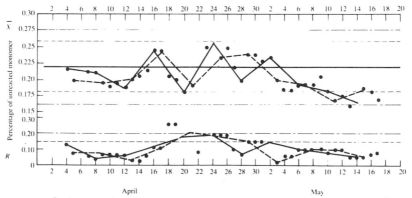

Figure 8.9 *Superimposed mean and range charts (unreacted monomer)*

fact it is entirely where we would expect to find it. The range, on the other hand, is in the warning zone and, as in the case of mean and range charts, we must take another sample. The repeat range also falls in the warning zone. An examination of the basic data shows that this is due to the presence of the single observation of 0.35 per cent on 23 April and the fact that again the two plotted range points are not independent. We shall have to take further samples to reach an independent point which will then tell us whether there is a clear signal that there is trouble with the range or not. The 23 April result can be seen to have had an effect on the range for four results and then it dropped out of the series and the range returned to the usual level. So the next independent point is after $(4 + 1) = 5$ consecutive points on the moving mean and moving range charts; in general $n + 1$ points. We have no indication from the charts of an assignable cause while the points were plotting in the warning zone and should not look for an assignable cause until we reach an action situation. In this case we see that after four points the results drop out of the warning zone, so this was one of those statistical events which will occur on average 1 in 40 times. There is no cause for any action. Compare the conclusion from the individual chart – panic – do something.

It is quite likely that the process will not be allowed to stay in the warning zone for the full $(n + 1)$ observations before either:

1   it is seen that the reason for being in the warning zone is associated with one isolated result. In this case, one may be somewhat reassured but sampling must be continued until the event is seen to be significant – either as an action or as a false alarm, or
2   in the case where there has been a real change of either the mean or the range, the chances of finding another high or low result are increased, and two or three high or low results will give rise to an action signal.

In practice, when observing warning signals in moving mean and moving range charts, one is often faced with limited analytical facilities so, instead of taking a repeat sample immediately, one sometimes adopts the alternative of increasing the frequency of sampling. For example, where one sample is examined per shift, the frequency is increased to four per shift or the interval between samples is reduced from eight hours to two hours. This is only continued until the signal has been confirmed and either requires action, or is seen as a false alarm – then the usual frequency of testing is reinstated. As already mentioned, the use of mean and range charts often demonstrates that the period of inherent stability within a process is longer than the period between samples, which has been established by experience. This means that one result of the use of charting is to reduce the frequency of sampling required to maintain process control. Clearly, there will be warning signals requiring repeat sampling and the analytical facilities must allow for the handling of these additional samples.

Returning to the plot of the results. Once can again see that at 29 April there was no justification for any action or repeat sampling.

Finally, it may be noted from Figure 8.7 that there was a tendency for the average values to decline during May. The rule for the mean chart is that trends are to be taken as 'warnings' when there are five points in a rising or falling series and as 'actions' when there are seven. For the moving mean and moving range charts this rule is modified to read five independent and seven independent points respectively. Five independent points are achieved after $(4n + 1)$ plots and seven after $(6n + 1)$. In our case where $n = 4$, the number of points in the downward trends during May is always less than 17 – so there is no significant trending.

It is worth pointing out that in the example of unreacted monomer, used to illustrate moving mean and moving range charts, adjustments would not be made to the process to increase the impurity if points fell below the lower action and warning lines, as was the case in the middle of May. Rather an 'action' situation should initiate an investigation into the assignable causes of the improvement. The use of any control charts in this way can be of value equal to, if not greater than, the usual search for problems.

The process overall

If the complete picture of Figure 8.7 is examined, rather than considering the values as they are plotted daily, it can be seen that the moving mean and moving range charts may be split into three distinctive periods:

- Beginning to mid April
- Mid April to early May
- Early to mid May

Clearly, a dramatic change in the variability of the process took place in the middle of April and continued until the end of the month. This is shown by the general rise in the level of the values in the range chart and the more erratic plotting on the mean chart.

An investigation to discover the cause(s) of such a change is required. In this particular example, it was found to be due to a change in supplier of feedstock material, following a shut down for maintenance work at the usual supplier's plant. When that supplier came back on stream in early May, not only did the variation in the impurity, unreacted monomer, return to normal, but its average level fell until on 13 May an action signal was given. Presumably this would have led to an investigation into the reasons for the low result, in order that this desirable situation might be repeated and maintained.

This type of 'map-reading' of control charts, integrated into a good quality management system, is an indispensable part of SPC.

Summary of moving mean and range charts

They are alternative ways of grouping results to observe changes in accuracy and precision.

They are particularly suited to industrial processes in which results become available infrequently. This is often a consequence of either lengthy, difficult, costly or destructive analysis in continuous processes or product analyses in batch manufacture.

Do not use moving mean and moving range charts for the analysis of historic data because it will give you four (or n) times the amount of work and no more information. The exception to this rule arises where only moving mean and moving range charts are in use and four times the work load is, in fact, a different computer program. If moving mean and moving range charts are to be used in all cases within an organization, their use for historic data keeps all the rules identical and is a useful standardization.

The rules for moving mean and moving range charts are the same as for mean and range charts except that there is a need to understand and allow for non-independent results. It is not, however, a good idea to mix mean and range with moving mean and moving range charts in the same operating units, as the different rules for interpretation of repeat results in the warning zone may cause confusion.

Exponentially weighted moving average (EWMA)

In mean and range control charts, the decision signal obtained depends largely on the last point plotted. In the use of moving mean charts some authors have questioned the reasonableness of giving equal importance to the most recent observation. The exponentially weighted moving average (EWMA) is a statistic which gives less and less weight to data as it gets older and older. A point plotted on an EWMA chart can be given varying degrees of 'memory' – short to long – by varying the weighting functions.

The formula for the EWMA to be plotted is:

$$EWMA = y_{t-1} + \lambda e_t$$

where y_{t-1} = previous EWMA value plotted at time $t - 1$

λ = constant between 0 and 1 which determines the length of memory of the EWMA

e_t = observed change of time t, = $y_t - y_{t-1}$ and y_t = observed value at time t

The choice of λ has to be left to the judgement of the quality control specialist, the smaller the value of λ, the greater the influence of the historical data. A value of $\lambda = 0.2$ is usually preferred.

The control lines for the EWMA chart are somewhat more complex to calculate than those for conventional mean or moving mean charts:

$$\text{Action lines} = T \pm 3 \sqrt{\left(\frac{\lambda}{2-\lambda}\right)} \sigma$$

where T is the target value.

Further λ terms can be added to the EWMA equation which are sometimes called the 'proportional', 'integral' and 'differential' terms in the process control engineer's basic proportional, integral, differential, or 'PID' control equation. (Reference: Hunter, J.S., (1986), *Journal of Quality Technology*, Vol. 18, pp. 203–10).

The EWMA has been used by some organizations, particularly in the process industries, as the basis of new 'control/performance chart' systems. Great care must be taken when using these systems since they do not show changes in variability very well, and the basis for weighting data is often either questionable or arbitrary.

A general approach for infrequent data

The following approach is recommended for those engaged in the control of batch and continuous processes, from which data is infrequently available.

1 Start by examining existing data. Data on material supplies together with in process and product parameters is usually available on file. If at least fifty consecutive data points are available and subgrouping is rational, submit them to the normal mean and range analysis. This should establish if the process is 'in statistical control' – both the means and the ranges of the grouped data are satisfactory when examined for unexpected results or trends (see Chapter 6, page 118). If the process is in control then all the observed variations of both means and ranges will appear to be due to random causes and the measure of this randomness can be made by calculating the standard

deviation. From this analysis the process capability can be assessed, possibly in the form of a calculated *Cpk* index.

2   This use of existing data will either confirm that the control method in use is effective or that there is a potential to improve it by making the distinction between accuracy and precision – by using either mean and range or moving mean and moving range charting for process control. If this type of analysis confirms that the control method is satisfactory, then pass on to the next available set of recorded data. It is not a good idea to seek solutions to problems which do not exist! Later, improved control can be sought as a part of the never-ending improvement approach.

3   Experience in the process industries shows that, because adjustments are often made, either manually or by instruments, after the observation of isolated results, processes are often not 'in control'. Even where processes appear to be in statistical control, frequent adjustment to the process parameters can sometimes show up as a series of seemingly random events. With this type of behaviour, the cusum technique is often useful in revealing what is actually going on (see Chapter 12).

4   If batches of product are produced only occasionally with parameters outside the specified limits, experience shows that this can occur as a result of the process not being quite capable, i.e. the degree of random variation and the sensitivity of detecting change combine in such a way that some out of specification material is inevitable. With, say, 1 per cent of the product outside the specified limits there is only a 1 per cent chance of it being observed. Knowledge of the process capability obtained from the historic data will show whether this is indeed the explanation of the occasional bad result.

5   In all cases when non-conforming product is obtained there is a need not only to dispose of it by finding a suitable less critical customer, blending if off, retreating it, scrapping it, etc., but also to investigate why the non-conformance occurred. It is only by tenaciously enquiring about all unexpected results that the move towards better control and never-ending improvement can be achieved. But start with the big steps first by locating the major contributors to non-conformance.

8.4   Control charts for standard deviation

Range charts are commonly used to control the precision or spread of processes. Ideally, a chart for standard deviation (σ) should be used but, because of the difficulties associated with calculating and understanding standard deviation, sample range is often substituted.

Significant advances in computing technology have led to the availability of cheap, pocket electronic calculators with a standard deviation key. Using such calculators, experiments in Japan have shown that the time required to calculate sample range is greater than that for sigma, and the number of miscalculations is greater when using the former statistic. The conclusions of this work were that mean and standard deviation charts provide a simpler and better method of process control for variables than mean and range charts, when a calculator with standard deviation key is available.

The standard deviation chart is very similar to the range chart (see Chapter 6). The estimated standard deviation (s_i) for each sample is calculated, plotted and compared to predetermined limits:

$$s_i = \sqrt{\sum_{i=1}^{n}(x_i - \bar{x})^2 / (n-1)}$$

Those using calculators for this computation must use the s or σ_{n-1} key and not the σ_n key. As we have seen in Chapter 5, the sample standard deviation calculated using the 'n' formula will tend to underestimate the standard deviation of the whole process, and it is the value of s or σ_{n-1} which is plotted on a standard deviation chart. The bias in the sample standard deviation is allowed for in the factors used to find the control chart limits.

Statistical theory allows the calculation of a series of constants (c_n) which enables the estimation of the process standard deviation (σ) from the average of the sample standard deviations ($\bar{s}$). The latter is the simple arithmetic mean of the sample standard deviations and provides the central line on the standard deviation control chart.

$$\bar{s} = \sum_{i=1}^{k} s_i/k$$

where

$\bar{s}$	= average of the sample standard deviations
s_i	= estimated standard deviation of sample i
k	= number of samples

The relationship between σ and s is given by the simple ratio:

$$\sigma = \bar{s}c_n$$

where

σ	= estimated process standard deviation
c_n	= a constant, dependent on sample size. Values for c_n appear in Appendix E.

The control limits on the standard deviation chart, like those on the range chart are asymmetrical, in this case about the average of the sample standard deviations ($\bar{s}$). The table in Appendix E provides four constants $B'_{.001}$, $B'_{.025}$, $B'_{.975}$ and $B'_{.999}$ which may be used to calculate the control limits for a standard deviation chart from $\bar{s}$. The table also gives the constants $B_{.001}$, $B_{.025}$, $B_{.975}$, $B_{.999}$ which are used to find the warning and action lines from the estimated process standard deviation, σ. The control chart limits for the control chart are calculated as follows:

Upper action line at $B'_{.001}\bar{s}$ or $B_{.001}\sigma$
Upper warning line at $B'_{.025}\bar{s}$ or $B_{.025}\sigma$
Lower warning line at $B'_{.975}\bar{s}$ or $B_{.975}\sigma$
Lower action line at $B'_{.999}\bar{s}$ or $B_{.999}\sigma$

An example should help to clarify the design and use of the sigma chart. Let us re-examine the steel rod cutting process which we met in Chapter 5, and for which we designed mean and range charts in Chapter 6. The data has been reproduced in Table 8.3 together with the standard deviation (s_i) for each sample of size 4. The next step in the design of a sigma chart is the calculation of the average sample standard deviation ($\bar{s}$). Hence:

$$\bar{s} = \frac{4.43 + 8.76 + 5.44 \ldots 7.42}{25}$$

$$= 4.75 \text{ mm}$$

The estimated process standard deviation (σ) may now be found. From Appendix E for a sample size, $n = 4$, $c_n = 1.085$ and

$$\sigma = 4.75 \times 1.085 = 5.15 \text{ mm}$$

This is very close to the value obtained from the mean range:

$$\sigma = \bar{R}/d_n = 10.8/2.059 = 5.25 \text{ mm}$$

The control limits may now be calculated using either σ and the B constants from Appendix E or $\bar{s}$ and the B' constants:

$$\text{Upper action line } B'_{.001}\bar{s} = 2.52 \times 4.75$$
$$\text{or } B_{.001}\sigma = 2.32 \times 5.15$$
$$= 11.96 \text{ mm}$$
$$\text{Upper warning line } B'_{.001}\bar{s} = 1.91 \times 4.75$$
$$\text{or } B_{.001}\sigma = 1.76 \times 5.15$$
$$= 9.07 \text{ mm}$$

Table 8.3 *100 steel rod lengths as twenty-five samples of size 4*

Sample number	Rod lengths (mm) (i)	(ii)	(iii)	(iv)	Sample mean (mm)	Sample range (mm)	Sample standard deviation (mm)
1	144	146	154	146	147.50	10	4.43
2	151	150	134	153	147.00	19	8.76
3	145	139	143	152	144.75	13	5.44
4	154	146	152	148	150.00	8	3.65
5	157	153	155	157	155.50	4	1.91
6	157	150	145	147	149.75	12	5.25
7	149	144	137	155	146.25	18	7.63
8	141	147	149	155	148.00	14	5.77
9	158	150	149	156	153.25	9	4.43
10	145	148	152	154	149.75	9	4.03
11	151	150	154	153	152.00	4	1.83
12	155	145	152	148	150.00	10	4.40
13	152	146	152	142	148.00	10	4.90
14	144	160	150	149	150.75	16	6.70
15	150	146	148	157	150.25	11	4.79
16	147	144	148	149	147.00	5	2.16
17	155	150	153	148	151.50	7	3.11
18	157	148	149	153	151.75	9	4.11
19	153	155	149	151	152.00	6	2.58
20	155	142	150	150	149.25	13	5.38
21	146	156	148	160	152.50	14	6.61
22	152	147	158	154	152.75	11	4.57
23	143	156	151	151	150.25	13	5.38
24	151	152	157	149	152.25	8	3.40
25	154	140	157	151	150.50	17	7.42

$$\text{Lower warning line } B'_{.975}\bar{s} = 0.29 \times 4.75$$
$$\text{or } B_{.975}\sigma = 0.27 \times 5.15$$
$$= 1.38 \text{ mm}$$
$$\text{Lower action line } B'_{.999}\bar{s} = 0.10 \times 4.75$$
$$\text{or } B_{.999}\sigma = 0.09 \times 5.15$$
$$= 0.47 \text{ mm}$$

Figure 8.10 shows control charts for sample standard deviation and range plotted using the data from Table 8.3. The range chart is, of course, exactly the same as that shown in Figure 6.8. The charts are very similar and either of them may be used to control the dispersion of the process, together with the mean chart to control process average.

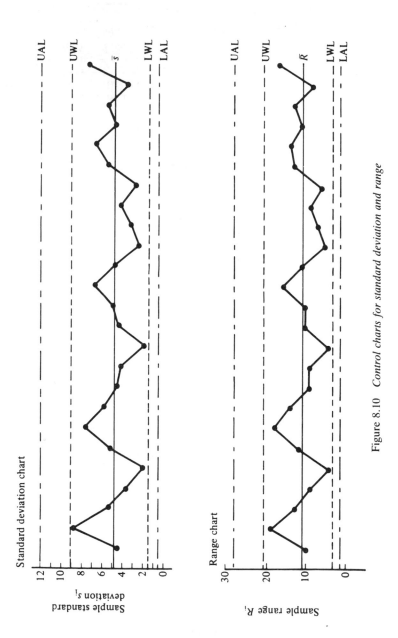

Figure 8.10 *Control charts for standard deviation and range*

If the standard deviation chart is to be used to control spread, then it may be more convenient to calculate the mean chart control limits from either the average sample standard deviation ($\bar{s}$) or the estimated process standard devication (σ). The formulae are:

Action lines at:

$$\bar{\bar{X}} \pm A_1\sigma$$
$$\text{or } \bar{\bar{X}} \pm A_3\bar{s}$$

Warning lines at:

$$\bar{\bar{X}} \pm 2/3A_1\sigma$$
$$\text{or } \bar{\bar{X}} \pm 2/3A_3\bar{s}$$

It may be recalled from Chapter 6, that the action lines on the mean chart are set at:

$$\bar{\bar{X}} \pm 3\sigma/\sqrt{n}$$

hence, the constant, A_1 must have the value:

$$A_1 = 3/\sqrt{n}$$

which for a sample size of four:

$$A_1 = 3/\sqrt{4} = 1.5$$

Similarly:

$$2/3\ A_1 = 2/\sqrt{n} \quad \text{and for } n = 4,$$
$$2/3\ A_1 = 2/\sqrt{4} = 1.0$$

In the same way the values for the A_3 constants may be found from the fact that:

$$\sigma = \bar{s}c_n$$

Hence, the action lines on the mean chart will be placed at:

$$\bar{\bar{X}} \pm 3\bar{s}c_n/\sqrt{n}$$

therefore,

$$A_3 = 3c_n/\sqrt{n}$$

which for a sample size of four:

$$A_3 = 3 \times 1.085/\sqrt{4} = 1.63$$

Similarly:

$$2/3\ A_3 = 2 \times c_n/\sqrt{n} \quad \text{and for } n = 4,$$
$$2/3\ A_3 = 2 \times 1.085/\sqrt{4} = 1.09$$

The constants A_1, $2/3A_1$, A_3 and $2/3A_3$ for sample sizes $n = 2$ to $n = 12$ have been calculated and appear in Appendix B.

Using the data on lengths of steel rods in Table 8.3, we may now calculate the action and warning limits for the mean chart:

$$\overline{\overline{X}} = 150.1 \text{ mm}$$

σ	$=$	5.15 mm	$\bar{s}$	$= 4.75$ mm
A_1	$=$	1.5	A_3	$= 1.63$
$2/3A_1$	$=$	1.0	$2/3A_3$	$= 1.09$

Action lines at:

$$150.1 \pm (1.5 \times 5.15)$$

or $\quad 150.1 \pm (1.63 \times 4.75)$

$= \quad 157.8$ mm and 142.4 mm

Warning lines at:

$$150.1 \pm (1.0 \times 5.15)$$

or $\quad 150.1 \pm (1.09 \times 4.75)$

$= \quad 155.3$ mm and 145.0 mm

These values are very close to those obtained from the mean range, $\overline{R}$ in Chapter 6:

Action lines at 158.2 mm and 142.0 mm

Warning lines at 155.2 mm and 145.0 mm

Chapter highlights

- SPC is based on basic principles which apply to all types of processes, including those in which isolated or infrequent data is available, as well as continuous processes – only the timescales differ.
- The use of individuals charts ignores the distinction between accuracy and precision. Individuals charts and range charts based on a sample of two are in common use, but their interpretation must be carefully managed. Mean charts with varying sample sizes require careful explanation to process operators.
- The median and the mid-range may be used as measures of central tendency and control charts using these measures are in use. The methods of setting up such control charts are similar to those for mean

charts. In the multi-vari chart, the specification tolerances are used as control limits.

- Data must be grouped together to allow the distinction between accuracy and precision to be made. When new data is available only infrequently it may be grouped into moving means and moving ranges. The interpretation of an individuals chart usually leaves something to be desired when compared to a moving mean and moving range chart, if subgrouping is rational.

- The method of setting up moving mean and moving range charts is similar to that for $\overline{X}$ and R charts. The rules for the moving $\overline{X}$ and moving R charts, however, differ from those for conventional $\overline{X}$ and R charts and care must be taken in their use.

- Under some circumstances, the latest data point may require weighting to give a lower importance to older data and then use can be made of an exponentially weighted moving average (EWMA) chart.

- The standard deviation is always a better assessment of the spread of a distribution than the range. The range is often more convenient and more understandable. Above sample sizes of twelve, the range ceases to be a good measure of spread and standard deviations must be used.

- Standard deviation charts may be derived from both estimated standard deviations for samples and sample ranges. Standard deviation charts and range charts, when compared, show little difference in controlling variability.

9 Random variation and its management

9.1 Introduction

The previous four chapters have been devoted to the subject of variables – things which are measured and not counted. The charting techniques discussed so far have been aimed at the control (or management) of the accuracy of the process but only the monitoring of the precision. There is no simple control loop for the management of the randon sources of variation, no simple knob to turn or simple procedure to follow, which will result in a decrease in the amount of the random variation. The use of mean and range control charts for control leads to less frequent adjustments of the mean; to building on the inherent stability of the process, and to a calmer and more stable operation.

The impact of the sample size in reducing the total variations to a limiting value, equal to the random component, is included in the discussion of process capability and process control in Chapter 7. There will be numerous occasions on which the introduction of mean and range charting, with a suitably chosen sample size, will suffice to reduce the total variations to a level where it is possible to demonstrate capability to meet the requirements of the customer. Where this is not possible, the investigation and reduction of the random component of the variations becomes essential. Even where capability of meeting the requirements exists, the never-ending improvement philosophy requires the random component to be investigated in order to ensure that the process stays ahead of the requirements. The subject of variation is vast and includes the analysis of variances, the design of experiments and parts of the so-called Taguchi methods. The treatment here will be to discuss variation and its management based on and around the data which conventional SPC techniques provide.

9.2 The components of capability

The assessment of the capability of a process involves the use of three components:

- The specification
- The accuracy
- The precision

Clearly all three components are potentially capable of being varied. The capability of a process is determined by the comparison of the specification with the combination of the inherent random variations and the ability of the control technique to detect changes of both accuracy and precision.

If a process is not capable there may be a possibility of varying the specification. Such a decision is only possible if the process is in statistical control. It would be in the hands of a level of management other than that

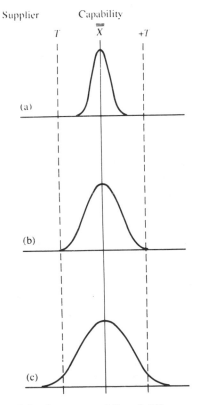

Figure 9.1 *Process capability of different suppliers*

at which the hour to hour, shift to shift, batch to batch responsibility for process and, hence, product control lies.

Charting enables this lower level of management to control the accuracy by noting when action is required and then following a predetermined procedure for adjustment. The process operator needs a 'help line' to pull when he receives an 'action on the range signal'. Some other level of management must then investigate and deal with the problem. An action signal on the range may be either good news or bad news – the range chart will indicate when this occurred and this should be a good clue to the possible contributory causes – look for the things which were changing at that time.

Figure 9.1 illustrates how reducing the random variations increases the capability of a process. It may enable a process in statistical control to be so improved that it reaches capability. It may also lead to a revision of the specification being proposed by the supplier and result in less variation being carried forward into the next process. The commercial advantages of this option are considerable.

9.3 The components of variability

What do we know about the random variations?

- They may be estimated by the range
- They are measured by the standard deviation
- They are made up of a number of elements or contributory causes

The causes of random variations will include, for example, measurement errors, varying methods of sample taking, the variations introduced in the raw materials used, the various methods of operation and control, the plant, the machinery, the equipment, the skills of operators, the environment, the planning, the product and so on. When the random variations are assessed in a process capability study it is the totality of all these contributory factors which is being estimated. So past, present and future data generated during normal process operation will keep giving information about the totality of the random variations and, if range or standard deviation charting is in use, the moments in time when there was an assignable change to these sources of variation.

Of course, the only way to reduce this totality of random causes is to establish the nature of the components, estimate their contributions and then tackle each identified source of random variation with a view to reducing its contribution. The range or standard deviation charts will assist in indicating when unusual events occurred, and both 'cause and effect' and 'Pareto' analysis can help to determine the likely causes and rank them for action.

The magnitude of the contribution of any individual component to the total random variation has to be estimated, either from existing data or by experimentation. For example, the contribution of the measurement method can be assessed by repeating a measurement on the same sample and using this data to calculate an estimated standard deviation. Testing often destroys the sample, in which case the combined contribution from the measurement, the sample taking and the sample preparation procedure may be estimated by taking a sample, large enough to be divided up for multiple tests, but in which the variation is thought to be relatively low. In practice this can be achieved by selecting a small area of product and making the assumption that the variations over such a small area will be relatively small by comparison with the total variations normally observed; or by taking a small volume in which it is reasonable to assume that the variations will be small; or by careful blending of a sample. The parameter, either a variable or an attribute, can then be measured on the sample ten, or better twenty, times. With this data both the mean and standard deviation of this family of results can be estimated. This will estimate the standard deviation of the measurement and/or sampling component of the total random variations. It is important to emphasize that this gives only an estimate of this standard deviation. (See Chapter 5, Section 5.3 for the detailed procedure for estimating the standard deviation of a family of results.)

9.4 The addition of independent components of variation

Given an estimate of the degree of variation due to the measurement method, the next question is: 'How much does it contribute to the total variations?' For the examples suggested above, the act of sampling and the act of measurement do not influence the other sources of random variations which make up the total – they are independent.

Not everything adds together in an arithmetic way, like $1 + 2 = 3$. You cannot add the lengths of two sides of a triangle to give the length of the third; you cannot add the probabilities of consecutive events (if the probability of heads when tossing a coin is 1/2. The probability of two consecutive heads is not $1/2 + 1/2$), and you cannot add square roots ($\sqrt{3}$ added to $\sqrt{4}$ does not give $\sqrt{7}$). Because a standard deviation is a square root, standard deviations do not add together arithmetically, neither do ranges or tolerances. But the square of the standard deviation, the variance, may be added arithmetically. Variations do not add through the standard deviations, they add through the variances.

The formula for the variance is:

$$\sigma^2 = \sum \frac{(x_i - \bar{x}^2}{(n-1)}$$

The addition of variances for populations which are independent and normally distributed may be expressed in a mathematical form as follows:

$$\sigma^2_{(total)} = \sigma^2_a + \sigma^2_b + \sigma^2_c + \ldots$$

This means that the total variance, or the standard deviation (of the totality of random variations) squared, is equal to the sum of a number of independent component variances, labelled a, b, c etc. The estimate of the measurement component of the total random variations can now be put into this formula to see how large or small a part it plays in the total.

This 'adding through the squares' has some very important consequences which can be illustrated by an example.

The research chemists have shown that the temperature in a new process must be controlled to $\pm 1°C$. This is the total tolerance which is allowed. The process operator or manager has to decide what tolerance he can accept in the thermometer, which will be used to assess the temperature for control purposes. He has the following additional information:

Thermometer manufacturer's specified tolerance °C	Relative price unspecified units
± 1	1
± 0.5	3
± 0.3	30
± 0.1	50
± 0.03	500

The manufacturer's specified tolerance will reflect the precision of the thermometer and hence be a function of its standard deviation. A $\pm 0.5°C$ thermometer will be capable of recording results to intervals smaller than $0.5°C$ but each result will be subject to an error related to the tolerance of the thermometer. The choice of thermometer will reflect the compromise between the desire for precision and the price of such precision. How can the decision be made rationally?

The total variance of the temperature allowed by the research chemist must be greater than or equal to the sum of the variance introduced by the lack of precision of the thermometer and the variances introduced by all the other sources of variation. This may be expressed as an equation:

$$\text{or } \sigma_{(total)}^2 \geqslant \sigma_{(therm)}^2 + \sigma_{(the\ rest)}^2 \text{ [Equation 1]}$$

Let us equate the maximum tolerance allowed in a given parameter to three times the standard deviation for that parameter (this is only approximately true, but if the approximation is maintained throughout all the following calculations, it has no effect on the resultant conclusions). So:

$$T = 3\sigma$$

and $\quad T_{(total)} = 1°C = 3\sigma_{(total)}$

or $\quad\quad \sigma_{(total)} = 1/3°C$

hence $\quad \sigma_{(total)}^2 = (1/3)^2 = 1/9 = 0.1111$

This has enabled the tolerance imposed on the process to be expressed as a variance, so the left-hand side of Equation 1 has been estimated.

Similarly for the thermometer, we may again approximate and write

$$T_{(Therm)} = 3\sigma_{(Therm)}$$

For the cheapest thermometer the tolerance is 1°C, so:

$$\sigma_{(therm)} = 1/3°C$$

hence $\quad \sigma_{(therm)}^2 = (1/3)^2 = 1/9 = 0.111$

The tolerance of this thermometer is now also expressed as a variance. If this value for variance is substituted in the right-hand side of Equation 1, we see that the variance available to 'the rest' may be zero. The result is not surprising; a thermometer with a tolerance of ±1°C is not capable of controlling a process for which the tolerance (or specification) is ±1°C.

One should avoid the temptation to dismiss this result as trivial, for the authors' experience includes numerous examples of parameters for which the test method is hopelessly inadequate, when compared with the specification. These examples fall into two broad categories. The first is where a rough and ready method of measurement is given to process operators, often backed up by a more precise method available to the laboratory or gauge room. In this case it is important to establish that the method used by the process operators is not *too* rough and ready. The second category is where the customer has tightened the specification beyond the precision of any known method of measurement – if this is recognized the solution lies in asking the customer, or whoever set the specification, to nominate the method of measuring to that precision.

Having rejected the ±1°C thermometer, the calculation can be repeated for the ±0.5°C thermometer. Again:

$$T_{(\text{therm})} = 3\sigma_{(\text{therm})}$$

so

$$\sigma_{(\text{therm})} = 0.5/3°C$$

and

$$\sigma_{(\text{therm})}^2 = (0.5/3)^2 = 0.0278$$

From Equation 1, slightly modified, we have:

$$0.1111 - 0.0278 \geqslant \sigma_{(\text{the rest})}^2$$

so

$$0.0833 \geqslant \sigma_{(\text{the rest})}^2$$

hence

$$\sqrt{0.0833} = 0.2886 \geqslant \sigma_{(\text{the rest})}$$

but

$$T_{(\text{the rest})} = 3\sigma_{(\text{the rest})}$$

so

$$T_{(\text{the rest})} \geqslant 0.886°C$$

This means that the remaining amount of tolerance available to 'the rest' of the possible sources of variation within the process, after allowing for the lack of precision within the thermometer, will be at least 86.6 per cent or more of the total permitted tolerance. If one uses a thermometer, for which the error can be expressed as a tolerance of ±0.5°C, to control a process in which one seeks to maintain a total variation which is within a tolerance of ±1°C, then the thermometer takes up some part of the total tolerance but leaves at least 86.6 per cent of it available for 'the rest' of the factors which will contribute to the process variation. A lack of precision, or margin of error of 0.5°C, in a thermometer still allows it to be used for control, even when the imposed tolerance is 1°C. Clearly a ±0.5°C tolerance thermometer is cheap and not too bad for control of this process.

The calculation can then be repeated for all the thermometers and gives the results shown in Table 9.1. Not surprisingly, this shows that the ±0.3°C thermometer is able to leave more room for the rest of the variations than the ±0.5°C thermometer. As the tolerance of the thermometers decreases, the amount of variation left available to the rest also decreases but much less rapidly. It will not be worth a lift of ten times in price to obtain the trivial advance in available tolerance given by buying the 0.03°C tolerance thermometer, instead of the 0.1°C tolerance thermometer.

Table 9.1   *The impact of a thermometer tolerance on the rest*

Tolerance thermometer (°C)	$\sigma_{(\text{therm})}^2$	$\sigma_{(\text{the rest})}^2$	Tolerance 'the rest'
1°	0.1111	0	0
0.5°	0.0278	0.0833	0.866
0.3°	0.0100	0.1011	0.954
0.1°	0.0011	0.1100	0.996
0.03°	0.0001	0.1110	0.9995

Sometimes the concept of adding through squares is not easily assimilated. One may have the feeling in this example, that one has eaten half a cake and then been tricked into believing that 86.6 per cent is still left. There is a simple explanation for this, using the language of chance and probability. It is that, given two sources of random variation, the probability of observing an extreme value from one source (for example at three standard deviations from its mean) and, at the same time by chance, observing a similarly extreme value from the other source is very remote. If the probabilities of the first and second events are both 1 in 1000, then the probability of the two occurring simultaneously is 1 in (1000×1000) i.e. one in a million. This probability of the combined event is considerably further away from the mean than three standard deviations, so the distribution of the probabilities of the combined event is much narrower – the impact of the 0.5°C thermometer on the total tolerance is less than 0.5°C.

Some readers may find the geometric model shown in Figure 9.2 of assistance. In this model of the thermometer problem, the total tolerance is represented by the radius and is always of constant length. This radius is made up of two components – the tolerance of the thermometer shown on the vertical axis in Figure 9.2 and the tolerance of 'the rest' on the horizontal axis. If one examines a situation in which the tolerance of the thermometer and 'the rest' are equal then, using Pythagoras' theorem, they will equal $1/\sqrt{2}T_{(total)}$ and not $1/2\ T_{(total)}$. Figure 9.2 also shows the 86.6 per cent result when $T_{(therm)}$ is 1/2 of $T_{(total)}$. Of course, when $T_{(therm)}$ equals $T_{(total)}$, the $T_{(the\ rest)}$ component is zero.

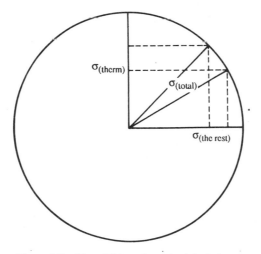

Figure 9.2 *The addition of standard deviations*

From this example we may conclude that:

- The contribution of any one source of variation to the total variations is added through a squares rule and so its individual influence is attenuated. If this were not so we would not be able to run processes involving multiple sources of variation and over which we seek to exercise control of several parameters simultaneously.
- Chasing the precision of measurement often makes no useful contribution to increasing knowledge or improving control. Yet laboratories and gauge rooms are full of expensive degrees of precision which are not always required by, nor indeed helpful to, their owners.
- A seemingly rough measurement (like the ±0.5°C thermometer) is surprisingly good.

To illustrate the consequences of the squares rule we examined the relative 'usefulness' of thermometers. In the wider area of the practical measurement of parameters, one clear message is *'do not chase precision'*. It is seldom the ability to measure something with increased precision which leads to an improvement in the capability of a process. It is clearly necessary to know how large a part of the total variance is contributed by the variance of the measuring and sample methods. Once this contribution has been found to be acceptable it is pointless to repeat its assessment except on an auditing basis.

On the other hand, the ±0.5°C thermometer example illustrates how a relatively rough measure can be used for control. Areas for the practical application of this conclusion are to be found, for example, when there are long delays in obtaining the precise measurement of a parameter. Rough and quick methods of assessment then become necessary, if not vital, for the control of a process. There are many examples of such rough measures which include:

- A long narrow 'V'-shaped gauge can be used to measure shaft diameters to quite high degrees of precision.
- No airfield is complete without its windsock, used to assess the direction, the speed and possible gusting of the wind.
- Viscosity can be assessed by measuring the pressure drop down a pipe, or by measuring the amps taken by the stirrer of a liquid.
- pH, litmus and albumen test paper.
- Touch, sight, smell and other senses of experienced operators etc.
- 'Ringing' of metal components to test their physical properties.
- The use of ultrasonics and gamma rays for measurements of moisture content and thickness.
- Melt indices as measures of molecular weight.

The message here is do not dismiss rough and ready methods of measurement, assess their variances and then decide on their potential for use. In almost all cases it is better to have a rough estimate quickly, and the possibility of acting on it, rather than a more precise result later. If the problem is that the assessment is too late, look for a rough and quick method. Measure its variance and compare it with the total variance available to see if it is acceptable (equivalent to the 0.5°C thermometer). It will also be necessary to establish the correlation between the two methods of assessment (see Section 9.5) in order to calibrate the rough and quick method from the precise one.

9.5 The addition of dependent components of variation

The reader will appreciate that the way in which dependent components of variation add together is a much more complex subject than the addition of independent components of variation. In chemical processes, for example, the random variations of temperature may give rise to related variations in other parameters such as pressure, conversion rate, reactivity of a catalyst, heat transfer rates etc. Each of these parameters may in turn be the subject of both independent and other dependent random variations. The philosophy, science and mathematics of this subject are included in 'the analysis of variances' and 'the design of experiments'. These are subjects which are outside the scope of this book. The further discussion of the addition of variances will therefore be limited to the way in which some understanding of this subject can be obtained by using data made available during both process capability studies and control charting for variables.

In any manufacturing process where SPC charting techniques are in use, it is normal to aim to keep all the variables in statistical control and to adjust them at the first significant indication of an assignable cause. This means that if a parameter of the process 'A', is interdependent with another parameter of the process 'B', it is normal to seek to keep both parameters in statistical control, during normal operation. If the variations in both A and B are entirely random there will not be a correlation between them and a scatter diagram will show its absence – the scatter diagram obtained by plotting values of A against simultaneous values of B will be a bunch of random points. Under normal process operation A will be the subject of assignable causes of variation which, if B is dependent on A, will 'drag' B along with the assignable causes of change in A. The corresponding point on a scatter diagram during such a period should be 'dragged' out of the random bunch of points. If individual values of A and B are plotted on a scatter diagram the scatter will reflect both the random component of the variations as well as any assignable components.

As in the case of the mean and range charts, this scatter can be 'calmed down' by plotting the mean values of A against the mean values of B when the random components of both variables will be attenuated. At the same time this procedure is made logical since, if there are correlations between A and B, a drift in the mean value of A must result in a drift in the mean value of B, whereas a random variation in A may not give rise to a correlated and identifiable variation in B.

We have used the formula

$$\sigma_{(total)}^2 = \sigma_a^2 + \sigma_b^2 + \sigma_c^2 + + +$$

and recognized that this is only true when a, b, c, etc. are independent and the random components of the variations are normally distributed.

We also recognize that, under normal manufacturing conditions, efforts are made to keep all the sources of variation under control and the total variations as small as possible. This must be so if both the customer is to be satisfied and the business is to be profitable. So, as a first approximation, one can assume that the relationship between the various sources of variation and the total variation are linear over the relatively small range which will occur during controlled operation of a manufacturing process. In addition, if a correlation is sought between the means of samples for which the sample size is four or more, the Central Limit Theorem tells us that the means will be distributed normally irrespective of the distribution of the population from which they are derived. So we may write:

$$SE_{(total)}^2 = ASE_a^2 + BSE_b^2 + CSE_c^2 + + K \, [\text{Equation 2}]$$

where SE_a^2 is the variance arising from parameter A, and A B C . . . and K are constants.

A, B, C . . . will be 1 when the sources of variation are independent. K will be negative, if in adding together the component variances some of the individual sources of variation have not been included.

What does this relationship mean? Clearly, the question is important because, in many processes, the sources of variation are not all independent of each other. If this relationship is a good approximation to the truth, and all the constants and standard errors in the equation are known, we will have total knowedge of the process, within the range of variation which is normally present. This would enable one to exercise excellent process control and probably initiate some very good process research. Like 'zero defects', this state of total knowledge will not be achieved overnight but it will only be achieved by accepting it as a target.

It may take several years to assess even the major components of the total variance, by a series of successive approximations or estimates of the values of the constants and the values of the standard errors.

If one of the causes or components of the total random variations has been identified, it may be possible to assess the amount by which it varies randomly and estimate its standard deviation and variance. This could require experimentation, but a production plant is not the recommended place for experiments. Sometimes the information is available from existing process data. For example, suppose that the temperature and pressure within a polymerization reactor are the subject of regular records, which also include the melt index of the resultant polymer. The data can be analysed to measure the process capability of all three parameters and obtain estimates of the means, ranges, standard deviations, standard errors and variances. In order to look for possible correlation between these three parameters, we may start with the values of the means assessed at the same times and plot these, one against another, on scatter diagrams – of temperature against pressure, temperature against melt index and pressure against melt index.

We already know from Chapter 4 that the eye is a fairly good first judge of whether a scatter diagram suggests that a correlation exists and, if a correlation looks likely, a second judgement can be made with the aid of a simple test. In plotting mean values on a scatter diagram it is important to make the best possible use of all the data. A direct plotting of means of one parameter against the means of another parameter, assessed at the same time, would give a diagram in which the time element would be lost. A simple remedy is to plot the points on the scatter diagram as letters of the alphabet or numbers – an example is shown in Figure 9.3. The mean of the first sample of Factor I is plotted against the mean of the first and simultaneous sample of Factor II – this is plotted as the letter 'A'. Then the second samples are plotted as the letter 'B' etc.

Of course, the plot in this form enables us to check whether the letters themselves are scattered about or whether there are distinct families – early letters in one group and later ones in another group. If such families are present it would be right to view the possible correlation with some suspicion. The presence of a 'flier' lying outside either a correlating or a non-correlating group will be immediately identifiable with the period when such behaviour occurred.

To assist in the judgement of whether a correlation seems likely or not, a line equal in length to four standard errors (4SEs) of the means of Factor I has been added horizontally in Figure 9.3. A similar line equal to four standard errors of the means of Factor II has been added vertically (the process capability or control chart data will include a knowledge of the standard errors of the means). In a band which is 4SEs wide, one will

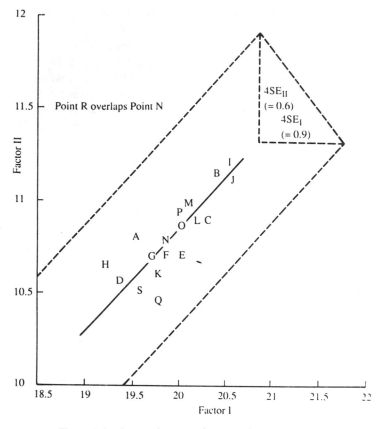

Figure 9.3   *Scatter diagram of means of Factors I and II*

expect to find 95 per cent of the means of Factor I and similarly for Factor II. The random variations of the means of the two Factors I and II will add through their variances. Pythagoras' theorem of right-angled triangles tells us that the hypotenuse squared is equal to the sum of the squares on the other two sides – so the length of the hypotenuse corresponds to 4SEs of the combined variance of the two parameters.

If there is a correlation, one may expect to find (with a 95 per cent degree of confidence) that the points on the scatter diagram will be scattered about a correlation line and lie within a 4SEs' bandwidth. The slope of the correlation line is related to the constants in Equation 2.

The data needed to assess capability and to control processes can be assembled into scatter diagrams to show whether correlations exist. Using the above method the chances of a false conclusion will be remote. Once it has been established that a correlation exists, the input data may be presented for a more thorough analysis to ascertain the precise characteristics of the correlation.

The above discussion of the use of scatter diagrams is based on the assumption that data is available from normal process operating conditions, during which there will have been some component of both random and non-random variations of the means of the control parameters and that, in addition, parameters within the process may have been interrelated so that the measured values will show correlations. The whole emphasis, when using mean and range charts for control, is that the process should be left alone and only adjusted when there is clear evidence that an adjustment of the mean values is required – so it follows logically that the 'philosophy' of SPC does not encourage experimentation on an operating plant.

It also follows that the range of variation within any SPC controlled process will be as small as can possibly be maintained. It is not

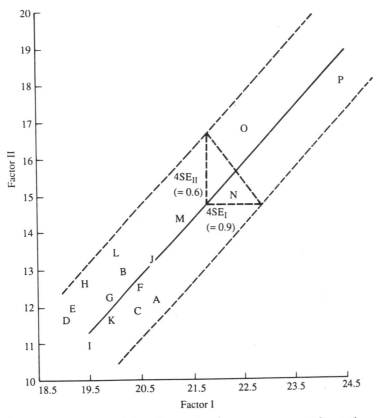

Figure 9.4   *Correlation shown when a process moves out of control*

uncommon, however, for things to go wrong, and under these conditions the immediate concern is to re-establish steady running. Figure 9.4 illustrates what will happen to the scatter diagram when the control of Factor I is lost. As I moves to higher values it will drag II along with it, but only if the two factors are interrelated. Under such conditions, one may have a rare sight of the correlations normally hidden during steady operation. So loss of control presents a special situation in which the chances of finding correlations will tend to be increased. The important conclusion is that when unsteady conditions occur, record all the data and analyse it to see if it yields otherwise elusive information.

A scatter diagram of means is easily constructed – it is also easy to interpret. When there is evidence of a correlation, assistance may be needed to determine the degree of confidence with which it is associated. Assistance should not be needed to answer questions such as: 'What does the existence of a correlation mean' and 'What should be done about it?' The answers to these questions will depend on the process involved, as well as the nature of the correlation, and are beyond the scope of this book. They should not be beyond the scope of the operators and managers of the process.

9.6   Blending and mixing

Blending and mixing are operations often carried out with the intention of reducing the amount of variation present. Mixing is the operation of putting two or more differing components together so that the members of one component are diffused among those of the others. Mixing can be of various types.

It is clear that no amount of mixing of mechanical or electrical components will result in any change in the physical dimensions or characteristics of the members of the populations being mixed and, in the absence of any real mixing, a discussion of its influence on variation is a nonsense.

There is a less well-defined area in which mixing may seem to have occurred. For example, mixing carbon black and powdered chalk will give rise to a grey appearance but the carbon and chalk remain distinct within the mixture. In this case there is no true blending or homogenizing of one component with another and the characteristics of the components remain unchanged, but a limit is imposed by the sensitivity of the eye, and the mixture 'appears' to be grey. Another example of the insensitivity of measurement causing a mixture to appear to be present is the blending of particles of low molecular weight polymer with high molecular weight polymer, when the estimate of the molecular weight of the mixture may appear to have changed. Unless some chemical reaction has occurred, however, the blend will continue to consist of two components. This can have important consequences in the polymer industry, where the assessment of the average molecular weight can fail to indicate the way in which the polymer will perform. Readers who are familiar with the process industries will know of many similar situations.

Blending and mixing of components which are miscible and capable of being homogenized is also possible and, in this case, the performance characteristics of the components are changed. So, before seeking the statistical interpretation of any results obtained from blending or mixing, one has to ask common sense questions in order to establish if blending is occurring. Where true blending occurs, the components are miscible. Each component has its individual mean and standard deviation for each of its measurable parameters. Consider the process illustrated by the flowchart shown in Figure 9.5. A common feedstock has one particular parameter of known mean and known standard deviation. The feedstock is fed to a series of reaction vessels set in parallel. During the reaction, additional sources of variation add to the variance of the output of each reactor and, for any given reactor, the variance in a given output parameter will be given by an equation of the type:

$$\sigma_{(\text{output})}^2 = \sigma_{(\text{input})}^2 + \sigma_{(\text{process})}^2$$

Under ideal circumstances the process variances and means of all four reactors will be equal, and it follows from this equation that all the output variances and means will also be equal. If this ideal condition is followed

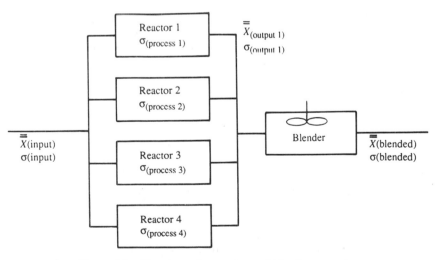

Figure 9.5 *Flowchart of a reaction and blending operation*

by perfect blending the resultant standard deviation of the blender output will be given by the formula:

$$\sigma_{(\text{blended})} = \frac{\sigma_{(\text{output})}}{\sqrt{n}}$$

where n is the number of identical components blended.

(The reader will recognize this formula as that used to derive the standard error of the means from the standard deviation of the initial population). This formula can be used to determine a value of n from estimates of the standard deviations of the blender inputs and output. This estimate of n reflects the effective number of blendings. In practice, a theoretical number of blendings is normally known from the number of inputs, or the number of batches being blended, or the ratio of blender volume and dwell time in continuous blending. The comparison of the theoretical value of n and that determined from the input and output standard deviations is a measure of the efficiency of the blending and can be used to assess and monitor the performance of blenders.

When blending is used to achieve greater uniformity we know that:

- There is a requirement that blending presents a real and not just an apparent opportunity to homogenize the components.
- Since perfect blending is not always possible, the standard deviation of the blended output will be reduced when the mean of the individual component inputs equal each other.
- The variation of the blended output is a function of the square root of the number of blendings carried out.
- The effectiveness of the blending can be estimated by comparing the theoretical value of the number of blendings with the value obtained by comparing the standard deviations of the inputs with the standard deviation of the outputs.

Clearly this subject becomes considerably more complex when either or both the means and the standard deviations of the input components are not identical. This simple discussion of blending offers an opportunity to assess its effectiveness using information which should be available if the process is being controlled by SPC techniques. In all cases, increasing the number of blendings is effective on a square-root rule. This means, for example, that if the number of batches blended together is increased from four to five, the cost of blending will rise by a factor related to 1 in 4 or 25 per cent, whereas the effectiveness of the additional blending in reducing the variation will be in the ratio of $\sqrt{5}/\sqrt{4}$ or about 12 per cent. A 12 per cent reduction in the input variation may cost less and will always be potentially of much greater benefit since it must result from improved process control, whereas blending is often simply a method of coping with unacceptable levels of process control and consequential variation.

Chapter highlights

- Whereas most processes are associated with a known procedure to correct assignable changes of the process mean, or accuracy, none have a simple procedure for adjustment of the amount of random scatter, or precision. Assignable changes of the precision are located in time by range control charts and may, therefore, be identified and eliminated.
- The precision of a process needs to be managed as well as monitored. It is, however, easier to manage accuracy than to manage precision.
- In practice, the precision of a process results from the sum of multiple individual sources of random variation, and can only be reduced by identifying and addressing the individual source.
- The capability of a process is a function of the specification, the total random variations and the sensitivity of the techniques of control. While increased capability can sometimes be obtained by increasing the sample size used in mean and range charting, often such improvements are not sufficient to satisfy the customer's requirements and it is necessary to investigate the make-up of the components of the random variations.
- When a component of the random variations is identified it may be possible to estimate its magnitude, and, through its *variance*, to assess the size of its contribution to the total.
- The addition of variances is a 'square' rule, which means that the impact of any individual component on the total of the random variations is attenuated. This also leads to important conclusions concerning measurement, where neither chasing precision nor ignoring 'rough and ready' methods of measurement are recommended.
- During process operation, experiments designed to provoke variation in the parameters used for control are not recommended; but during normal operation both random and assignable variations do occur. Data collected during normal operation can be used in scatter diagrams of mean values to reveal significant correlations between the various parameters being measured.
- Knowledge of correlations between measured parameters leads to greater understanding and the potential for improving the control of a process.
- Additional opportunitites for seeking correlations between measured

parameters exist when data is recorded during periods of accidental loss of control of a process.

- Blending is an operation used to reduce variation. Data available from charting can be used to assess the efficiency of blending. Blending efficiency is typically related to only the square root of the number of blendings, so it is often less expensive to decrease the variations in the inputs to a blender rather than increase the number of blendings.

10 Managing out-of-control processes

10.1 Introduction – process types and variability

Statistical process control is based on a number of basic principles which apply to the various methods available for doing work. Methods of doing work may be classified as:

- *Project* – such as the one-off design and construction of buildings, specialized plant, haute-couture dresswear, the creation of works of art, the preparation of a purchase contract, etc.
- *Job* – when a series of processes are used to convert inputs to outputs using the multiple skills and methods of one operator. This includes the work of artisans and is often the method of manufacture or operation used when a project or service is first offered to fulfil a limited demand. Job also includes one-off 'batch manufacture' as used in various process industries, the diagnosis and treatment of illness and disease, equipment servicing, setting up a mortgage facility etc.
- *Batch* – which describes work when a limited set of processes are repeated to produce the same outputs which are collected into a batch, before being passed on to the next conversion process. This includes various types of mass production, such as mechanical and electrical component manufacture, checking out at supermarkets, data inputting in accounting, and other services. Work in batches is typically associated with the need for sampling.
- *Flow* – where each operator carries out only a limited number of processes on a batch of size one. As the inputs from one process pass through an operator and become the inputs for the next process, there is minimum delay between the various process steps. This method of working is used in the mass assembly of cars, electronic equipment, household goods, making-up of textile garments, serving of meals in restaurants, mail sorting, telephone exchange manning, etc.

- *Continuous* – in which a continuous flow of inputs is transformed into an equally continuous flow of outputs. This includes the manufacture of various chemical-based products, where stability of operation and economy of scale are achieved by continuity, and electricity generation and hospital services, where the demand determines the need for continuity of supply.

While the various types of work may be defined discretely, actual work often includes a mixture of types. For example, in the construction of a building, the whole of the work is a project: concrete mixing, brick laying and electrical installation are all jobs; timbers, cable conduit, and fastenings will have been made in batches; fuse boxes, telephones and window frames may have been assembled in flow; and drawn steel rod manufacture and water supplies are continuous processes.

Within all these possible types of process, the basic principles underlying statistical process control apply with varying degrees of relevance. The basic principles are:

- All processes are inherently reproducible.
- Within the limits of reproducibility there is variation.
- Within the limits of reproducibility there is stability.
- Causes of variation are of two types – random and assignable.
- The control procedures add to the total variation.
- The common objective is perfection.
- SPC techniques contribute to the management of objectives, variation, stability and reproducibility.

The authors have met, and continue to meet, process operators and managers who are convinced that their processes are uniquely different, and even some who believe that they succeed in what are essentially miraculous achievements. The first step towards effective management of variables may be the need to accept that all processes are covered by these same basic principles and hence a knowledge of statistics will assist in their management. Project work lies at one extreme of the types of work where the amount of data available for analysis may seem to be too small for the application of SPC. No one starts to design or execute project work without some previous experience which they seek to incorporate in their current work. If, during previous project work, the problems which arise, their causes, and the solutions adopted, were not adequately recorded, analysed and made known, the same old problems will recur and result in the wasteful need for repeated remedial action. The authors have found numerous examples of project work in which the bank of previous experience is not formally recorded, and the same mistakes keep being repeated; SPC embraces the recording and use of such data.

Moving through the range of processes, from project to continuous, the possibilities for data collection tend to increase. Indeed in mass production of batches, as well as flow and continuous processes, there is often so much data available that the maladies of 'data diarrhoea' and 'information constipation' often co-exist.

10.2 The evaluation of actual process control procedures

For no matter what process, each time it starts or continues, the whole recipe of components and the methods of operation are reproduced in the expectation that the output will be similar to previous work and lie within predetermined specification limits. Stability of the output is sought by either holding conditions constant or by seeking to change them in a way which will hold the output constant (the output may be the product to be manufactured, the service to be rendered or a parameter important to the product or process control). Non-conformance is recognized when the outputs are not what was expected, it is *not* limited to those cases where the output is unacceptable, or difficult to dispose of.

The key to justifying existing procedures is to undertake process capability studies. Variable data on raw material supplies, process and product parameters is often available on file. If at least fifty consecutive data points are available, submit them to the normal mean and range analysis by grouping them into samples of size four or more. Examine the charts and establish if the process is 'in statistical control' (both the means and the ranges of the grouped data are satisfactory when examined for unexpected results or trends). In all cases this analysis will give a value for the mean range from which the standard deviation of the process may be estimated. The estimated standard deviation may be used in conjunction with the specification to calculate a process capability index, *Cpk* (see Chapter 7).

The existing data is unlikely to have been collected during a period limited to that associated with the inherent stability of the process. So this assessment of statistical control, along with the estimates of both the standard deviation and the process capability index, are based on data collected over periods when the conditions of operation and the control procedures may have been changing. It is also possible that the data is not exactly consecutive and, in the case of raw materials, there may have been periods of either eliminated or unavailable data. All of this needs to be taken into account when reviewing the results of the analysis which may then proceed by using the flowchart in Figure 10.1 The reader is recommended to follow this flowchart through in detail, taking note of the following observations concerning some of the more critical steps:

1   Start with an array of existing data in chronological order.
2   First examine the range chart and then the mean chart.
3   Investigate capability only when statistical control has been established.
4   The investigation can be assisted at various stages by the use of cusum charts (see Chapter 12).
5   The assessments of the random component of variation includes the possibility of random-type control behaviour.
6   The flowchart leads to a number of possible ends as follows:

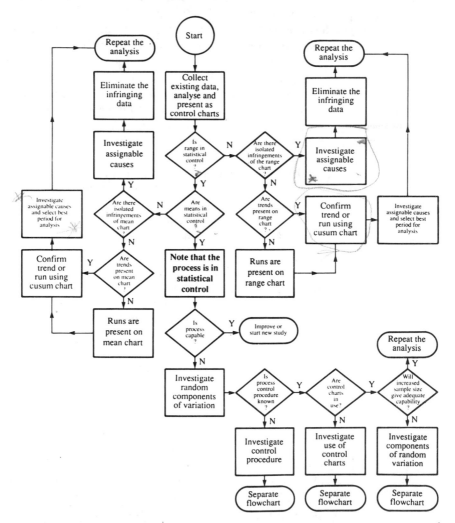

Figure 10.1   *Flowchart for review of capability analysis*

- The recognition that the process is both in statistical control and capable. In this case either consider further possible improvements or move to the analysis of other available data.
- The identification of isolated infringements of either the mean or the range chart and the elimination of the isolated data before repeating a full analysis. Reprocessing the data after this elimination will lead to new estimates of the state of control and the measures of capability. Before eliminating such data and repeating the analysis it is worth glancing at the results not so far examined. This may reveal, for example, that the control procedure is chasing the means around quite violently, in which case repeating the whole analysis before considering the use of SPC charting techniques for control is probably of little value. Alternatively the *Cpk* may be very high, say above 3, in which case it may be expedient to leave unchallenged a seemingly capable process which is not in statistical control and give greater priority to the study of other parameters.
- The recognition that further work is required before a useful analysis can be repeated. This further work may be the investigation of the control procedures normally in use; or the possibility of introducing capability studies and control charting as the normal control procedure; or a detailed investigation of the random causes of variation which are seen to be limiting the capability of the process; or the use of an increased sample size to achieve the required increase in capability.

The authors have often encountered unwillingness to even attempt the above analysis of available data, the typical objections being:

- We make one-off batches using a complex route and no batch can be the same as another.
- We run a continuous process.
- We seek to change the parameter not to hold it constant.
- Experience has shown that it is better to make it, test it and then correct it based on the test results – this works in most cases.
- Our plant is fully automated with computerized control.

For all processes the constant objective is to reproduce earlier experience. Without the distinction between accuracy and precision being made, the process will be adjusted by changing the mean value, both when this is necessary and when it is not. Similarly with parameters which are being made to change, the difference between the observed value at any moment in time and the target value for that time will again be a compound of both the random variations which determine the precision of the process and the mean value for which control is sought.

Computers can be programmed to rules which distinguish accuracy and precision. Often computer-controlled plant is programmed to change parameters more frequently than can be achieved manually, and, because this may lead to excessive hunting, an arbitrary degree of damping is sometimes introduced. Reprogramming the computer is relatively easy and proving that this is necessary is one of the results to be expected from the analysis of existing data as suggested above.

10.3   The analysis of existing data – some examples

Variable data from process inputs

A supplier of a raw material, biscuits in this case, has been asked to give more detailed information about the product. The supplier normally quotes the 'lot sample result' of the moisture content of the biscuits and has been requested to make available the moisture content data collected during manufacture. This is shown in Table 10.1 from which one may note that the specification is 4.9 per cent maximum and 3.0 per cent minimum, that the 'lot sample result' is given as 4.3 per cent, and that the data is in fact a series of results taken every three hours over a period of twelve days. What does the data in this table tell us? There is no record of out-of-tolerance results although there are quite a number of results at the upper specification limit of 4.9 per cent.

Table 10.1   *Analysis of percentage moisture content*
Limits: 4.9 per cent maximum
          3.0 per cent minimum

Day	Samples taken at:							
	03.00	06.00	09.00	12.00	15.00	18.00	21.00	24.00
1	4.6	4.2	4.7	4.3	4.1	3.9	4.3	4.4
2	4.4	4.5	4.5	4.4	4.2	4.1	3.8	4.0
3	4.9	4.4	4.2	4.3	4.1	4.0	4.3	4.2
4	3.8	4.4	4.5	3.9	4.4	4.0	4.3	3.9
5	4.6	4.0	4.1	4.1	4.5	4.1	4.4	4.5
6	4.1	4.3	4.2	4.4	4.3	4.5	4.1	4.3
7	4.8	4.7	4.9	4.2	4.8	4.2	4.9	4.3
8	4.2	4.7	4.5	4.5	4.6	4.4	4.8	4.2
9	4.3	4.6	4.7	4.9	4.4	4.6	4.8	4.9
10	4.3	4.7	4.8	4.9	4.4	4.5	4.9	4.3
11	4.6	4.5	4.2	4.9	4.8	4.6	4.5	4.4
12	4.4	4.9	4.9	4.6	4.5	4.5	4.4	4.4

Lot sample: 4.3 per cent

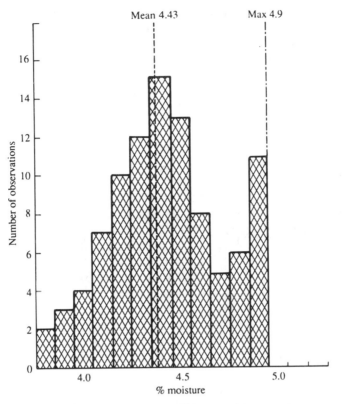

Figure 10.2 *Distribution of biscuit moisture contents*

The distribution of the results can be more closely examined by constructing a bar chart. This has already been shown in Figure 3.6 and is reproduced as Figure 10.2. It reveals that:

- There is no recorded result below 3.8 per cent or above 4.9 per cent.
- The most frequently observed result is 4.4 per cent and on either side of this value the number of observations falls away (the mean of all the results has been calculated at 4.43 per cent).
- The frequency of observation rises again at 4.7 per cent and there is then a sharp cut-off at 4.9 per cent, the upper specification limit.

How can these observations be interpreted? Are the suppliers cheating, and, when they find a result of 5.0 per cent or more, do they 'adjust' it by 'rounding down' or 'taking a repeat sample' or by 'adjusting the process and then taking a repeat sample, which is the subject of the record now available'? The distribution before 'adjustment' might have been the one in Figure 10.3. Displacing, revising or replacing eight of the observations

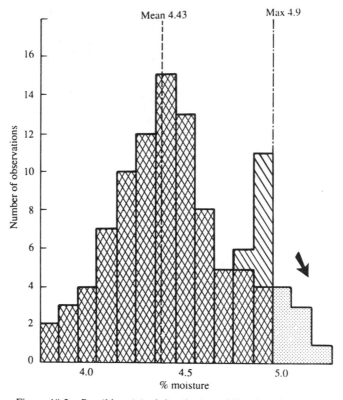

Figure 10.3   *Possible original distribution of biscuit moisture contents*

gives a more probable distribution. If this represents the truth then 8 in 96 results would have been above the specified maximum, i.e. about 8 per cent off-specification biscuits. There is an incentive to do this, since selling water instead of biscuit is a way to higher profits, but what does it do to customer confidence? Is the distribution bi-modal and have they been selecting those or results which just lie inside the specification? We may have our suspicions but we do not have a clear demonstration of what was going on.

The original data was time related so we can see somewhat more clearly what was happening if we plot out all the results against time as in Figure 10.4. This shows that the moisture results are constantly scattered about. It suggests that there was relatively high scatter on days 1, 2 and 3 and less on days 4, 5 and 6. During the first six days there was only one 'flier' at 4.9 per cent. From day 7, although the scatter was less, several results (nine in all) bounced along the top limit of the specification.

So far we have used very simple statistical and graphical methods to analyse the data, with a view to seeking justification of the procedure

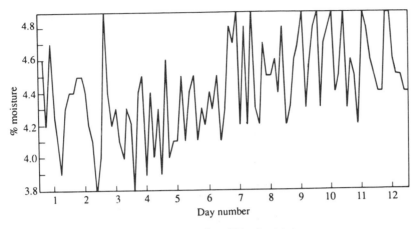

Figure 10.4 *Individuals plot of biscuit moisture contents*

used to control moisture content. Now we may group consecutive results together and plot mean and range charts which are shown in Figure 10.5, from which we may see that:

- Contrary to the earlier conclusion, based on the plot of the individual results, the range chart suggests that the scatter was steady throughout the period. The standard deviation is 0.24 per cent. Some part of this random variation may have been induced by the control procedure in use.
- With a standard deviation of 0.24 per cent and a tolerance zone of (4.9 − 3.0) = 1.9 per cent the potential capability, using the existing control procedures, has a Cp of $2T/6\sigma = 1.9/1.44 = 1.32$. This is a

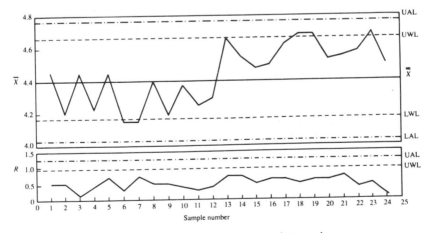

Figure 10.5 *Biscuit data – mean and range chart*

Cp_{delivery} and, since it is greater than one, it may reflect a potential to control the moisture content at a satisfactory level of capability.

- The error associated with the moisture content of any individual biscuit is ± 2 standard deviations (at the 95 per cent confidence level) or ± 0.48 per cent.
- The mean values were centred around 4.3 per cent up to sample 12 (which is the latter part of day 6) when they increased to, and remained at, around 4.6 per cent. During both these periods there were action signals arising from consecutive points lying in the warning zones, as well as from seven consecutive points lying on one side of the mean.

Following the flowchart in Figure 10.1 we may conclude that the range is in control but that the means are not. It is reasonable to assume from these charts that, if the supplier used control charting techniques, his $Cp_{\text{production}}$ would be greater than 1.32 and might even exceed 2, at which level it would reflect a process of high capability (given that $n = 4$). Presenting these results to the supplier, asking them what happened at the twelfth sample (end of day 6) and why, and suggesting the use of control charting, would almost certainly meet with a positive response. The data has yielded considerable information about the supplier's process as well as a meaningful plan for action.

Starting to use control charting techniques on raw materials is a good idea both because the variations present in raw materials will pass into a process and be reflected in the outputs, and because it is sometimes easier to make a supplier conform than it is to enforce conformance on oneself. There is another side to this coin, which says that this analysis of suppliers can be performed by your customers on your products.

Variables data from within the process

Figure 10.6 shows a mean and range chart constructed from a recorded parameter within a chemical distillation process – the record stretches over a period of approximately six weeks, during which numerous changes were made to the inputs and process conditions, some manually, some by automated instruments, and probably some unwittingly. Not all the changes were recorded. The parameter was recorded every six hours and the charts had been set up by simply grouping the basic data into consecutive daily samples of four data points, from which the means and the ranges were calculated.

At first sight one is tempted to conclude that this demonstrates that the process was not in statistical control and probably never would be. It certainly looks a mess!

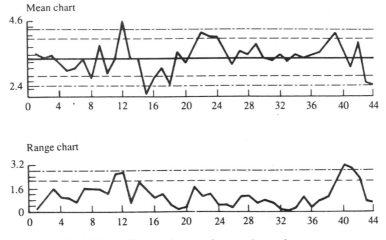

Figure 10.6   *Mean and range charts – chemical parameter*

If we follow the procedure outlined in the flowchart of Figure 10.1 we see by first examining the range chart, that there was absence of statistical control, both at the beginning and at the end of the period. During a long period, of about twenty days, the range remained reasonably steady, however, showing a potential for the random component of the variations to remain stable. Closer examination of the periods of out-of-control behaviour on the range chart shows that these were associated with a period during which the mean values were changing quite violently. These mean values are made up by grouping four results taken over a period encompassing eighteen hours. If the mean of the process was swinging during such a period the difference between the first and last result in a group could be large, giving rise to unusually high values of the range. It may be reasonable to assume that the out-of-control ranges observed at both the beginning and the end of the period under review were due to the swings in the means.

There is a period between the twenty-sixth day of the first month and the seventh of the second month during which the process range and the process mean were both relatively steady – this is the period which shows the potential of the process and its control. The available data may seem to be short, but it covers some thirteen days during which fifty-two recordings were made. So this is the period of highest stability of behaviour which demands closer examination. Figure 10.7 shows the mean and range chart for this period. The mean range based on this limited period is now 0.61 instead of the previous estimate of 1.09. The exclusion of the periods of high variability has decreased the estimate of the average range and, hence, the standard deviation. Since the means

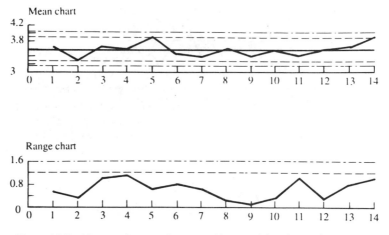

Figure 10.7   *Mean and range charts – stable period for chemical parameter*

were also in statistical control during this period, it is meaningful to calculate the process capability index. The specification was 3.5 ± 2.0 so the Cp is 2.2 and the Cpk is 1.71. This is a process which, if properly centred and managed, would be one of high capability.

The apparent mess reflected in the original mean and range chart is now seen to include the encouraging news that over a period of thirteen days the process was capable. Equally clearly, during a large part of the period under review, there were numerous changes resulting from assignable causes, all of which need to be investigated.

10.5   The use of control charts for trouble shooting

In the preceding section and some previous chapters we have looked at mean and range charts with a view to understanding the process behaviour which they reflect. It will be useful to reverse this process and define the ways in which the process may vary and how this will be reflected in the charts. We have already considered the case of a process in statistical control in Chapter 6. When any of the rules for statistical control are infringed, the process is seen to be subjected to assignable causes of change. These assignable causes may take three basic forms:

- Changes of the process mean, with no change in the scatter, spread or standard deviation.
- Changes of the process spread (scatter or standard deviation) with no change of the mean.
- Changes in both the population mean and standard deviation.

The manner of the change may also vary as follows:

- A sustained shift.
- A drift or trend – including cyclical behaviour.
- Frequent and irregular shifts.

All the possible combinations are illustrated in the charts shown in Figures 10.9 to 10.14. The examples used for illustration were taken from a tablet-making process which was known to be in statistical control and capable, when assessed during a period of one hour, and in which samples of five tablets were collected each half-hour, for normal process control. In all cases the data is plotted on the mean and range charts, set up as a result of the process capability study, for which the standard deviation was 4.5 mg and the process mean was 300 mg.

Sustained shift of process mean – Figure 10.8

The process varied as in (a) but, of course, no one can actually know that the changes were exactly as shown. Our best estimate of the real behaviour can be obtained by reading the mean and range charts. These show, (c), that throughout the whole period the range was in control. The mean, (b), remained in control until time 3 when an action signal immediately followed the large actual change of the process mean. This change was 10 mg, equivalent to 2.2 standard deviations of the total

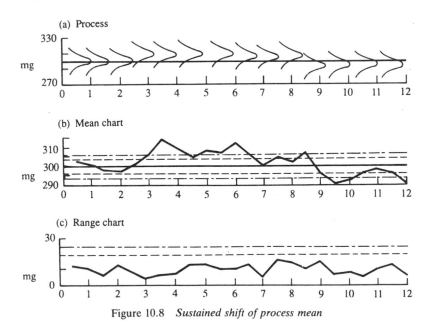

Figure 10.8   *Sustained shift of process mean*

population, and 5.0 standard errors of the mean. At time 6, when the process mean fell to 305 mg, action and warning signals continued on the mean chart and the process mean can be seen to average about the 305 mg level. This relatively small change is noticeably less distinct than the following more marked fall to 295 mg, which is again immediately detected by the mean chart.

The mean chart clearly responds, and immediately, to changes of 2 standard deviations but is less explicit when the changes are smaller.

Drift of process mean – Figure 10.9

When the process varies as shown in Figure 10.9(a), the mean and range charts ((b) and (c) respectively) respond as expected. The range chart shows an in-control situation throughout the period, while the mean chart starts with low mean action and warning signals, moves on to a period of no abnormalities, and then on to a period in which high action and warning signals occur. The whole plot of means reflects a continuously rising trend in the means.

Frequent irregular shifts of process mean – Figure 10.10

As in the previous two examples, the changes shown in Figure 10.10(a)

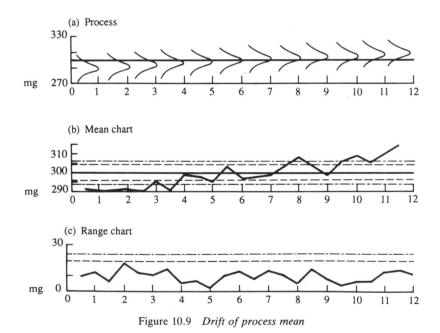

Figure 10.9   *Drift of process mean*

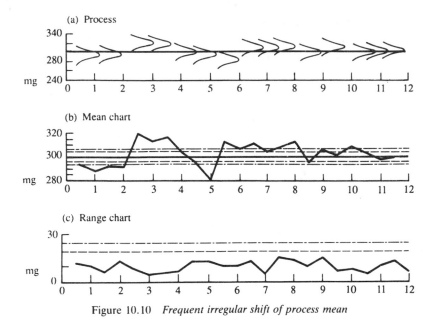

Figure 10.10 *Frequent irregular shift of process mean*

include a steady standard deviation which is reflected in an in-control situation shown in the range chart (c). The changes of the process mean, by amounts exceeding 5 standard errors of the means, are clearly shown up in the highly erratic mean chart. The mean chart cannot totally reveal the process behaviour because it will always include the scatter due to sampling, but its picture of events is a good reflection of the reality. Certainly the mean chart shows erratic behaviour of the mean throughout the whole period.

Sustained shift of process standard deviation – Figure 10.11

Again the actual process changes are shown in Figure 10.11(a). The initial period, up to time 2.5, was of in-control behaviour for both the mean and the range. This behaviour seems to have been maintained throughout the following period up to time 6, although in fact the standard deviation declined in the second period – while this is reflected in the range chart it does not show up as a significant event, because the change was too small. The following period is particularly interesting. Figure 10.11(a) shows that the standard deviation increased to about twice the value on which the charts were based i.e. to about 9 mg. This is reflected in action signals on the range chart, as would be expected. It is also reflected in the mean chart, because, during this period of increased standard deviation, the standard error of the mean is also increased. This increase shows up as

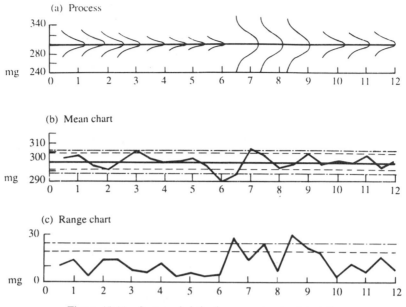

Figure 10.11   *Sustained shift of process standard deviation*

increased scatter of the sample mean results which, of course, remain scattered about the stable process mean but now by an amount which is sufficient to give rise to apparent warning and action signals on the mean chart.

The discipline of examining the range chart first, and, if the range is not in control treating any later observations on the corresponding mean chart with caution, is clearly demonstrated. At the end of the period under review the standard deviation returns to a value consistent with that obtained during the process capability study and the range chart and mean chart both again appear to be in control.

The importance of reading the range chart *before* reading the mean chart, and interpreting the mean chart with caution, if the range is not in control, is clearly demonstrated.

Drift of process standard deviation – Figure 10.12

Figure 10.12(a) shows that the standard deviation declined steadily during this period from an initial bell width of about 30 mg to a final bell width of only 10 mg (the charts are based on a standard deviation of 4.5 mg). The range chart (c) clearly reflects this steady decline in the standard deviation, with some action and warning signals being followed by seemingly normal behaviour and a trend, towards the end, to an

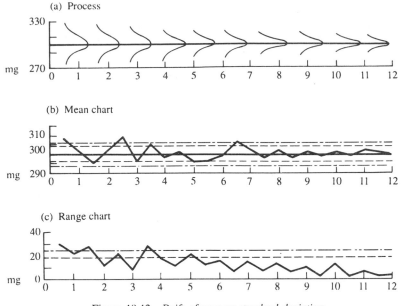

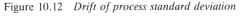

Figure 10.12 *Drift of process standard deviation*

average range which is clearly below the value observed when the chart was set up (10 mg). Once again the mean chart reflects this behaviour of the range, even though Figure 10.12(a) clearly shows that during the period the mean was absolutely constant. The mean chart is entirely consistent with a declining standard deviation and a steady, and on target mean, even though there were warning and action signals on the mean chart during the period of abnormally high standard deviation.

Frequent irregular shift of process standard deviation – Figure 10.13

The highly variable behaviour of the standard deviation depicted in Figure 10.13(a) gives a somewhat confused picture on the mean and range charts. Clearly the range was not in control and most, but not all, the periods of abnormally high standard deviation are picked up by action signals on the range chart. The range chart does include periods of 4.5 mg standard deviation during which the mean chart shows in-control behaviour suggesting that at those times both the mean and the range were in control. But, as noted above, when the range chart is not in control, the signals from the mean chart are no longer clear and, in this case, action and warning signals on the mean chart are due to the increased standard deviation, and not to any abnormal behaviour of the mean.

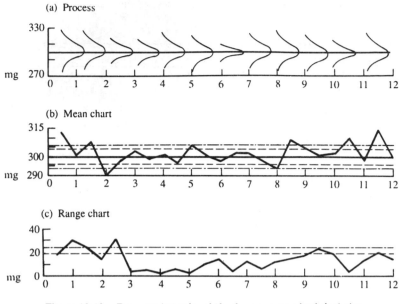

(a) Process

(b) Mean chart

(c) Range chart

Figure 10.13   *Frequent irregular shift of process standard deviation*

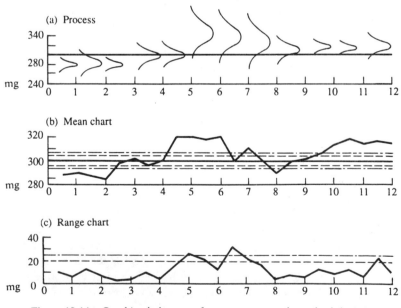

(a) Process

(b) Mean chart

(c) Range chart

Figure 10.14   *Combined changes of process mean and standard deviation*

Combined changes of process mean and standard deviation – Figure 10.14

The process behaviour shown in Figure 10.14(a) is totally out of control, with various and frequent changes of both the process mean and range. The examination of the mean and range charts also suggests a highly unstable process, in which the range appears to be in control up to time 4.5, but during the first part of this period the mean was not on target. The period from time 4.5 to 8 is out of control on both the mean and the range charts. It is not possible from the charts to be certain of what is happening, although the charts are entirely consistent with the process behaviour shown in (a). While the range appears to return to an in-control situation after time 8, the mean is clearly above target from time 9.5 onwards. Once more the highly erratic behaviour of the process is reflected in the highly variable mean and range charts.

*   *   *

All the above examples demonstrate how the mean and range charts enable one to have some indication of the changes that are actually occurring within a process. One never knows exactly what is happening within a process because the pictures which we have of the process performance are clouded, in part, by the constant presence of random variations. The mean and range charts give a very good picture, but not one which is completely 'in focus'. The 'focus' can often be improved by obtaining additional information.

It is the authors' experience that the correct reading of control charts is achieved fairly quickly, once proper initial training of the process operators or chart reviewer has been completed. It is also their experience that issuing instructions on how to plot the charts, and hoping that proper interpretation will follow, is a highly dangerous way of introducing SPC.

10.5   Assignable causes

Variability cannot be ignored, it must be recognized and managed. When a process is found to be out of control, the first action must be to investigate the responsible assignable causes. In some cases, this may require the charting of parameters other than the product parameters. For example, it may be that the viscosity of a chemical product is directly affected by the pressure in a reactor vessel, which in turn may be directly affected and controlled by the reaction temperature. A control chart for pressure, with the recorded changes in temperature, could well be the first step in breaking into the complex relationship involved. The important point is to ensure that all the adjustments to the process are recorded and that the relevant data is charted.

There can be no compromise on processes which are shown to be out of control. The simple device of changing the charting method and/or the control limits will not bring the process into control. A proper process investigation must take place.

Control charts indicate when assignable causes occur. There are numerous possible causes for processes being out of control. It is extremely difficult, even dangerous, to try and find an association between types of causes and patterns shown on control charts. There are clearly many causes which could give rise to different patterns in differing industries and conditions. While the causes may be identified by examination of the process outputs, they will be found only in the direct and indirect process inputs and within the process itself.

It may be useful to list some of the general types of assignable causes.

1 People

- Fatigue, illness, state of health.
- Lack of training/novices.
- Inadequate supervision, lack of discipline.
- Lack of knowledge of the requirements, both general and specific.
- Lack of motivation, attitudes, die-hard approaches.
- Changes/improvements in skills, uncontrolled experimentation.
- Rotation of shifts, changes within teams.

2 Plant/machines

- Rotation of machines.
- Differences in test or measuring devices.
- Lack of adequate, scheduled, preventative maintenance.
- Badly designed equipment.
- Worn machinery or tools.
- Gradual deterioration of plant/machines.

3 Process/methods

- Unsuited techniques of production and/or test.
- Untried/new production processes.
- Changes in inspection/test methods.

4 Materials

- Merging or mixing of batches, parts, components.
- Subassemblies, intermediates, etc.
- Accumulation of waste products or impurities.
- Lack of homogeneity.
- Changes of sources of supply.

5   *Environment*

- Gradual deterioration in conditions.
- Seasonal, daily, weekly changes.
- Variations in temperature and humidity which affect the process.
- Variations of noise or dust which affect the operator.

It should be clear from this non-exhaustive list of the major areas that an intimate knowledge of the process is essential for the effective use of control charts in trouble shooting and process improvement programmes. The control chart, when used carefully and with proper understanding, tells us when to look for the source of trouble, and only does this when the probability of there not being an assignable cause is extremely low (one can afford to be wrong 1 in 1000 times). This is the simple and easy part of an improvement programme. The bulk of the work in realizing improvements is associated with finding where to look and which causes are operating. When the improvement has been made, the control chart plays a second role in assessing its effectiveness.

Chapter highlights

- Methods of running processes range from project, through job, batch and flow to continuous. All types of process are inherently stable, have random and assignable causes of variation present, have control procedures which add to the variation, have a common objective of perfection and SPC techniques apply. Failure to recognize, investigate and eliminate assignable causes of variation means that they go on needlessly adding to the total variation.
- Non-conformance occurs when the outputs from a process are not what was expected – it is not limited to outputs which are unacceptable or difficult to dispose of.
- Evaluating process capability from existing data leads to the identification of any major problem areas, and introduces the concepts of variation, stability and their management to process operators and managers.
- The assessment of process capability from existing data collected during the normal running of a process, ideally requires a minimum of fifty data. The two steps in the assessment are, first examine the state of statistical control, and then compare the precision of the process with the specification.
- Any assignable variations of either the mean or the range made evident by the analysis of existing data must be investigated and eliminated.

- Where data exists, carry out the procedure for judging process capability before questioning the relevance of charting to any given process. The time for discussion of the relevance and meaning of any data is after its analysis.
- Assessment of process capability is illustrated in Section 10.3 by examples which include measurements of variables on a raw material and a parameter within a manufacturing process. These examples illustrate how information can be exposed or released from data by mean and range charting. Data often contains information which only analysis can reveal. The information leads to identified requirements for action. Action includes more meaningful discussions with suppliers, process operators/managers and customers.
- The actual behaviour of a process is never totally known. The constant presence of random variations will always mask the absolute truth. The interpretation of mean and range charts can be assisted by examining their likely behaviour for assumed changes to a process. All the common changes and trends of accuracy and precision are illustrated by examples in Section 10.4.
- Variability must be managed. Assignable causes must be recognized and their causes investigated – no compromise is possible. The assignable cause may be found in the direct and indirect inputs to a process or the process itself. Assignable causes are often recognized in process outputs; they cannot have their origin there.
- Section 10.5 includes a non-exhaustive list of the possible origins of assignable causes of variation to both accuracy and precision.

11   Process control by attributes

11.1   Underlying concepts

The quality of many products and services is dependent upon characteristics which cannot be measured as variables. These are called attributes and may be counted, having been judged simply as either present or absent, conforming or non-conforming, acceptable or defective. Such properties as bubbles of air in a windscreen, the general acceptance of a paint surface, accidents, the particles of contamination in a sample of polymer, clerical errors in an invoice, and the number of telephone calls, are all attribute parameters. It is clearly not possible to use the methods of measurement and control for variables described in Chapters 5 to 10 when addressing the problem of attributes.

An advantage of attributes is that they are in general more quickly assessed, so often variables are converted to attributes for assessment. But, as we shall see, attributes are not so sensitive a measure as variables and, therefore, detection of small changes is less sensitive.

The statistical behaviour of attribute data is different from that of variable data and this must be taken into account when designing process control systems for attributes. To identify which type of data distribution we are dealing with, we must know something about the product or service form and the attribute under consideration. The following types of attribute lead to the use of different types of control chart, which are based on different statistical distributions:

1   *Conforming or non-conforming units*, each of which can be wholly described as failing or not failing, acceptable or defective, present or not present, etc., e.g. ball-bearings, invoices, workers, respectively.
2   *Non-conformities*, which may be used to describe a product or service, e.g. number of defects, errors, sales calls, faults, truck deliveries.

Hence, a defective is an item or 'unit' which contains one or more flaws, errors, faults, or defects. A defect is an individual flaw, error, or fault.

When we examine a fixed sample of the first type of attribute, for example 100 ball-bearings or invoices, we can state how many are defective or non-conforming. We shall then very quickly be able to work out how many are acceptable or conforming. So in this case, if two ball-bearings or invoices are classified as unacceptable or defective, ninety-eight will be acceptable. This is different from the second type of attribute. If we examine a product such as a windscreen and find four defects – scratches or bubbles – we are not able to make any statements about how many scratches/bubbles are not present. This type of defect data is similar to the number of goals scored in a football match. We can only report the number of goals scored. We are unable to report how many were not.

The two types of attribute data lead to the use of two types of control chart:

1   Number of non-conforming units (or defectives) chart.
2   Number of non-conformities (or defects) chart.

These are each further split into two charts, one for the situation in which the sample size (number of units, or length or volume examined or inspected) is constant, and one for samples of varying size. Hence, the collection of charts for attributes becomes:

1   (a)   Number of non-conforming units (defectives) (np) chart – for constant sample size.
    (b)   Proportion of non-conforming units (defectives) (p) chart – for samples of varying size.
2   (a)   Number of non-conformities (defects) (c) chart – for samples of same size every time.
    (b)   Number of non-conformities (defects) per unit (u) chart – for varying sample size.

The specification

Process control can be exercised using these simple charts on which the number or proportion of units, or the number of incidents or incidents per unit are plotted. Before commencing to do this, however, it is absolutely vital to clarify what constitutes a defective, non-conformance, defect or error, etc. No process control system can survive the heated arguments which will surround badly defined non-conformances. It is evident that in the study of attribute data, there will be several degrees of imperfection. The description of attributes, such as defects and errors, is a subject in its own right, but it is clear that a scratch on a paintwork or table top surface may range from a deep gouge to a slight mark, hardly visible to the naked eye; the consequences of accidents may range from death or severe injury to mild inconvenience. To ensure the smooth control of a process using attribute data, it is often necessary to provide representative samples, photographs, or other objective evidence to support the decision maker. Ideally a sample of an acceptable product and one that is just not acceptable should be provided. These will allow the attention and effort to be concentrated on improving the process rather than debating the issues surrounding the severity of non-conformances.

Attribute process capability and its improvement

When a process has been shown to be in statistical control, the average level of events, errors, defects per unit, or whatever, will represent the capability of the process when compared with the specification. As with variables, to improve process capability requires a systematic investigation of the whole process system – not just a diagnostic examination of particular apparent causes of lack of control. This places demands upon management to direct action towards improving such contributing factors as:

● Operator performance, training and knowledge.
● Equipment performance, reliability and maintenance.
● Material suitability, conformance and grade.
● Methods, procedures and their consistent usage.

A philosophy of never-ending improvement is always necessary to make inroads into process capability improvement, whether it is when using variables or attribute data. It is often difficult, however, to make progress in process improvement programmes when only relatively insensitive attribute data is being used. One often finds that some form of alternative variable data is available or can be obtained with a little effort and expense. The extra cost associated with providing data in the form of measurements may well be trivial compared with the savings that can be derived from reducing process variability.

11.2 Charts for number of defectives or non-conforming units (*np*)

Consider a process which is producing ball-bearings, 10 per cent of which are defective – p, the proportion of defects is 0.1. If we take a sample of one ball from the process, the chance or probability of finding a defective is 0.1 or p. Similarly, the probability of finding a non-defective ball-bearing is 0.90 or $(1-p)$. For convenience we will use the letter q

instead of $(1-p)$ and add these two probabilities together:

$$p + q = 0.1 + 0.9 = 1.0$$

A total of unity means that we have present all the possibilities, since the sum of the probabilities of all the possible events must be one. This is clearly logical in the case of taking a sample of one ball-bearing for there are only two possibilities – finding a defective or finding a non-defective.

If we increase the sample size to two ball-bearings, the probability of finding two defectives in the sample becomes:

$$p \times p = 0.1 \times 0.1 = 0.01 = p^2$$

This is one of the first laws of probability – the *multiplication law*. When two or more events are required to follow consecutively, the probability of them all happening is the product of their individual probabilities. In other words, for A *and* B to happen, multiply the individual probabilities p_A and p_B.

We may take our sample of two balls and find zero defectives. What is the probability of this occurrence?

$$q \times q = 0.9 \times 0.9 = 0.81 = q^2$$

Let us add the probabilities of the events so far considered:

Two defectives – probability		0.01 (p^2)
Zero defectives – probability		0.81 (q^2)
	Total	0.82

Since the total probability of all possible events must be one, it is quite obvious that we have not considered all the possibilities. There remains, of course, the chance of picking out one defective followed by one non-defective. The probability of this occurrence is:

$$p \times q = 0.1 \times 0.9 = 0.09 = pq$$

However, the single defective may occur in the second ball-bearing:

$$q \times p = 0.9 \times 0.1 = 0.09 = qp$$

This brings us to a second law of probability – the *addition law*. If an event may occur by a number of alternative ways, the probability of the event is the sum of the probabilities of the individual occurrences. That is, for A *or* B to happen, add the probabilities p_A and p_B. So the probability of finding one defective in a sample of size two from this process is:

$$pq + qp = 0.09 + 0.09 = 0.18 = 2pq$$

Now, adding the probabilities:

Two defectives – probability 0.01 (p^2)
One defective – probability 0.18 ($2pq$)
No defectives – probability 0.81 (q^2)
Total probability 1.00

So when taking a sample of two from this process, we can calculate the probabilities of finding one, two or zero defectives in the sample. Those who are familiar with simple algebra will recognize that the expression:

$$p^2 + 2pq + q^2 = 1$$

is an expansion of:

$$(p + q)^2 = 1$$

and this is called the *binomial* expression. It may be written in a general way:

$$(p + q)^n = 1$$

where:

n = sample size (number of units)
p = proportion of defectives or 'non-conforming units' in the population from which the sample is drawn
q = proportion of non-defectives or 'conforming units' in the population = $(1 - p)$

To reinforce our understanding of the binomial expression, look at what happens when we take a sample of size four:

$$n = 4$$
$$(p + q)^4 = 1$$

expands to:

$$p^4 \quad + \quad 4p^3q \quad + \quad 6p^2q^2 \quad + \quad 4pq^3 \quad + \quad q^4$$

Probability of four defectives in the sample	Probability of three defectives	Probability of two defectives	Probability of one defective	Probability of zero defectives

The mathematician represents the probability of finding x defectives in a sample of size n when the proportion present is p as:

$$P(x) \left(\frac{n}{x} \right) p^x (1-p)^{(n-x)}$$

where:

$$\binom{n}{x} = \frac{n!}{(n-x)!x!} \text{ and } n! \text{ is } 1 \times 2 \times 3 \times 4 \times \ldots \times n$$

For example, the probability $P(2)$ of finding two defectives in a sample of size five taken from a process producing 10 per cent defectives ($p = 0.1$) may be calculated:

$$n = 5$$
$$x = 2$$
$$p = 0.1$$

$$P(2) = \frac{5!}{(5-2)!2!} \times 0.1^2 \times 0.9^3$$

$$= \frac{5 \times 4 \times 3 \times 2 \times 1}{(3 \times 2 \times 1) \times (2 \times 1)} \times 0.1 \times 0.1 \times 0.9 \times 0.9 \times 0.9$$

$$= 10 \times 0.01 \times 0.729 = 0.0729$$

This means that, on average, about seven out of one hundred samples of size 5 ball-bearings, taken from the process will have two defectives in them. The *average* number of defectives present in a sample of five will be 0.5.

It may be possible at this stage for the reader to see how this may be useful in the design of process control charts for number of defectives or classified units. If we can calculate the probability of exceeding a certain number of defectives in a sample, we shall be able to draw action and warning lines on charts, similar to those designed for variables in Chapter 6.

To use the probability theory we have considered so far we must know the proportion defective being produced by the process. This may be discovered by taking a reasonable number of samples – say twenty to fifty – over a 'stable' period, and recording the number of defectives or non-conforming units in each. Table 11.1 lists the number of defectives

Table 11.1   *Number of defectives found in samples of 100 ballpoint pen cartridges*

2	2	2	2	1
4	3	4	1	3
1	0	2	5	0
0	3	1	3	2
0	1	6	0	1
4	2	0	2	2
5	3	3	2	0
3	1	1	1	4
2	2	2	3	2
3	1	1	1	1

Table 11.2

Number of defectives in sample	Tally chart (number of samples with that number of defectives)	Frequency
0	⊞ 11	7
1	⊞ ⊞ 111	13
2	⊞ ⊞ 1111	14
3	⊞ 1111	9
4	1111	4
5	11	2
6	1	1

found in fifty samples of size $n = 100$ taken every hour from a process producing ballpoint pen cartridges. These results may be grouped into the frequency distribution of Table 11.2 and shown as the histogram of Figure 11.1. This is clearly a different type of histogram from the symmetrical ones derived from variables data in Chapter 5.

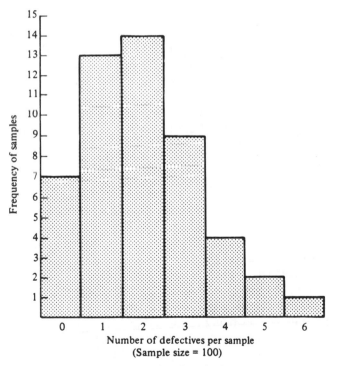

Figure 11.1   *Histogram of results from Table 11.1*

The average number of defectives per sample may be calculated by adding the number of defectives and dividing the total by the number of samples:

$$\frac{\text{Total number of defectives}}{\text{Number of samples}} = \frac{100}{50} = 2 \quad \begin{array}{l}\text{(Average number of}\\ \text{defectives per sample)}\end{array}$$

This value is $n\bar{p}$ – the sample size multiplied by the average proportion defective in the process. Hence, $\bar{p}$ may be calculated:

$$\bar{p} = n\bar{p}/n = 2/100 = 0.02 \text{ or 2 per cent}$$

Although the individual values of np do not make up a normal distribution, their distribution tells us something about the probability of events and it is saying that the probability of finding a sample defective number close to the mean of 2 is high and that, as we move away from this mean, the probabilities decrease. So we can see a marked similarity between this distribution and the normal distribution but they are not identical.

So, there is some similarity between attributes and variables – but can we have a range? In the case of variables we say that each sample observation is from a continuous distribution and that there is always some variation about a mean value. In this case we are saying that the percentage of defectives produced is stable at 2 per cent and that we are finding different answers in the samples, not because the number of defectives varies about a mean, but simply because we are looking only at samples. The 'scatter' of results is simply the reflection of sampling variation and not due to inherent variation within the process. There is no possibility of errors of 'accuracy' and 'precision'. Errors may occur in obtaining the sampling results, but these will be in integer values and may reflect human errors in counting, classifying, or sample taking. We may observe results which reflect a change in the proportion defective produced but we cannot observe results which reflect a change in the 'random scatter'.

Looking at Figure 11.1, we can see that at some point around five defectives per sample, results become less likely to occur and at around seven they are very unlikely. As with mean and range charts, we can argue that if we find, say, eight defectives in the sample, then there is a very small chance that the percentage defective being produced is still at 2 per cent and this odd result is simply due to the sampling. For a sample result of eight there is, of course, another explanation, which is that the percentage of defectives being produced has risen above 2 per cent.

We may use the binomial distribution to set action and warning lines for the so-called 'np process control chart', sometimes known in the USA as a pn chart.

Attribute control chart practice in the American automotive industry is to set only one set of outer limits at three standard deviations (3σ) either side of the average number defective (or non-conforming units). It will be found that use of the 3σ formula usually sets the limits at a probability of between 0.01 and 0.005, when $n\bar{p} = 2$ and similar values.

The standard deviation (σ) for a binomial distribution is given by the formula:

$$\sigma = \sqrt{n\bar{p}(1-\bar{p})}$$

We can show that this is true by calculating σ using the conventional formula:

$$\sigma = \sqrt{\frac{\Sigma f x_i^2}{\Sigma f} - \bar{x}^2}$$

and comparing the result with that obtained from the new formula. Table 11.3 shows the data from Tables 11.1 and 11.2 being used to give the value of $\Sigma f x^2$. The calculation of σ now becomes:

$$\sigma = \sqrt{(300/50) - 4}$$
$$= 1.41$$

Use of the simpler and quicker formula requiring knowledge of only n and np gives a similar result:

$$\sigma = \sqrt{100 \times 0.02 \times 0.98} = 1.4$$

Table 11.3

Number of defectives (x)	Frequency (f)	x^2	fx^2
0	7	0	0
1	13	1	13
2	14	4	56
3	9	9	81
4	4	16	64
5	2	25	50
6	1	36	36
	$\Sigma f = 50$		$\Sigma f x^2 = 300$

Now, the upper action line (UAL) or control limit (UCL) may be calculated:

$$\text{UAL (UCL)} = n\bar{p} + 3\sqrt{n\bar{p}(1-\bar{p})}$$
$$= 2 + 3\sqrt{100 \times 0.02 \times 0.98}$$
$$= 6.2, \text{ i.e. between 6 and 7}$$

This result is the same as that obtained by setting the upper action line at a probability of about 0.005 (one in 200) using binomial probability tables.

This formula offers a simple method of calculating the upper action line for the *np* chart, and a similar method may be employed to calculate the upper warning line. This will be set at two standard deviations above the average number defective:

$$\text{UWL} = n\bar{p} + 2\sqrt{n\bar{p}(1-\bar{p})}$$
$$= 2 + 2\sqrt{100 \times 0.02 \times 0.98}$$
$$= 4.8, \text{ i.e. between 4 and 5}$$

Again this gives the same result as that derived from using the binomial expression to set the warning line at about 0.05 probability (one in twenty).

It is not possible to find fractions of defectives in attribute sampling, so the presentation may be simplified by drawing the control lines between the whole numbers. The sample plots then indicate clearly when the limits have been crossed. In our sample, four defectives found in a sample indicates normal sampling variation, while five defectives gives a warning signal that another sample should be taken immediately because the process may have deteriorated. In control charts for attributes it is commonly found that only the upper limits are specified since we wish to detect an increase in defectives. Lower control lines may be useful, however, to indicate when a significant process improvement has occurred, or to indicate when suspicious results have been plotted. In the case under consideration, there are no lower action or warning lines, since it is expected that zero defectives will be periodically found in the samples of 100, when 2 per cent defectives are being generated by the process. This is shown by the negative values for $(np-3\sigma)$ and $(np-2\sigma)$. The use of warning limits is strongly recommended since their use improves the sensitivity of the charts and tells the 'operator' what to do when results *approach* the action limits – take another sample but do not act until there is a clear signal to do so. Runs and trends on attribute charts may be used, as with variable charting, to indicate the presence of assignable causes of change. The rules for runs and trends on control charts are given in Chapter 6.

Figure 11.2 is an *np* chart on which are plotted the data concerning the ballpoint pen cartridges from Table 11.1. Since all the samples contain less defectives than the action limit and only three out of fifty enter the warning zone, and none of these are consecutive, the process is considered to be in statistical control. We may, therefore, reasonably assume that the process is producing a constant level of 2 per cent defective, and the chart may be used to control the process. The method of interpretation of control charts for attributes is exactly the same as that described for mean and range charts in Chapter 6.

Figure 11.3 shows the effect of increases in the proportion of defective pen cartridges from 2 per cent through 3 per cent, 4 per cent, 5 per cent, 6 per cent to 8 per cent in steps. For each percentage defective, the run length to detection, that is the number of samples which needed to be taken before the action line is crossed following the increase in process defective, is given below:

Percentage process defective	*Run length to detection from Figure 11.3*
3	>10
4	9
5	4
6	3
8	1

Clearly, this type of chart is not as sensitive as mean and range charts for detecting changes in process defective. For this reason, the action and warning lines on attribute control charts are set at the higher probabilities of approximately one in two hundred (action) and approximately one in twenty (warning).

This lowering of the action and warning lines will obviously lead to the more rapid detection of a worsening process. It will also increase the number of incorrect action signals. Since inspection for attributes by, for example, using a go/no-go gauge, is usually less costly than the measurement of variables, an increase in the amount of resampling may be tolerated.

If the probability of an event is, say, 0.25, on average it will occur every fourth time, as the average run length (ARL) is simply the reciprocal of the probability. Hence, in the pen cartridge case, if the proportion defective is 3 per cent ($p = 0.03$), and the action line is set between 6 and 7, the probability of finding seven or more defectives may be calculated or derived from the binomial expansion as 0.0312 ($n = 100$). We can now work out the average run length to detection:

$$\text{ARL } (3\%) = 1/P(\geqslant 7) = 1/0.0312 = 32.1$$

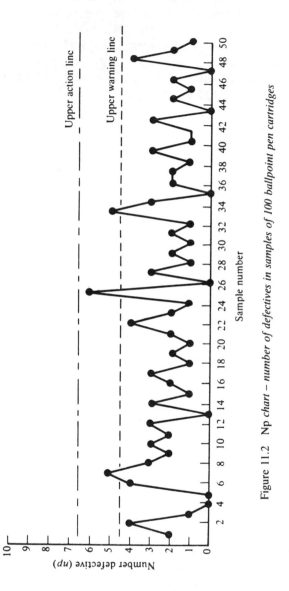

Figure 11.2   Np chart – number of defectives in samples of 100 ballpoint pen cartridges

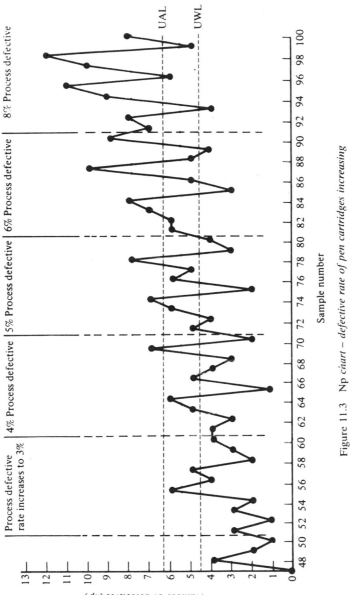

Figure 11.3 Np chart – defective rate of pen cartridges increasing

For a process producing 5 per cent defectives, the ARL for the same sample size and control chart is:

$$\text{ARL } (5\%) = 1/P(\geqslant 7) = 1/0.234 = 4.3$$

The conclusion from the run length values is that, given time, the 'np' chart will detect a change in the proportion of defectives being produced. If the change is an increase of approximately 50 per cent, the np chart will be very slow to detect it, on average. If the change is a decrease of 50 per cent, the chart will not detect it because, in the case of a process with 2 per cent defective, there are no lower limits. This is not true for all values of defective rate.

Before we move on to other types of attribute charts it is important to understand what we are doing when we use an attribute control chart. If we turn back to the control chart constructed in Figure 11.2, this is telling us that from this process there will be, on average, two defectives present in a sample of 100, and that we will not be surprised when we find zero, one, two, three and four; that five is not alarming but may spell trouble. If five defectives are observed in a sample, we should check by getting more information, which we do by taking another sample; and only when we reach seven or more in a sample of 100 can we be sufficiently certain to take remedial action. Even in this latter case, there is approximately a 1 in 200 chance that we shall have done the wrong thing. If we translate this type of information into everyday life, we find that when there are two incorrect invoices per hundred, or two defective printed circuit boards per batch, then this is the level to which we are accustomed, and we tend to congratulate ourselves when we find zero and attempt to change things when we find four. But the binomial distribution is telling us that both these values (zero and four) will occur occasionally and that neither result should cause surprise. There is a parallel between these charts and the mean and range charts in that they help us to:

- Stop people chasing shadows.
- Ensure that action is taken only when there is clear evidence that it is required.
- 'Calm down' the control mechanism.
- Increase the chances of acting only when there is trouble.
- Not precipitate trouble by reacting to normal sampling variation.

The 'np' chart is easy to maintain and its use and interpretation easily explained to an operator or manager, but they will need either a procedure or a 'help line' to decide what action is to be taken when action is indicated.

11.3   Charts for proportion defective or non-conforming (p)

In cases where it is not possible to maintain a constant sample size for attribute control, the *p* chart, or proportion defective or non-conforming chart may be used. It is, of course, possible and quite acceptable to use the *p* chart instead of the *np* chart even when the sample size is constant. However, plotting directly the number of defectives in each sample on to an *np* chart is simple and usually more convenient than having to calculate the proportion defective. The data required for the design of a *p* chart is identical to that for an *np* chart, both the sample size and the number of defectives needing to be observed.

An example will be used to explain the design and use of the *p* chart. Standard components used in the assembly of various types of electrical appliances were mixed during their manufacture. They were packed into small boxes of 144, stored and used by the customer on a 'first in, first out' basis, and issued at 102 per cent of the works order requirement. The assembly operators were asked to collect the defective components and count them at the end of each batch of assembled goods.

Table 11.4 shows the number issued and the number rejected. How can the performance of the supplier be assessed with this data? The number issued varies from 405 to 2360. For each issued 'sample', the proportion defective has been calculated:

$$p_i = x_i/n_i$$

where: p_i is the proportion defective in issued 'sample'*i*
x_i is the number of defectives in 'sample'*i*
n_i is the size (number of items) of the *i*th 'sample'

As with the *np* chart, the first step in the design of a *p* chart is calculation of the average proportion defective ($\bar{p}$):

$$\bar{p} = \sum_{i=1}^{k} x_i \Big/ \sum_{i=1}^{k} n_i$$

where *k* is the number of samples, and:

$\sum_{i=1}^{k} x_i$ is the total number of defective items

$\sum_{i=1}^{k} n_i$ is the total number of items inspected

For the components in question:

$$\bar{p} = 280/27{,}930 = 0.010$$

Table 11.4 *Results from the issue of components in varying numbers*

'Sample' number	Issue size	Number of rejects	Proportion defective
1	1135	10	0.009
2	1405	12	0.009
3	805	11	0.014
4	1240	16	0.013
5	1060	10	0.009
6	905	7	0.008
7	1345	22	0.016
8	980	10	0.010
9	1120	15	0.013
10	540	13	0.024
11	1130	16	0.014
12	990	9	0.009
13	1700	16	0.009
14	1275	14	0.011
15	1300	16	0.012
16	2360	12	0.005
17	1215	14	0.012
18	1250	5	0.004
19	1205	8	0.007
20	950	9	0.009
21	405	9	0.022
22	1080	6	0.006
23	1475	10	0.007
24	1060	10	0.009

Control chart limits

If a constant sample size is being checked, the p control chart limits would remain the same for each sample. When p charts are being used with samples of varying sizes, the standard deviation and control limits change with n, and unique limits should be calculated for each sample size. However, for practical purposes, an average sample size ($\bar{n}$) may be used to calculate action and warning lines. These have been found to be acceptable when the individual sample or lot sizes vary from $\bar{n}$ by no more than 25 per cent each way. For sample sizes outside this range, separate control limits must be calculated. There is no magic in this 25 per cent, it simply has been shown to work.

The next stage then in the calculation of control limits for the p chart with varying sample sizes is to determine the average sample size ($\bar{n}$) and the range 25 per cent either side:

$$\bar{n} = \sum_{i=1}^{k} n_i/k$$

Range of sample sizes with constant control chart limits =

$$\bar{n} \pm 0.25\,\bar{n}$$

For the components under consideration:

$$\bar{n} = 27{,}930/24 = 1{,}164$$

Permitted range of sample size:

$$= 1{,}164 \pm (0.25 \times 1{,}164)$$
$$= 873 \text{ to } 1455$$

For sample sizes within this range, the control chart limits may be calculated using a value of σ given by:

$$\sigma = \frac{\sqrt{\bar{p}\,(1-\bar{p})}}{\sqrt{\bar{n}}} = \frac{\sqrt{0.010 \times 0.99}}{\sqrt{1164}} = 0.003$$

Then, action limits $= \bar{p} \pm 3\sigma$
$= 0.01 \pm 3 \times 0.003$
$= 0.019 \text{ and } 0.001$

Warning limits $\qquad = \bar{p} \pm 2\sigma$
$= 0.01 \pm 2 \times 0.003$
$= 0.016 \text{ and } 0.004$

Control limits for sample numbers 3, 10, 13, 16 and 21 must be calculated individually as these fall outside the range 873 to 1455:

$$\text{Action limits} = \bar{p} \pm 3\ \sqrt{\bar{p}(1-\bar{p})}\,/\,\sqrt{n_i}$$
$$\text{Warning limits} = \bar{p} \pm 2\ \sqrt{\bar{p}(1-\bar{p})}\,/\,\sqrt{n_i}$$

Table 11.5 shows the detail of the calculations involved and the resulting action and warning lines. Figure 11.4 shows the *p* chart plotted with the varying action and warning lines. It is evident that the design, calculation, plotting, and interpretation of *p* charts are more complex than that associated with *np* charts.

The process involved in the delivery of the components is out of control. Clearly, the supplier has suffered some production problems during this period and some of the component deliveries are of doubtful quality. Complaints to the supplier after the delivery corresponding to sample 10 seemed to have a good effect until delivery 21 caused a warning signal. This type of control chart may improve substantially the dialogue and partnership between suppliers and customers.

Sample points falling below the lower action line also indicate a process which is out of control. Lower control limits are frequently omitted to avoid the need to explain to operating personnel why a very low

Table 11.5   *Calculation of p chart limits for sample sizes outside the range 873 to 1455*

General formulae:

Action limits $= \bar{p} \pm 3 \; \sqrt{\bar{p}(1-\bar{p})} \, / \, \sqrt{n}$

Warning limits $= \bar{p} \pm 2 \; \sqrt{\bar{p}(1-\bar{p})} \, / \, \sqrt{n}$

$\bar{p} = 0.010$

and $\sqrt{\bar{p}(1-\bar{p})} = 0.0995$

Sample number	Sample size	$\sqrt{\bar{p}(1-\bar{p})} \, / \, \sqrt{n}$	UAL	UWL	LWL	LAL
3	805	0.0035	0.021	0.017	0.003	neg. (i.e. 0)
10	540	0.0043	0.023	0.019	0.001	neg. (i.e. 0)
13	1700	0.0024	0.017	0.015	0.005	0.003
16	2360	0.0020	0.016	0.014	0.006	0.004
21	405	0.0049	0.025	0.020	neg. (i.e. 0)	neg. (i.e. 0)

proportion defectives is classed as being out of control. When the *p* chart is to be used by management, however, the lower limits are used to indicate when an investigation should be instigated to discover the cause of an unusually good performance. This may also indicate how it may be repeated. The lower control limits are given in Table 11.5. An examination of Figure 11.4 will show that none of the sample points fall below the lower action lines.

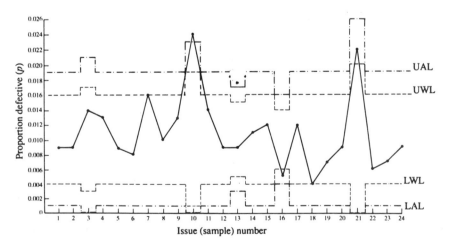

Figure 11.4   p *chart – for issued components*

11.4 Charts for number of defects or non-conformities (*c*)

The control charts for attributes considered so far have applied to cases in which a random sample of definite size is selected and examined in some way. In the process control of attributes, there are situations where the number of events, defects, errors, or non-conformances can be counted, but there is no information about the number of events, defects, or errors which are *not* present. Hence, there is the important distinction between defectives and defects already given in Section 11.1. So far we have considered defectives where each item is classified either as conforming or non-conforming (a defective). In this case we always know both the number of defective items in the sample examined and the number of non-defectives – this 'two sides of the coin' gives rise to the term binomial. In the case of defects, such as holes in a fabric or fisheyes in plastic film, we know the number of defects present but we do not know the number of non-defects present. Other examples of these include the number of imperfections on a painted door, errors in a typed document, the number of faults in a length of woven carpet and the number of sales calls made. In these cases the binomial distribution does not apply.

This type of problem is described by the Poisson distribution, named after the Frenchman who first derived it in the early nineteenth century. Because there is no fixed sample size when counting the number of events, defects, etc., theoretically the number could tail off to infinity. Any distribution which does this must include something of the *exponential distribution* and the constant *e*. This contains the element of fading away to nothing since its value is derived from the formula:

$$e + \frac{1}{0!} + \frac{1}{1!} + \frac{1}{2!} + \frac{1}{3!} + \frac{1}{4!} + \frac{1}{5!} + \ldots \frac{1}{\infty!}$$

If the reader cares to work this out, the value *e* = 2.7183 is obtained.

The equation for the Poisson distribution includes the value of *e* and looks rather formidable at first. The probability of observing *x* defects in a given unit is given by the equation:

$$P(x) = e^{-\bar{c}} (\bar{c}^x/x!)$$

where: *e* = exponential constant, 2.7183
 $\bar{c}$ = average number of defects per unit being produced by the process

The reader who would like to see a simple derivation of this formula should refer to the excellent book *Facts from Figures* by M.J. Moroney. (Pelican, 1983).

So the probability of finding three bubbles in a windscreen from a process which is producing them with an average of one bubble present is given by:

$$P(3) = e^{-1} \times \frac{1^3}{3 \times 2 \times 1}$$

$$= \frac{1}{2.7183} \times \frac{1}{6} = 0.0613$$

As with the *np* chart, it is not necessary to calculate probabilities in this way to determine control limits for the *c* chart. Once again the UAL (UCL) is set at three standard deviations above the average number of events, defects, errors, etc.

Let us consider an example in which, as for *np* charts, the sample is constant in number of units, or volume, or length, etc. In a polythene film process, the number of defects – fisheyes – on each identical length of film are being counted. Table 11.6 shows the number of fisheyes which have been found on inspecting fifty lengths, randomly selected, over a twenty-four-hour period. The total number of defects is 159 and, therefore, the average number of defects, $\bar{c}$ is given by:

$$\bar{c} = \sum_{i=1}^{k} c_i / k$$

where:

c_i is the number of defects on the *i*th unit
k is the number of units examined.

In this example:

$$\bar{c} = 159/50 = 3.2$$

Table 11.6 *Number of fisheyes in identical pieces of polyethelene film (10 square metres)*

4	2	6	3	6
2	4	1	4	3
1	3	5	5	1
3	0	2	1	3
2	6	3	2	2
4	2	4	0	4
1	4	3	4	2
5	1	5	3	1
3	3	4	2	5
7	3	2	8	3

The standard deviation of a Poisson distribution is very simply the square root of the process average. Hence, in the case of defects.

$$\sigma = \sqrt{\bar{c}}$$

and for our polythene process

$$\sigma = \sqrt{3.2} = 1.79$$

The UAL (UCL) may now be calculated:

$$\text{UAL (UCL)} = \bar{c} + 3\sqrt{\bar{c}}$$
$$= 3.2 + 3\sqrt{3.2}$$
$$= 8.57 \text{ i.e. between 8 and 9}$$

This sets the UAL at approximately 0.005 probability, using a Poisson distribution. In the same way, an upper warning line may be calculated at approximately 0.05 probability:

$$\text{UWL} = \bar{c} + 3\sqrt{\bar{c}}$$
$$= 3.2 + 2\sqrt{3.2}$$
$$= 6.78 \text{ i.e. between 6 and 7}$$

Figure 11.5 which is a plot of the fifty polythene film inspection results used to design the c chart, shows that the process is in statistical control, with an average of 3.2 defects on each length. If this chart is now used to control the process, we may examine what happens over the next twenty-five lengths, taken over a period of twelve hours. Figure 11.6 is the c chart plot of the results. As previously we have to recognize that, if on average, there are 3.2 defects present in a sample of constant size then we shall expect to find some sample defect values frequently and others hardly at all. The picture tells us that all was running normally until sample 9, which shows eight defects on the unit being inspected; this signals a warning and another sample is taken immediately. Sample 10 shows that the process has drifted out of control and results in an investigation to find the assignable cause. In this case, the film extruder filter was suspected of being blocked and so it was cleaned. An immediate resample after restart of the process shows the process to be back in control. It continues to remain in that state for at least the next fourteen samples.

As with all types of control chart, an improvement in quality and productivity is often observed after the introduction of the c chart. The confidence of having a good control system, which derives as much from

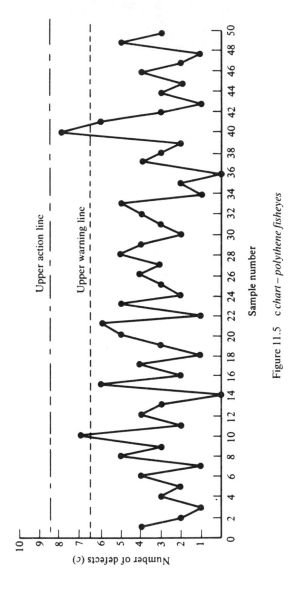

Figure 11.5   *c chart – polythene fisheyes*

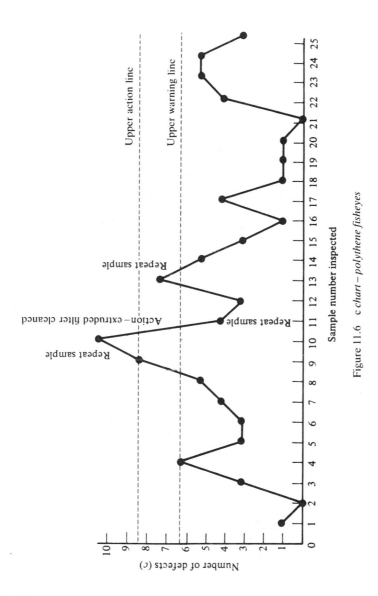

Figure 11.6 *c chart – polythene fisheyes*

knowing when to leave the process alone as when to make adjustments, leads to more stable processes, less variation, and less interruptions from unnecessary alterations.

11.5   Charts for number of defects or non-conformities per unit (*u*)

We saw in the previous section how the *c* chart applies to the number of events, defects, or errors in a constant size of sample, such as a table, a length of cloth, the hull of a boat, a specific volume, a windscreen, an invoice, or a time period. The length of pieces of material, volume or time, for instance, may vary. At other times, it may be desirable to continue examination until a defect is found and then note the sample size. If, for example, the average value of *c* in the polythene film process had fallen to 0.5, the values plotted on the chart would be mostly 0 and 1, with an occasional 2. Control of such a process by a whole number *c* chart would be nebulous.

The *u* chart is suitable for controlling this type of process, as it measures the number of events, defects, or non-conformities per unit or time period, and the 'sample' size can be allowed to vary. In the case of inspection of cloth or other surfaces, the area examined may be allowed to vary and the *u* chart will show the number of defects per unit area, e.g. per square metre. The statistical theory behind the *u* chart is very similar to that for the *c* chart.

An example of *u* chart design should help to explain the usefulness of the technique. Consider a process which is producing cooling fan blades. Each blade has no less than seventeen measurements made on it. If manually plotted mean and range charts are introduced to control this process, the production operators may spend all their time plotting graphs! Examination of previous inspection records shows that the number of blades examined during each batch manufactured has varied from batch to batch. In this case we shall not examine the variables data in detail, although that would be essential at some stage to establish that the process was capable of achieving the specified requirements. Instead we shall use the data in attribute form. This conversion is quite easily achieved by comparing each measurement made with the design tolerance and deciding whether it is inside or outside specification – acceptable or not acceptable. Table 11.7 shows the number of measurements outside specification in each sample from twenty different batches. The sample sizes vary from batch to batch. The number of 'defects' per unit in each case has also been calculated.

The design of the *u* chart is similar to the design of the *p* chart for

Table 11.7 *Number of defects (measurements
out of specification) on cooling fan blades*

Batch number	Number of blades in sample	Number of defects in sample	Number of defects per unit (u)
1	85	40	0.47
2	88	82	0.96
3	92	95	1.03
4	83	78	0.94
5	78	125	1.60
6	75	50	0.67
7	80	105	1.31
8	72	35	0.49
9	80	72	0.90
10	92	85	0.92
11	75	68	0.91
12	81	77	0.95
13	43	75	1.74
14	80	46	0.58
15	125	120	0.96
16	120	105	0.88
17	155	250	1.61
18	81	152	1.88
19	45	17	0.38
20	50	43	0.86

proportion defective. The control lines will vary for each sample size, but for practical purposes may be kept constant if sample sizes remain with 25 per cent either side of the average sample size, $\bar{n}$. From Table 11.7, $\bar{n}$ may be calculated:

$$\bar{n} = \sum_{i=1}^{k} n_i/k$$

where:

n_i = the size of the ith sample
k = the number of samples
$\bar{n}$ = 1680/20 = 84

25 per cent either side of this value gives the range 63 to 105. The first twelve batches are within this range.

As for the p chart, it is necessary to calculate the process average defect rate. In this case we introduce the symbol u:

$$\bar{u} = \text{process average defects per unit}$$

$$= \frac{\text{total number of defects}}{\text{total sample inspected}}$$

$$= \sum_{i=1}^{k} x_i \Big/ \sum_{i=1}^{k} n_i$$

where x_i = the number of defects in sample i.

Hence, $\bar{u}$ = 1720/1680 = 1.02 measurements per blade outside specification.

The control chart limits for the first twelve batches may now be set at 3 and 2 standard deviations from the process average. The defects found per unit will follow a Poisson distribution, the standard deviation σ of which is the square root of the process average. Hence, for the first twelve batches:

$$\text{Action limits} = \bar{u} + 3 \sqrt{\bar{u}} / \sqrt{\bar{n}}$$

$$= 1.02 \pm (3 \sqrt{1.02} / \sqrt{84})$$

$$= 1.02 \pm 0.33 = 1.35 \text{ and } 0.69$$

$$\text{Warning limits} = u \pm 2 \sqrt{\bar{u}} / \sqrt{\bar{n}}$$

$$= 1.02 \pm (2 \sqrt{1.02} / \sqrt{84})$$

$$= 1.02 \pm 0.22 = 1.24 \text{ and } 0.80$$

Control chart limits for batches 13, 15, 16, 17, 19 and 20 are shown in Table 11.8.

Table 11.8   *Calculation of u control chart limits for sample sizes outside the range 63–105*

General formulae:

$$\text{Action lines} = \bar{u} \pm 3 \sqrt{\bar{u}} / \sqrt{n}$$

$$\text{Warning lines} = \bar{u} \pm 2 \sqrt{\bar{u}} / \sqrt{n}$$

$$\bar{u} = 1.02$$

$$\text{and } \sqrt{\bar{u}} = 1.01$$

Sample number	Sample size	$\sqrt{\bar{u}} / \sqrt{n}$	UAL	UWL	LWL	LAL
13	43	0.154	1.48	1.33	0.71	0.56
15	125	0.090	1.29	1.20	0.84	0.75
16	120	0.092	1.30	1.20	0.84	0.74
17	155	0.081	1.26	1.18	0.86	0.78
19	45	0.151	1.47	1.32	0.72	0.57
20	50	0.143	1.45	1.31	0.72	0.59

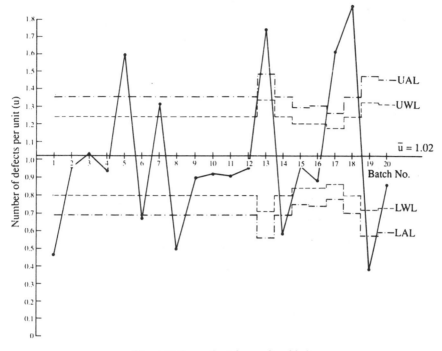

Figure 11.7   u *chart for cooling blades*

Figure 11.7 shows the defects per unit results from Table 11.7 plotted on the u chart. There are several points above the upper action line and the process is clearly not in statistical control. In this situation it would be prudent, before proceeding to detailed and possibly expensive investigations, to do the following:

1 Ensure that the data was valid:
   - Were the measurements carried out correctly?
   - Was a calibrated micrometer with known precision used?
   - Has the measurement system changed?
   - Has the data been edited or altered in any way?
   - Is the data from a process of acceptable capability?
2 Ensure that all calculations and graphs are correct:
   - Are the values of u and n correct?
   - Have the sample defects per unit values been determined correctly?
   - Have the u chart limits been calculated accurately and drawn precisely on the chart?
   - Have the sample results been plotted correctly?

If after examination of these points an out-of-control condition is still demonstrated, the process must be investigated to determine the assignable causes. When discovered, the causes must be corrected and, if possible, prevented from recurring. The problem-solving techniques, such as the Pareto and cause and effect analyses described in Chapter 4 should be helpful. Assuming that the major assignable causes have been identified and corrected, the initial control chart limits may be recalculated, excluding the points associated with these causes. The remainder of the historical data may then be checked against the revised limits to ensure that no other points indicate the presence of further assignable causes.

In cases such as the measurement of cooling fan blades, the preliminary study with historical data may be complicated by the passage of time and confounded by causes or symptoms that come and go and whose presence was not recorded. All that can be recommended is that the analysis be carried out as systematically as possible. If the historical data show or, following correction of assignable causes, can be made to show a performance consistent with the trial control limits, the charts may be extended and used for future control of the process.

A summary table

Table 11.9 shows a summary of all four attribute control charts in common use. Approximations of the binomial distribution by the Poisson distribution, and of both of these distributions by the normal distribution, may be useful in the design of process control systems for attributes. These are given in Appendix J.

11.6 Managing specifications based on subjective assessments

In a number of industries, attributes are the subject of regular subjective decisions. Deciding whether a defect is acceptable or unacceptable is often the source of endless debate. Such debate frequently results from a flexible specification with its associated commercial convenience. It may allow customers to reject merchandise when delivery is embarrassing, and to accept any quality of goods when quick delivery has top priority. Examples of such attributes include scratches, flash-free mouldings, clean surfaces, dirty marks, on or off-shade materials, seam puckering, readable copy, squareness, straightness, levelness, uniformity of

Table 11.9 *Attribute data – control charts*

What is measured	Chart name	Attribute charted	Centre line	Warning limits	Action or control limits	Comments
Number of defectives in sample of constant size n	'np' chart or 'pn' chart	np – number of defectives in sample of size n	$n\bar{p}$	$n\bar{p} \pm 2\sqrt{n\bar{p}(1-\bar{p})}$	$n\bar{p} \pm 3\sqrt{n\bar{p}(1-\bar{p})}$	n = sample size p = proportion defective $\bar{p}$ = average of p
Proportion defective in a sample of variable size	'p' chart	p – the ratio of defectives to sample size	$\bar{p}$	$\bar{p} \pm 2\sqrt{\dfrac{\bar{p}(1-\bar{p})}{\bar{n}}}$	$\bar{p} \pm 3\sqrt{\dfrac{\bar{p}(1-\bar{p})}{\bar{n}}}$	$\bar{n}$ = average sample size $\bar{p}$ = average value of p
Number of defects/flaws in sample of constant size	'c' chart	c – number of defects/flaws in sample of constant size	$\bar{c}$	$\bar{c} \pm 2\sqrt{\bar{c}}$	$\bar{c} \pm 3\sqrt{\bar{c}}$	$\bar{c}$ = average number of defects/flaws in sample of constant size
Average number flaws/defects in sample of variable size	'u' chart	u – the ratio of defects to sample size	$\bar{u}$	$\bar{u} \pm 2\sqrt{\dfrac{\bar{u}}{\bar{n}}}$*	$\bar{u} \pm 3\sqrt{\dfrac{\bar{u}}{\bar{n}}}$*	u = defects/flaws per sample $\bar{u}$ = average value of u n = sample size $\bar{n}$ = average value of n

* Only valid when n is in zone $\bar{n} \pm 25$ per cent

appearance, clarity, freshness, flavour, smoothness, softness, shininess, etc.

Various techniques are used to attempt to manage these subjective decisions. These include the submission of a preproduction or operation sample for acceptance. While this can *limit* the area of debate, experience in the textile, mechanical engineering, food, capital equipment, construction, banking, and numerous other industries shows that it does not go far enough in defining the requirements with precision and, hence, does not provide a sound basis on which to monitor or manage the transformation processes involved.

A first step towards containing the subjectivity of the assessment is to provide two samples – one which is just outside the limits of acceptability and the other which is just acceptable. Faced with two such samples, the room for debate becomes strictly limited and the monitoring of quality performance should become less subjective and, hence, more reliable. The sensitivity of attribute charting to change is known to be less than that of variable charting and converting attribute assessment into variable measurement will always present the opportunity for a greater sensitivity of detection and correction. In a number of the cases cited above, it is possible to find instruments and methods of measuring the specified parameter, but both the cost of the instruments and the time taken to measure may render this route unacceptable.

The basic requirement for treating any parameter or characteristic as a variable is simply that an assessment of the characteristic can be made on some predetermined and calibrated scale. It is often possible to record a fault on a scale varying, for example, through: none, very slight, slight, marked, very marked, to appalling. If a reference sample at each gradation exists, the degree of subjectivity during assessment can be measured against the scale. Ideally, any such scale will include gradations of acceptability as well as unacceptability.

With a minimum of six gradations within the scale, both mean and range charts are effective as methods of monitoring performance and judging capability. The higher the number of grades, the more valid the use of mean and range charting becomes. The setting up of a scale with gradations from appalling to very good, will probably not reflect a series of linear steps in the characteristic which one seeks to measure by a subjective method. This need be of no concern, provided that the mean and range or moving mean and moving range charts being used are based on sample sizes of four or more.

How is the scale set up? How is its validity demonstrated to a customer? Assuming that output, or product, is available for joint examination, a minimum of thirty-two examples is required. These will vary across the width of the proposed scale in order to make use of a simple 'better/worse' decision. The method of establishing the scale is as follows:

1 Take a set of thirty-two examples of output, which vary from the extremes of excellent to appalling. Pick up any two examples in the set and make a better/worse decision. (A choice must be made in each case.) Place the 'better' example on the *left* and the 'worse' example on the *right*. Take another two examples, decide on 'better' or 'worse' and place the examples in the appropriate piles. This will produce a 'better' pile of sixteen examples on the left and a 'worse' pile of sixteen samples on the right.

2 The first pile of sixteen 'better' examples is then subjected to a similar treatment to give two sets of eight examples, again one 'better' on the left-hand side and one 'worse' on the right-hand side. The sixteen 'worse' examples from the first sort are similarly resorted to give two sets of eight and the 'better' of these sorts is placed on top of the 'worse' of the previous second sort of sixteen. So one ends up with three piles of eight, sixteen and eight representing a 'better and better again' assessment, a 'better first time/worse second time or worse first time/better second time' assessment, and a 'worse both times' assessment.

3 The procedure is then repeated as illustrated below:

Number of sort

				32				
First	Better			16	16			Worse
Second			8	16		8		
Third		4		12	12		4	
Fourth		?	8	12		8	2	
Fifth	1	5		10	10		5	· 1
Scale	1	2		3	4		5	6

to give 6 piles going from best to worst.

If the sorting has involved numerous close decisions, there will be a fair chance that examination of the resultant six piles will reflect a gradation of the defect but that the number of discernable gradations may be less than six. In this case, a more discerning assessment is required. Examination of the output or product under different lighting conditions, viewed from different angles, etc., may be used to facilitate a scale of at least six gradations. If this is not done, the method will not be successful, although the mean chart may be useful to detect trends, given only four gradations of the defect.

If, during the sorting, there have not been many close decisions, the resultant piles will be found to have approximately the same grade of defect present in all the examples within one pile, and there will be a distinct difference between adjacent piles. If the examination of each pile suggests that more than six gradations are possible, the piles of five and ten can be subdivided to expand the number of gradations. The increased number of gradations will result in a scale of greater sensitivity to change.

The 'better/worse' decision is readily accepted by any assessor, and the above technique has been at the source of establishing a scale for better control of processes and better definition of a customer's requirements in numerous industries.

At worst, the use of this technique will result in only one detectable grade of defect, when further research will be necessary before improved control can follow. The next worse possibility is that two grades will be identified and, if these are classified as the barely acceptable and the just unacceptable, the room for debate will have been contained. If four gradations are found, the use of mean charts is possible and should assist in future control of the process mean. If six gradations are found, the scale can be used for control by mean and range charting, provided that observations against the scale are grouped into samples of at least four. The greater the number of gradations defined on such a scale, the more sensitive the mean and range chart will become.

Chapter highlights

- Attributes, things which are counted and are generally more quickly assessed than variables, are often used to determine quality. These require different control methods from those used for variables.
- Attributes may appear as numbers of non-conforming or defective units, or as numbers of non-conformities or defects. In the examination of samples of attribute data, control charts may be further categorized into those for constant sample size and those for varying sample size. Hence, there are charts for:

number defective (non-conforming)	np
proportion defective (non-conforming)	p
number of defects (non-conformities)	c
number of defects (non-conformities) per unit	u.

- It is vital, as always, to define attribute specifications. The process capabilities may then be determined from the average level of defectives or defects measured. Improvements in the latter require investigation of the whole process system. Never-ending improvement applies equally well to attributes, and variables should be introduced where possible to ensure this.

● Control charts for number (np) and proportion (p) defective are based on the binomial distribution. Control charts for number of defects (c) and number of defects per unit (u) are based on the Poisson distribution.

● A simplified method of calculating control chart limits for attributes is available, based on an estimation of the standard deviation, σ.

● The concept of processes being in and out of statistical control applies to attributes. Attribute charts are not as sensitive as variable control charts for detecting changes in non-conforming processes. Attribute control chart performance may be measured, using the average run length (ARL) to detection.

● np and c charts use constant sample sizes and, therefore, the control limits remain the same for each sample. For p and u charts, the sample size (n) varies and the control limits vary with n. In practice, an 'average sample size' ($\bar{n}$) may be used in most cases.

● Specifications based on subjective, attribute type, assessments can be difficult to manage. The subjectivity may be contained by the use of preproduction/operation samples, and the provision of good and bad examples of output.

● Attributes may be 'converted' into variables, which may then be plotted on $\bar{X}$ and R charts, by the creation of a predetermined, calibrated scale, containing six gradations, e.g. from appalling to very good. The method of doing so requires a set of thirty-two examples of output and a series of better/worse decisions to be made. The greater the number of gradations on the scale, the more sensitive the $\bar{X}$ and R will be.

12 Cumulative sum charts

12.1 Introduction

The control charts so far discussed for both variables and attributes are based on those originally proposed by Shewhart during the 1920s. They were designed to allow a process operator to collect new information about the operation of a process and make an immediate control decision. The rules for the Shewhart charts also include guidelines for the detection of trends and runs, but, again a decision is made as each point is plotted.

One of the basic hypotheses on which these charts are based is that processes are inherently stable. It follows that one should not react to all observations of variation but only those which have a high and known probability of being associated with assignable causes. Inherent stability, coupled with the inevitability of change, is the common property of all processes. The inherent stability of processes is also the hypothesis which lies behind the cumulative sum techniques in which, as for Pareto analysis, additional information can be extracted from data by cumulating the results.

The *cusum* technique was developed in Britain during the 1950s and found its first applications in sales forecasting. It is a powerful tool for management which, when used in conjunction with control charting, leads to improved management of processes, improved control of outputs and the more efficient use of resources.

12.2 The detection of trends and runs – attributes

The advantages of cumulating the data available when Shewhart charting is in use can be illustrated by an example using attributes. Samples of fifty ($n = 50$) lithographic plates from a production process are being taken every half-hour and examined for imperfections – burrs at the edges,

Table 12.1   *Number of defective lithographic plates in samples of size fifty*

Sample number	Number defective	Sample number	Number defective	Sample number	Number defective	Sample number	Number defective
1	1	11	3	21	2	31	1
2	4	12	4	22	1	32	4
3	3	13	2	23	2	33	1
4	5	14	3	24	4	34	3
5	4	15	7	25	1	35	1
6	3	16	3	26	2	36	5
7	6	17	5	27	5	37	5
8	3	18	1	28	0	38	2
9	2	19	3	29	5	39	3
10	5	20	3	30	2	40	4

marks, scratches, squareness – each plate is classified as accepted or rejected. Table 12.1 shows the results following the inspection of forty samples, each of fifty plates. Looking at the number of defectives alone will not give the reader a clear picture of the behaviour of the process. Figure 12.1 is the '*np*' chart for this data. The control limits have been set at the observed mean value of *np* (3.1) plus two standard deviations, for the warning limit, and plus three standard deviations, for the action limit, using the formula for the standard deviation of a binomial distribution:

$$\sigma = \sqrt{n\bar{p}(1-\bar{p})}$$

where:

σ = the standard deviation
n = the sample size
and $\bar{p}$ = the average proportion defective present

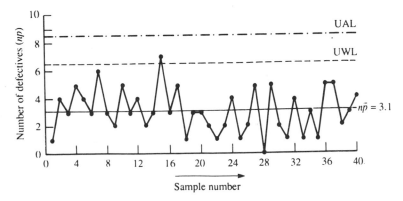

Figure 12.1   *Number defective chart*

This gives a standard deviation of 1.73 and sets the warning limit at 6.5 and the action limit at 8.5. Inspection of the '*np*' chart in Figure 12.1 reveals a process which is in statistical control since none of the points lie outside the action limit and only one of the forty points lies in a warning zone. In the absence of the warning and action limits it would not be possible to judge whether the amount of scatter present was consistent with random variations arising from the process and the sampling.

Closer examination of this presentation reveals:

- A run of points below the average value of '*np*' (3.1), starting at sample 18 but only continuing for six consecutive points, so no significant evidence of a run.
- No upward or downward trends of seven or more consecutive points.
- Some suggestion that prior to about sample 16 the results show a tendency to be above the average.

In Figure 12.2 the same data are plotted on a cusum chart. The calculations necessary to achieve this are simple and are shown in Table 12.2. A target value, convenient and close to the average value of '*np*', in this case 3.0, has been selected. This target value has been subtracted from each sample results and the residue cumulated to give the cusum score '*Sr*' for each sample.

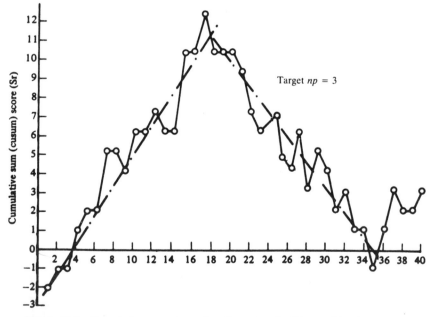

Figure 12.2   *Cumulative sum chart of attribute data for lithographic plate process*

Table 12.2 *Cumulative sum vaslues of data from Table 12.1*
Target *np= 3*

Sample number	Number defective	Cusum score. Sr	Sample number	Number defective	Cusum score. Sr
1	−2	−2	21	−1	+9
2	1	−1	22	−2	+7
3	0	−1	23	−1	+6
4	2	+1	24	1	+7
5	1	+2	25	−2	+5
6	0	+2	26	−1	+4
7	3	+5	27	2	+6
8	0	+5	28	−3	+3
9	−1	+4	29	2	+5
10	2	+6	30	−1	+4
11	0	+6	31	−2	+2
12	1	+7	32	1	+3
13	−1	+6	33	−2	+1
14	0	+6	34	0	+1
15	4	+10	35	−2	−1
16	0	+10	36	2	+1
17	2	+12	37	2	+3
18	−2	+10	38	−1	+2
19	0	+10	39	0	+2
20	0	+10	40	1	+3

The calculation of the cusum score, *Sr*, is simple and may be represented by the formula:

$$Sr = \sum_{i=1}^{i=r} (x_i - t)$$

where:

Sr is the cusum score of the *r*th value
x_i is the *i*th observation or result
t is called the target value.

The values of '*Sr*' are plotted on the cusum chart which shows dramatically that there was a difference between the behaviour of the process during the two halves of the period under review. In addition to joining up the points of the cusum plot, trend lines for the initial and final periods have been added. During the initial period, the cusum plot has a positive (upward) slope, indicating that, on average, the values being observed were greater than the target value of 3.0, so that the cumulative sum of the differences between the target value and the actual observations tended to increase continuously. This behaviour changed at the point of intersection of the two trend lines, i.e. at sample 18, when the slope of the cusum became negative (downwards) because the observations then tended, on average, to be below the target value and the cumulative sum of their differences from the target value tended to decline continuously.

The trend lines on the cusum plot may be used to estimate the values of '*np*' during these two periods of stable, but off-target output. The trend line during the initial period started at a value for sample 1 of about -1 and reached a value of about $+11$ at sample 18. There was a rise of $+12$ over an interval of seventeen samples or an average difference from the target value of $12/17 = 0.7$. Since the target value was 3 and the differences from the target were positive, the estimated average value of '*np*' during this period was $(3.0 + 0.7) = 3.7$. Similarly the slope then declined from about $+11$ at sample 18 to zero at sample 35: a decline of eleven in an interval of seventeen observations or an average of $11/17 = 0.6$ per observation. So during this period the average value of '*np*' was $(3.0 - 0.6) = 2.4$. Was there a further change of behaviour over the last few observations?

We shall deal with a method of adding known degrees of confidence to the conclusions reached, when reading a cusum chart for attributes, later in Section 12.5. For the moment it suffices to note that, during a stable period of the observed parameter, the initial values of '*Sr*' contain information from only a limited number of observations. As we move along the cusum plot the degree of confidence we may have in any trend will increase with each new observation. This means that when adding the trend lines to cusum plots 'by eye', we must give more emphasis or weight to the later points on the trend line. The trend lines shown in Figure 12.2 are clearly more influenced by the later observations in the trends than by the early observations. In the absence of a more refined technique, use can be made of a rough rule, that trends in cusum charts for attributes may be assumed to exist when they are associated with a family of at least five consecutive points. This rule might suggest that at the end of the period of observations on lithographic plates there is a family of five points scattered about a trend line which is approximately horizontal, implying that the observations differ from the target, on average, by zero, or that the average value of '*np*' is equal to the target of 3.0. It might also be possible to suggest that there is a family of the last six points which show an upward trend. A few more observations plotted on the cusum chart would allow the change, if any, to be identified. One can then determine with greater confidence, both the magnitude of the change in the average value of '*np*' and the time at which the change most probably occurred. Armed with this information the search for an assignable cause can be directed towards the time when it most probably took place.

It will be evident from the discussion so far that, in examining cusum charts, the search is for changes of slope as indicators of changes of behaviour between stable periods of process operation. Charts which rely on the search for changes of slope must be used with considerable care, since they are easily capable of misinterpretation. Unlike most graphs, the actual value of the points plotted, the cusum scores, are of no special significance; it is the slope of the family or families of plotted points which indicates their relevance to a target value. The use of cusum plots at the point of operation exposes them to misinterpretation which can only be overcome by careful training.

The rules for the interpretation of cusum plots may be summarized as follows:

- The cusum slope is *upwards*, the observations are *above* the target.
- The cusum slope is *downwards*, the observations are *below* the target.
- The cusum slope is *horizontal*, the observations are *on target*.
- The cusum slope *changes*, the observations are *changing*.
- The *absolute value* of the cusum score has *little meaning*.

The interpretation of the lithographic plate data, when reading from both the control chart (Figure 12.1) and the cusum chart (Figure 12.2), is as follows:

- The '*np*' chart shows some slight evidence of a change in behaviour between the first and second halves of the period under review, when the process was in statistical control, about a mean '*np*' of 3.1, even though sample 17 fell within a warning zone.
- The cusum chart shows clear evidence of an abrupt change from an 'above target of 3' period to a 'below target of 3' period at about sample 18 (i.e. from an average '*np*' of 3.8 to 2.4), and the possibility of a second change (from 2.4 to about 3.0) at about sample 35/36.

This combination of charts gives a very full description of the process behaviour and indicates those moments in time when changes occurred, or may have occurred, and when a search for assignable causes should be initiated. If the assignable causes present in the process of manufacturing lithographic plates could be identified and eliminated, the least success to be anticipated would be the achievement, over longer periods, of an average '*np*' of 2.4.

The full 'picture' is derived from the data being presented in both charts. The control chart picks up the short-term isolated high results more readily than the cusum, and the cusum is more sensitive to smaller changes maintained over long periods. This split of characteristics is consistent with the requirements for both short-term control, using the control charts, and longer term analysis, aimed at identifying trends and runs in order to improve process capability. In general, the control chart is ideal for use at the points and moments of control, while the combination of the control chart and the associated cusum chart is best suited to reviews of process performance and development, which typically lie outside the hour-to-hour responsibility for control.

12.3   Cusum charts – variables

The interpretation of cusum charts is concerned with the assessment of the gradient or slope of graphs. Careful design of the charts is, therefore, necessary to ensure that the appropriate sensitivity to change is catered for.

The choice of the value of 't' varies with the application. In the attribute example considered above, 't' was chosen as the nearest integer to the mean of the number of defectives present. When used to monitor forecasting performance, 't' would be the forecast for the rth period for which the actual achievement was x_r. In the manufacture of pharmaceutical tablets, 't' could be the target weight at the mid-point of the specification tolerance band or, alternatively, the average weight of tablets actually produced. Choosing an integer or round number may ease the calculation. Choosing a value for 't' which is significantly different from the actual average of the results being studied will give relatively high, positive or negative, slopes which could lead to the irritating result that the plotted graph runs off the graph paper!

Since we are interested in the slope of a cusum plot, the cusum chart design must be concerned with vertical and horizontal scales. The design of conventional mean and range charts for variables includes setting control limits at certain distances from the process mean. These correspond to two and three standard errors of the means (SEs). We may make use of these known limits in the design of the vertical and horizontal scales of a cusum chart for variables.

When reviewing sample means presented in a cusum plot, a major change, such as a change of 2SE, should show clearly, yet not so obviously that the cusum plot varies wildly during periods of normal sampling variation. This requirement may be met by arranging the scales such that a shift in sample mean of 2SE is represented on the chart by an approximate 45° slope. This is illustrated in Figure 12.3. It requires that the distance along the horizontal axis which represents one sample plot is approximately equal to the distance on the vertical axis represented by 2SEs. An example will clarify the explanation.

In Chapter 6, a process manufacturing steel rods was examined. Mean rod lengths from twenty-five samples, each made up of four rods, had the

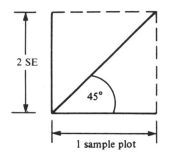

2 SE

45°

1 sample plot

Figure 12.3   *Slope of cusum chart for a change of 2SE in sample mean*

following characteristics:

Grand or process mean length $\overline{\overline{X}}$ = 150.1 mm
Mean sample range $\overline{R}$ =   10.8 mm

The process standard deviation was estimated using the formula:

$$\sigma = \overline{R}/d_n$$

(d_n for a sample size of four, is 2.059)

Hence, σ = 10.8/2.059 = 5.25 mm

This value may in turn be used to calculate the standard error of the means using the formula:

$$SE = \sigma/\sqrt{n} = 5.25/\sqrt{4} = 2.625 \text{ mm}$$
$$\text{and } 2SE = 2\times 2.625 = 5.25 \text{ mm}$$

The vertical and horizontal scales may now be set for the cusum chart. If the distance on the horizontal scale between two consecutive plots is to be one unit, say 1 cm, then 2SE on the vertical scale will need also to be about 1 unit or 1 cm. In the steel rod process 2SE = 5.25. No one would be happy plotting a graph which required a scale of 1 cm equivalent to 5.25 mm, so it is necessary to round up or round down. Clearly a scale of 1 cm/5 mm is the nearest and the most convenient option. Having designed the cusum chart for variables, it is usual to provide a key showing the slopes which correspond to being on target and changes of 2 and 3 SEs (Figure 12.4).

The cusum chart may now be used to interpret data. Table 12.3 shows the sample means from thirty groups of four steel rods, which were used in plotting the mean chart of Figure 6.11 (Chapter 6), now reproduced as Figure 12.5(a). In Table 12.3 the process average of 150.1 mm has been selected as the target value and subtracted from each of the sample results. The cusum score is also shown. The latter has been plotted in

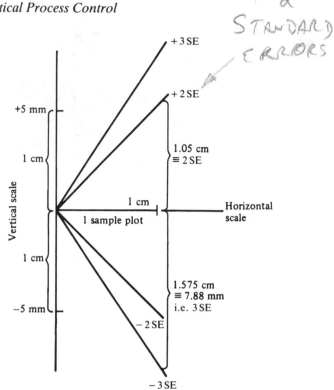

Figure 12.4   *Scale key for cusum plot*

Figure 12.5(b), so that the Shewhart and cusum charts may be read together.

If the reader compares the charts, certain features will be immediately apparent. First, an examination of sample plots 11 and 12 on both charts demonstrates that the mean chart more readily identifies large changes in the process mean. This is by virtue of the sharp peak on the mean chart which is not evident on the cusum chart unless the slopes of the chart and the key are compared. This is again illustrated at samples 13, 20, 21, 27 and 29. Second, the zero slope or horizontal line on the cusum plot between samples 12 and 13 shows what happens when a process observation is perfectly on target. The cusum score of sample 13 is still quite high, but its absolute value does not concern us. Our interest is only in the slope of the cusum chart at this point.

The care necessary when interpreting cusum charts is shown again at sample 21. On the mean chart there is a clear indication that the process has been overcorrected between samples 20 and 21 and that the mean length of the rods is significantly too short. On the cusum plot the negative slope between plots 20 and 21 indicates the same effect, and,

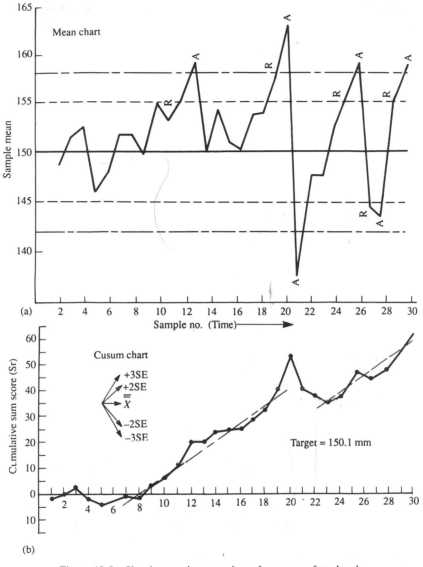

Figure 12.5 *Shewhart and cusum charts for means of steel rods*

even though the cusum score remains high at over 40 mm, there is a need to increase the lengths of the rods. The addition of the trend lines to the cusum plot enables us to see that up to about sample 8 the process was running on target but from this point it maintained a value steadily above the target value. At about sample 11, and again at sample 18, the mean chart and the slope of the cusum plot begin to change. While the mean

Table 12.3　*Cusum values of sample means (n= 4) for steel rod cutting process*

Sample number	Sample mean, $\bar{x}$ (mm)	$(\bar{x} - t)$ mm (t= 150.1 mm)	Sr
1	148.50	−1.60	−1.60
2	151.50	1.40	−0.20
3	152.50	2.40	2.20
4	146.00	−4.10	−1.90
5	147.75	−2.35	−4.25
6	151.75	1.65	−2.60
7	151.75	1.65	−0.95
8	149.50	−0.60	−1.55
9	154.75	4.65	3.10
10	153.00	2.90	6.00
11	155.00	4.90	10.90
12	159.00	8.90	19.80
13	150.00	−0.10	19.70
14	154.25	4.15	23.85
15	151.00	0.90	24.75
16	150.25	0.15	24.90
17	153.75	3.65	28.55
18	154.00	3.90	32.45
19	157.75	7.65	40.10
20	163.00	12.90	53.00
21	137.50	−12.60	40.40
22	147.50	−2.60	37.80
23	147.50	−2.60	35.20
24	152.50	2.40	37.60
25	155.50	5.40	43.00
26	159.00	8.90	51.90
27	144.50	−5.60	46.30
28	153.75	3.65	49.95
29	155.00	4.90	54.85
30	158.50	8.40	63.25

chart shows a very erratic behaviour after sample 19, the cusum indicates that from sample 23 the variations were steady about a value above target.

The rough rule used for the cusum plots of attributes, which only recognizes the possibility of a change when it is associated with a family of five consecutive points, does not apply to cusum charts of means and range values for sample sizes of greater than four. The reason is simply that the information contained in two such consecutive plots comes from eight separate individual observations. The degree of confidence in two such points is clearly comparable with at least five individual results. It is for this reason that, when using variables, it is recommended that cusum plots be made of means and ranges, which will show potential trends and

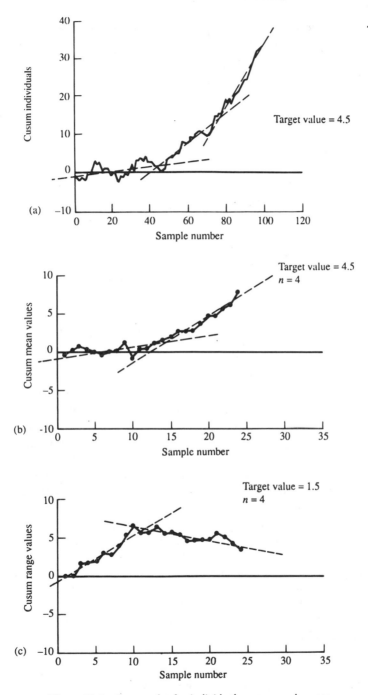

Figure 12.6 *Cusum plot for individuals, means and ranges*

runs in the accuracy and the precision of the process. The use of individual results for cusum plots will be a mixture of both random and non-random variations and be less enlightening. This is illustrated by the cusum plots in Figures 12.6(a), (b) and (c).

Although the method of cumulating differences and plotting them has a great impact on the management of the outputs of processes used in the manufacture of artefacts, it also provides monitors in such areas as:

- Forecasting performance – actual versus forecast sales.
- Detection of small but maintained changes of absenteeism, output, foreign exchange rates, share values, delivery delays, creditors, debtors, order intake, sales performance etc.

12.4   Decision procedures

As in the case of Shewhart charts, the interpretation of the cusum chart must be contained within an appropriate framework, otherwise the significance of the interpretation, or the degree of confidence one may have in the conclusion, adjustment and/or change to the process, will not be known. The largely subjective treatment adopted so far, while helpful, is not entirely satisfactory. A set of reasonably simple but objective decision rules is required to assist in identifying significant phenomena, where action is required, and segregating them from insignificant events, which should not lead to action. While several decision procedures for cusum charts have been developed, one of the earliest and simplest has found universal application and complements the use of Shewhart charts. This is the V-shaped mask first described by G.A. Barnard in 1959. It is a simple mask, often made from a transparent material, which can be placed over a cusum plot and used to distinguish between significant and insignificant events, the degree of significance, or the decision interval being selected, to reflect both the typical behaviour of a specific process and the degree of sensitivity of the cusum plot sought by the user. The design and use of V-masks is described in the following section of this chapter.

Other decision interval techniques are aimed at detecting change in one direction only; either increases or decreases. These require the target value for calculation to be chosen so that the chart will normally have a predetermined positive or negative slope and, when a predetermined absolute value of the cumulative sum is reached, a trend is recognized and the need for action is signalled. Two such charts may be used to monitor both upward and downward trends. The equivalence between this technique and the V-mask scheme is explained by K.W. Kemp in *Applied*

Statistics (1962), Volume II, in which the whole subject of the design of decision procedures for cusum charts and the explicit use of V-masks is given a thorough mathematical treatment.

12.5   The design and use of V-masks – attributes

Figure 12.7 shows a typical V-mask. The mask is placed over a cusum plot so that the line A0 is parallel with the horizontal axis, the vertex, 0, points forwards in the time series of the plot and the point A lies on the last cusum score plotted. A significant change is indicated by any part of the cusum plot being covered by either limb of the V-mask, as in Figure 12.7. This infringement of the V-mask is recognized as a significant event and should give rise to a search for the assignable cause. If all the points lie within the limbs of the V-mask, the process is assumed to be stable and any variations are classified as not significant and hence not requiring investigation or correction. The design of the V-mask obviously depends on the choice of the lead distance 'd' (measured in units of sample plots) and the angle of the mask 'θ'. These two parameters determine the geometry of the mask and need to be chosen to reflect the degree of significance sought.

Before proceeding to a discussion of the geometry and the decision intervals used in V-masks, it should be emphasized that:

- The V-mask is placed over the cusum plot horizontally, which means that the target value (t) must be chosen to reflect the fact that a

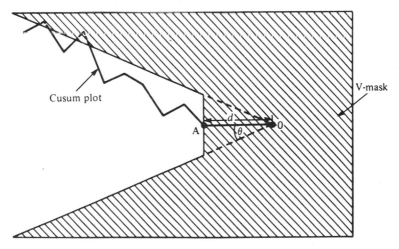

Figure 12.7   *V-mask for cusum chart*

horizontal plot of the cusum is the normal or expected behaviour from which abnormal changes are to be detected.

- Unlike the key of slopes (see for example, Figure 12.6), the V-mask is drawn backwards from the point under review. This merely reflects the fact that, in making a review of the cusum at any moment in the cusum score time series, the data available from all the previous scores is taken into account. Reversing the mask to point forwards will reveal the zone in which it is reasonable to make forecasts. Its V shape reflects the obvious fact that the further ahead one seeks to extrapolate past behaviour into the future, the wider the zone in which the forecast may fall.

The design of a V-mask for use with a cusum plot based on attributes can best be illustrated by an example. The cusum plot for defectives found in samples of fifty lithographic plates, discussed in Section 12.2 of this chapter, is shown in Figure 12.8 (the trend lines have been excluded). The reader will recall that Table 12.1 and the *np* chart shown in Figure 12.1 record the same data, which has a mean of 3.1, and a standard deviation of 1.73. The first step in the design of a suitable V-mask is to choose a decision interval which reflects the degree of sensitivity to change required. The choice is somewhat arbitrary. Let us follow an established

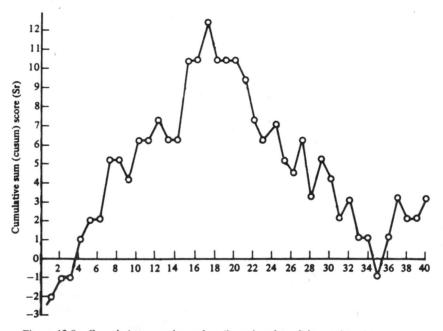

Figure 12.8   *Cumulative sum chart of attribute data from lithographic plate process*

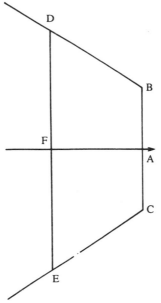

Figure 12.9   *V-mask for Figure 12.8*

practice of fixing the decision interval at two standard deviations i.e. 2×1.73 or 3.46. Figure 12.9 shows the resulting V-mask which is constructed as follows:

- Add vertically, above and below A, the points B and C, the distance AB and AC being equal to one decision interval or 3.46 units on the scale on which the cusum plot exists.
- Construct the line BC.
- Move from A horizontally to the point F, the distance AF being ten sample points on the existing cusum plot.
- Add vertically, above and below F, the points D and E, the distances FD and FE being equal to two decision intervals or 6.92 units on the cusum plot scale.
- Construct the lines BD and CE and extrapolate them if appropriate.
- Cut out the mask and, since the point A will overlay the existing cusum plot at the point of test, cut out a small nick to allow the point on the cusum plot to be seen.

The reader may find it useful to trace the V-mask in Figure 12.9 and from a piece of paper make an identical mask in order to move it around on the cusum plot shown in Figure 12.8. At sample 7 the V-mask shows that the process is above the target of 3.1 since the cusum plot infringes the lower line. The use of the mask enables this change to be identified after seven samples have been plotted on the cusum chart. Moving on through the plot with the V-mask shows that the infringement of the lower line continues and, at sample 22, a new infringement of the upper line occurs, so the process at this point is identified as running below the target of 3.1. This new infringement results from a change which is now positively identified. Trend lines added to the cusum plot after the twenty-second sample would allow the point of inflexion in the slope to be determined and an investigation of the assignable cause should concentrate enquiries to events occurring at that time.

In this example, the process changes were not detected until several data points reflecting the changed behaviour had been cumulated. The mean value of np during the first eighteen points was 3.8 and changed to 2.4 thereafter (see Section 12.2). The use of the V-mask on the cusum plot identified a change from the target of 0.7 units some six data points after the change, or an average run length to detection (ARL) of about 6. A change of 0.7 units in the mean value, in this case, is equivalent to 0.4 standard deviations. While a change of 0.4 standard deviations is detected by this V-mask with an ARL of 6, the same mask would require longer sustained runs or a larger ARL to detect smaller changes and would detect larger changes after shorter runs or a smaller ARL.

Separate V-masks have to be made for each parameter which is the subject of control, both because cusums for different parameters will require different scales and because each parameter will have its own standard deviation from which the decision interval is determined. The reader will appreciate that the use of the V-mask allows cusum decision making to be formalized. The choice of the decision interval in multiples of standard deviations also highlights a limitation for cusum charts based on attributes, namely that, since the standard deviation for both 'p' and 'u' is a function of the sample size a new V-mask has to be designed for each sample size outside a limited range of sample sizes. The cusum technique and the use of V-masks is straightforward for both 'np' (number of effected attributes in samples of constant size) and 'c' (number of effects in samples of constant size).

12.6   The design and use of V-masks – variables

As already indicated in Section 12.4, in the case of variables the cusum technique is most effectively used to detect changes in either the mean values or in the random scatter, assessed for example by the range. A V-mask enables the decision making to be formalized and once again a choice has to be made between the sensitivity of the mask to change and the probability of a false alarm. In the case of the mean values, the decision interval will clearly be related to the standard error of the means

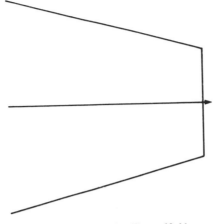

Figure 12.10 *V-mask for Figure 12.11*

(SE), which will be known from normal Shewhart charting. A key showing the slope of ± 2SE and ± 3SE has already been described as an aid to interpreting a cusum chart. There cannot be an absolute rule about the size of the decision interval for a V-mask because, for each case, the sensitivity to detection of a change has to be weighed against the probability of false alarms. A general rule, which is more fully justified in the following section of this chapter, is to start by using a decision interval of 5SE.

Figure 12.10 shows the detail of a V-mask using a decision interval of 5SE and is similar to that described above for a decision interval of 2 standard deviations for an '*np*' cusum plot. The cusum chart, shown previously as Figure 12.5(b), is repeated in Figure 12.11 but without either the key for slopes or the trend lines. Use of the 5SE V-mask first indicates a significant trend at sample 12 (reference back to the mean chart – Figure 12.5(a) – shows that this is the point at which the action line was crossed and an adjustment to the mean took place). A V-mask based on a decision interval of 2SEs would have detected an above average run at sample 11 and again at sample 17, whereas the 5SE V-mask would have detected the same run only at sample 19. The price for the more rapid detection of change by the 2SE mask is, of course, the increased probability of false alarms.

A cusum plot of ranges may be used to detect small trends or runs in the observed values of the range through a series of samples. In this case, the distribution of the range is already known to be skewed and not normal. The V-mask for the cusum chart of ranges should be based on a decision interval of two standard deviations of the range. The standard deviation of the range may be estimated by assuming a Poisson distribution and

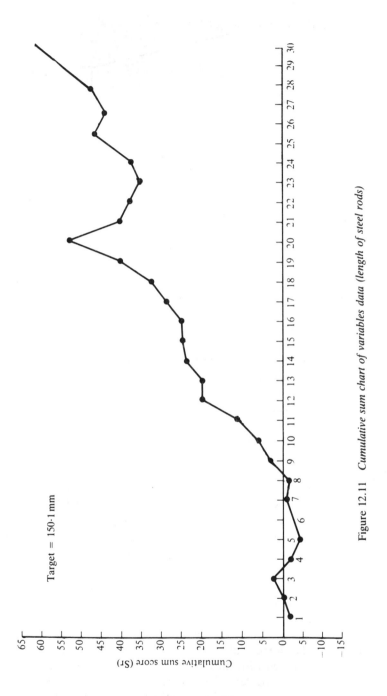

Figure 12.11 *Cumulative sum chart of variables data (length of steel rods)*

equating it to the square root of the average range, i.e.:

$$\sigma_R = \sqrt{\overline{R}}$$

where σ_R is the estimated standard deviation of the range

and $\overline{R}$ is the average range

12.7   Shewhart charts and cusums in combination

The above examples of cusum charts for both attributes and variables
have illustrated how they may be used in combination with Shewhart
charts, but the decision interval for the cusum charts was chosen
somewhat arbitrarily. While the choice of the decision interval may have
been capricious, the consequences of the choice were not. In all data
analysis techniques there is both a risk of not detecting a real event (this
can be expressed as the delay in detecting an event, or the ARL) and also
a risk of a false alarm. The greater the sensitivity to change demanded of
any control technique, the greater the probability of it giving rise to false
alarms. The choice which has to be made is concerned with the balance
between sensitivity to change and probability of false alarms. Assistance
in making the choice can be obtained by considering the consequences of
false alarms and then choosing the V-mask decision interval so that the
sensitivity of the cusum chart complements the sensitivity of the Shewhart
chart.

Examination of Figure 12.10 suggests that the sensitivity of the V-mask
based on 5SEs is such that, if there is a change in the process mean of
more than 5.5SEs between two successive sample means, the decision
line should be crossed – the event should be classed as significant. But
what is the degree of the significance? Figure 12.12 shows that when the
process mean is 5.5SEs above the target value, both the mean chart and
the cusum chart will probably detect the change but the risk of the signal
being false on the cusum chart is 0.5, because half of the distribution bell
is above the infringed V-mask, whereas the probability of a false alarm on
the mean chart is only that associated with a tail of the normal distribution
beyond 2.5 standard deviations (this is 0.006). The average run lengths to
detection are very similar but the risks of false alarms are not, and the
mean chart at this magnitude of change is the more reliable detector of
change. If we now consider the case where the change is smaller, say 1SE,
Figure 12.13 indicates an ARL on the cusum chart of 10, at which point
the risk of a false signal is again 0.5, but the probability of detection of this
change on the mean chart is only 0.0228 and the ARL will be the
reciprocal of this probability i.e. 43.9. If, or when, the mean chart shows
an action signal, however, the probability of it being a false alarm is only

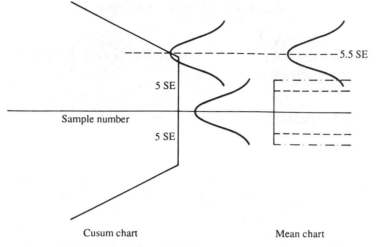

Figure 12.12   *Comparison of 5SE V-mask and mean chart sensitivity*

about 0.001. In this case the cusum chart is a superior detector of change by comparison with the mean chart. So the combination of a Shewhart mean chart and a mean cusum plot with a V-mask using a decision interval of 5SEs will cater for the detection, with high levels of confidence, of both rapid large and small sustained changes of a process mean.

In the case of Shewhart attribute charts, the probability of a false alarm

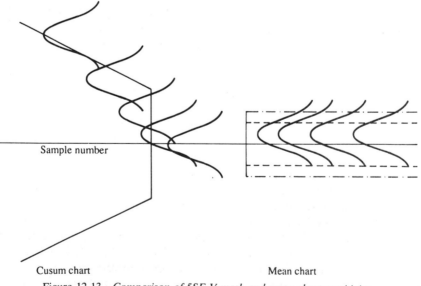

Figure 12.13   *Comparison of 5SE V-mask and mean chart sensitivity*

resulting from an action signal is somewhat reduced at 1 in 200, but at any infringement of the cusum V-mask the probability of a false alarm will always be much larger at about 0.5, as above.

In general, one may conclude that, while the cusum chart is more sensitive to the detection of small changes in the mean values of a parameter, the risk of it giving false alarms is greater than for the Shewhart charts. SPC thinking rests on the assumption that processes are stable in their performance and that assignable causes of change should only lead to interference with the process when the probability of false alarms is very small. The cusum chart has a characteristic which is not entirely consistent with this thinking and, were it to be used at the point of manufacture to review the status of the process after the addition of each new data point, false alarms and unjustified actions or investigations would occur more frequently than with the Shewhart charts. When reviewing the past performance of a process, the nature of the risk of action following a false alarm is changed because the action will be to investigate a possible assignable cause and *not* to change the process. When reviewing data as a cusum plot, it may show that after infringing the V-mask the process continued along the trend line, and as it did so the probability of a false alarm decreases. If the last point on the cusum plot indicates an infringement of the V-mask the reviewer of the process data is faced with the possibility of either taking the risk of investigating a possibly fictive assignable cause or waiting for more information to confirm or deny the existence of a change.

The reader may be interested in the operating characteristics shown in Figure 12.14 which illustrate the characteristics of the recommended 5SE and 2σ decision interval V-masks by comparison with the respective Shewhart charts. This shows the relationship between the ARL and the magnitude of change. Not surprisingly, it confirms that the cusum is quicker to detect small sustained changes, but, as already discussed, the probabilities of false alarms resulting from the two types of chart are seriously different.

Shewhart charts perform more reliably at the point of manufacture. When they are reviewed in combination with the cusum chart, the reviewer has the potential to see both large and small sustained changes or trends. There are two reasons for recommending the use of only Shewhart charts at the point of manufacture. The first is that the interpretation of cusums is not straightforward and process operators may have difficulties in handling the concept of monitoring performance by reference to the slope of a chart and by seeking to find significant changes of slope with a V-mask. The second is that the risk of false alarms arising from the decision rules for a cusum are not as readily acceptable as those arising from the decision rules for a Shewhart chart.

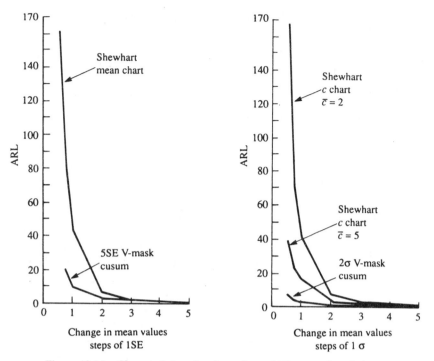

Figure 12.14   *Characteristics of various charts ARL – v – size of change*

12.8   Some examples of cusum and Shewhart charts

Figure 12.15 shows the fault rates from three parallel garment-making processes operating at the same time. What does this tell us? We may suspect periodic swings of all the processes and the fault rates coming closer together with time. The cusum charts shown in Figure 12.16 confirm the periodic swings and show that they have the same time period, so some external factor is probably affecting all three processes. The cusum charts also show that process 3 was the nearest to target – this can also be seen on the individuals chart but less obviously. In addition, process 4 was initially above target and process 5 even more so. Again, once this is pointed out, it can also be seen in Figure 12.15. After an initial separation of the cusum plots they remain parallel and the same distance apart. By referring to the individuals plot we see that this distance was close to zero. Reading the two charts together gives a very complete picture of the behaviour of the processes.

Figure 12.17(a) shows a noisy plot of a semi-continuous temperature recording with, perhaps, a gentle trend upward throughout the period. The addition of the cusum plot in Figure 12.17(b) suggests that there were

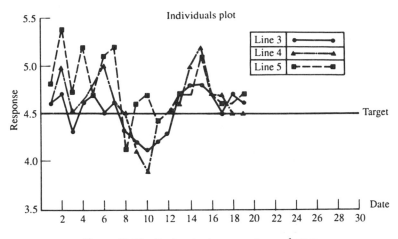

Figure 12.15 *Fault rates in garment manufacture*

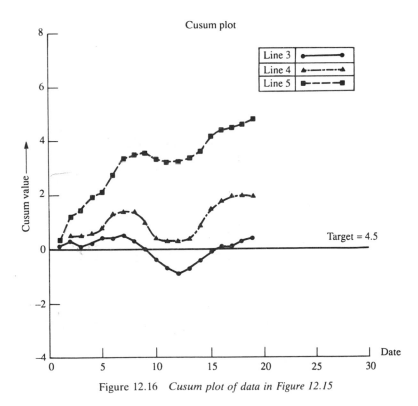

Figure 12.16 *Cusum plot of data in Figure 12.15*

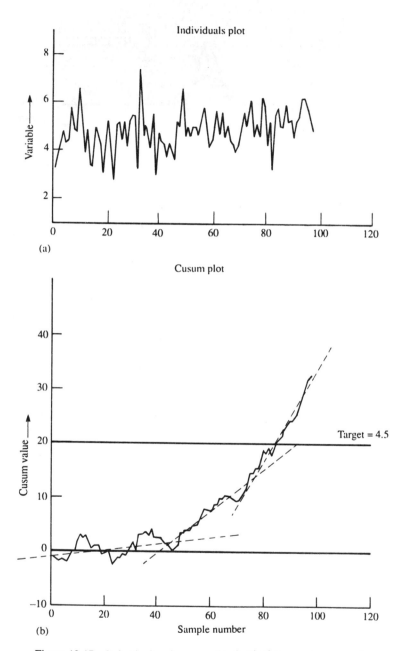

Figure 12.17   *Individual and cusum of individual temperature recordings*

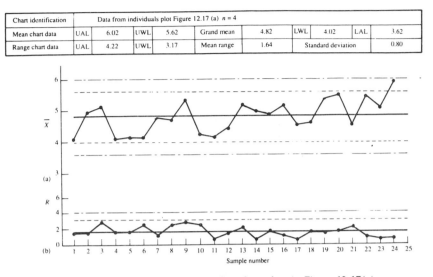

Chart identification	Data from individuals plot Figure 12.17 (a) *n* = 4									
Mean chart data	UAL	6.02	UWL	5.62	Grand mean	4.82	LWL	4.02	LAL	3.62
Range chart data	UAL	4.22	UWL	3.17	Mean range	1.64	Standard deviation			0.80

Figure 12.18 *Mean and range chart from data in Figure 12.17(a)*

two discrete changes in the parameter observed, at about samples 44 and 76, and not a gentle trend upwards (a gentle trend upwards would give a gentle and steady increase in the slope of the cusum plot). The time when these changes occurred can now be determined, and the search for assignable causes could be pursued around these discrete moments in time, but when looking at the individuals chart and the corresponding cusum plot, no attempt is made to distinguish between accuracy and precision and the resultant pictures and the potential conclusions will confuse the two different sources of variation. Greater clarity should result from making the distinction between accuracy and precision by grouping results and assessing the groups or samples by both their mean and range values.

Figures 12.18(a) and (b) show the mean and range charts for this data. From this we may conclude that both the means and the ranges were in statistical control although there is some suggestion of an upward trend of the means during the second half of the period. The cusum charts for both the means and the ranges are shown in Figures 12.19(a) and (b) which show one discrete change in the means at about observation number 56 and another discrete change in the range at about observation 48. The confused individuals charts give different and probably erroneous answers.

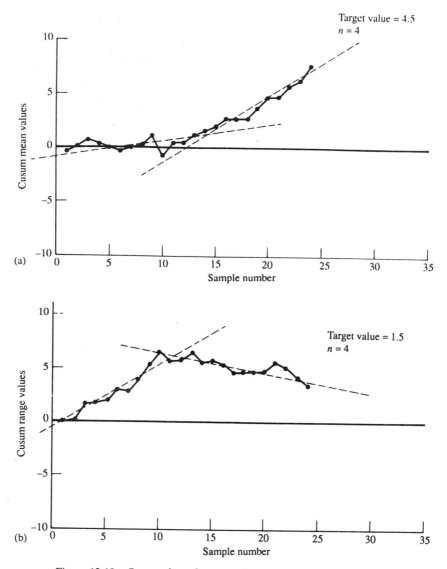

Figure 12.19 *Cusum plots of mean and range – temperature recording*

Chapter highlights

• Shewhart charts allow a decision to be made after each plot. While rules for trends and runs exist, they are only used infrequently. Since processes are inherently stable, cumulating process data can give longer-term information. The cusum technique is a method of analysis in which data is cumulated to give information about longer-term trends.

- Cusum charts for attributes are obtained by determining the difference between the values of individual observations and a target value, and cumulating these differences to give a cusum score which is then plotted.
- When a trend line, drawn through a cusum plot, is horizontal, it indicates that the observations were scattered around the target value; when the observed values are above the target value the slope of the cusum is positive; when the observed values lie below the target value the slope of the cusum plot is negative, and when the observed values are changing the slope of the cusum plot changes.
- The cusum technique can also be used for variables when some refinements may be added. These include making the distinction between accuracy and precision by grouping data to give means and ranges, predetermining the scale for plotting the cusum scores, choosing the target value and setting up a key of slopes corresponding to predetermined changes.
- The behaviour of a process can be described very fully by using the Shewhart and cusum charts in combination. The Shewhart charts are best used at the point of control, while the cusum chart is reserved for a later review of data. This distinction arises because of both practical difficulties in interpreting process behaviour from changes of slope within a plot, and the much greater risks, with cusum charts, of receiving a false alarm when adding the latest plot.
- Shewhart charts are more sensitive to rapid changes within a process, while the cusum is more sensitive to the detection of small sustained changes.
- Various decision procedures for the interpretation of cusum plots are possible including the use of V-masks.
- A V-mask has to be made for each parameter plotted as a cusum. Once prepared, the use of the mask distinguishes between insignificant and significant trends, the level of significance being specified in the design of the V-mask by its decision interval.
- Using the average run length to detection as a measure of the sensitivity of charting, the operating characteristics of Shewhart charts can be compared with those of cusums and V-masks of known decision interval.
- As in previous chapters, the use of cusum charts, alone and in combination with Shewhart charts, is illustrated by several examples.

13 Designing the process control system

13.1 SPC and the quality system

For successful SPC there must be an uncompromising commitment to quality, which must start with the most senior management and flow down through the organization. It is essential to set down a *quality policy* for implementation through a *documented management system*. Careful consideration must be given to this system as it forms the backbone of the quality skeleton. The objective of the system is to cause improvement of products and services through reduction of variation in the processes. The focus of the whole workforce from top to bottom should be on the processes and not the outputs. This approach makes it possible to control variation and, more importantly, to prevent non-conforming products and services, while steadily tightening standards.

The quality management system should apply to, and interact with, all activities of the organization. This begins with the identification of the requirements and ends with their satisfaction, at every transaction interface, both internally and externally. The activities involved may be classified in several ways – generally as processing, communicating, and controlling, but more usefully and specifically as:

1 Marketing
2 Market research
3 Design
4 Specifying
5 Development
6 Procurement
7 Process planning
8 Process development and assessment
9 Process operation and control
10 Product or service testing or checking
11 Packaging (if required)
12 Storage (if required)

13   Sales
14   Distribution/logistics
15   Installation/operation
16   Technical service
17   Maintenance/servicing

The impact of a good quality management system, such as one which meets the requirements of the international standard ISO 9000 series, is that of gradually reducing process variability to achieve continuous or never-ending improvement. The requirement to set down defined procedures for all aspects of an organization's operations, and to stick to them, will reduce the variations introduced by the numerous different ways often employed for doing things. Go into any factory without a good quality system and ask to see the operators' 'black book' of plant operation and settings. Of course, each shift has a different black book, each with slightly different settings and ways of operating the process. Is it any different in office work or for salespeople in the field? Do not be fooled by the perceived simplicity of a process into believing that there is only one way of operating it. There are an infinite variety of ways of carrying out the simplest of tasks – the authors recall seeing various course participants finding fourteen different methods for converting A4 size paper into A5 size (half A4) in a simulation of a production task. The ingenuity of human beings needs to be controlled if these causes of variation are not to multiply together to render processes completely incapable of consistency or repeatability.

The role of the quality system is to define and control process procedures and methods. Continual system audit and review will ensure that procedures are either followed or corrected, thus eliminating assignable or special causes of variation in materials, methods, equipment, information, etc. to ensure a: 'Could we do this job with more consistency' approach (Figure 13.1).

The task of measuring, inspecting or checking is taken by many to be the passive one of sorting out the good from the bad, when it should be an active part of the feedback system to prevent errors, defects, or non-conformance. Clearly any control system based on detection of poor quality by post-production/operation inspection or checking is unreliable, costly, wasteful and uneconomical. It must be replaced by the strategy of prevention, and the inspection must be used to check the system of transformation, not the product. Inputs, outputs and proc- esses need to be measured for effective quality management. The measurements monitor quality and may be used to determine the extent of improvements and deterioration in quality. Measurement may take the form of simple counting to produce attribute data, or it may involve

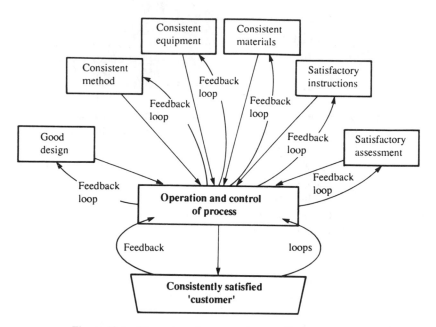

Figure 13.1   *The systematic approach to quality management*

more sophisticated methods to generate variable data. Processes operated without measurement and feedback are processes about which very little can be known. Conversely, if inputs and outputs can be measured and expressed in numbers, then something is known about the process and control is possible. The first stage in using measurement, as part of the process control system, is to identify precisely the activities, materials, equipment, etc., which will be measured. This enables everyone concerned with the process to be able to relate to the target values and the focus provided will encourage improvements.

For measurements to be used for quality improvement they must be accepted by the people involved with the process being measured. The simple self-measurement and plotting, or the 'how-am-I-doing' chart, will gain far more ground in this respect than a policing type of observation and reporting system which is imposed on the process and those who operate it. Similarly, results should not be used to illustrate how bad one operator or group is, unless their performance is 100 per cent under their own control. The emphasis in measuring and displaying data must always be on the assistance that can be given to correct a problem or remove obstacles preventing the process from meeting its requirements first time, every time.

Economic design of the system

Considerable attention has been given by several authors to the design of process control systems on economic grounds. For control charts to be used, the following parameters must be specified:

- Sample size, n
- Sampling frequency, h (the interval of time between samples)
- Control limits (set at $k\sigma$ from the process mean)

The selection of these values is effectively the design of the control chart, and traditionally charts have been designed considering only statistical concepts. The design of control charts, however, has economic consequences since the costs of the following are all affected by the selection of the control chart parameters:

- Taking samples
- Inspection
- Taking repeat samples and inspection after 'warning' indications
- Investigating 'action' signals, and any accompanying loss of production/operation
- Correcting assignable causes
- Defective products or services reaching the customer

The work carried out in this field has involved many mathematical manipulations which are outside the scope of this book, but the results were of a general nature. They are based on the fact that there is an optimum level of out-of-specification material to be produced, which is cheaper to produce than using more frequent sampling and larger sample sizes. This clearly contradicts the current thinking on SPC, expressed throughout this text.

The mathematics of economic design of control chart limits are irrelevant when the action to be taken is simply to investigate or 'adjust' the process at the right time. It has already been shown in Chapters 7–12 that the sample size to be used is influenced by the process capability, the performance of the control chart, and the desire to separate changes in accuracy and precision. Sampling frequency, as already stated, is a function of the inherent process stability which is, in turn, determined by the nature of the process itself.

13.2 Teamwork and process control

Teamwork will play a major role in any organization's efforts to make never-ending improvements. The need for teamwork can be seen in many human activities. In most organizations, problems and opportunities for improvement exist between departments. Seldom does a single department own all the means to solve a problem or bring about improvement alone.

Sub-optimization of a process seldom improves the total system performance. Most systems are complex, and input from all the relevant processes is required when changes or improvements are to be made. Teamwork throughout the organization is an essential part of the implementation of SPC. It is necessary in most organizations to move from a state of independence to one of interdependence, through the following stages:

Little sharing of ideas and information
Exchange of basic information
Exchange of basic ideas
Exchange of feelings and data
Elimination of fear
Trust
Open communication

Time

The communication becomes more open with each progressive step in a successful relationship. The point at which it increases dramatically is when trust is established. After this point, the barriers that have existed are gone and open communication will proceed. This is critical for never-ending improvement and problem solving, for it allows people to supply good data and all the facts without fear.

Teamwork brings diverse talents, experience, knowledge, and skills to any process situation. This allows a variety of problems that are beyond the technical competence of any one individual to be tackled. Teams can deal with problems which cross department and divisional boundaries. All of this is more satisfying and morale boosting for people than working alone.

A team will function effectively only if the results of its meetings are communicated and used. Someone should be responsible for taking minutes of meetings. These need not be formal, and may simply reflect decisions and action assignments. The minutes may be handwritten, copied, and delivered to the team members on the way out of the door. More formal sets of minutes might be drawn up after the meetings and sent to sponsors, administrators, supervisors, or others who need to know what happened. The purpose of minutes is to inform others of decisions made and list actions to be taken. Minutes are an important part of the communication chain with other people or teams involved in the whole process.

Quality improvement and 'Kaisen' teams

A quality improvement team (QIT) is a group of people with the appropriate knowledge, skills, and experience who are brought together specifically by management to tackle and solve a particular problem, usually on a project basis. They are cross-functional and often multi-disciplinary.

The 'task force' has long been a part of the culture of many organizations at the technological and managerial levels. But quality improvement teams go a step further – they expand the traditional definition of 'process' to include all production and operating systems. This includes paperwork, communication with other units, operating procedures and the process equipment itself. By taking this broader view all process problems can be addressed.

The management of quality improvement teams is outside the scope of this book and is dealt with in *Total Quality Management* (Oakland, Heinemann Professional Publishing, 1989). It is important, however, to stress here the role which SPC techniques themselves can play in the formation and work of teams.

The management in one company which was experiencing a 17 per cent error rate in its invoice generating process, decided to try to draw a flowchart of the process. Two people who were credited with knowledge of the process were charged with the task. They soon found that it was impossible to complete the flowchart, because they did not fully understand the process. Progressively five other people who were involved in the invoicing, had to be brought to the table in order that the chart could be finished to give a complete description of the process. This assembled group were kept together as the quality improvement team, since they were the only people who collectively could make improvements. Simple data collection methods, brainstorming, cause and effect, and Pareto analysis were then used, together with further flowcharting, to reduce the error rate to less than 1 per cent within just six months.

The flexibility of the cause and effect (C/E) diagram makes it a standard tool for problem-solving efforts throughout an organization. This simple tool can be applied in manufacturing, service, or administrative areas of the company and can be applied to a wide variety of problems from simple to very complex situations.

Again, the knowledge gained from the C/E diagram often comes from the method of construction, not just the completed diagram. A very effective way to develop the C/E diagram is with the use of a team representative of the various areas of expertise on the effect and processes being studied. The C/E diagram then acts as a collection point for the current knowledge of possible causes, from several areas of experience.

Brainstorming in a team is the most effective method of building the C/E diagram. This activity contributes greatly to the understanding, by all those involved, of a problem situation. The diagram becomes a focal point for the entire team and will help any team develop a course for corrective action.

Quality improvement teams usually find their way into an organization as problem solving groups. This is the first stage in the creation of problem prevention teams, which operate as common work groups and whose main objective is constant improvement of processes. Such groups may be part of a multi-skilled, flexible workforce, and include 'inspect and repair' tasks as part of the overall process. The so called 'Kaisen' team operates in this way to eliminate problems at the source by working together, and using very basic tools of SPC where appropriate, to create less and less opportunity for problems and reduce variability. Kaisen teams are usually provided with a 'help line' which, when 'pulled', attracts help from human, technical and material resources from outside the group. These are provided specifically for the purpose of eliminating problems and aiding process control.

13.3   Improvements in the process

To improve the quality of a process it is important to first recognize whether the process control is limited by the random or the assignable causes of variation. This will determine *who* is responsible for the specific improvement steps, *what resources* are required, and *which statistical tools* will be useful (see Figure 13.2). The comparison of actual product quality characteristics with the requirements (inspection) is not a basis for action on the process, since unacceptable product or service can result from either random or assignable causes. Product or service inspection is useful only to sort out good from bad and to set priorities on which processes to improve.

Any process left to natural forces will suffer from deterioration, wear and breakdown. Therefore, management must help people identify and prevent those natural causes through ongoing improvement of the processes they manage. The organization's culture must encourage communications throughout and promote a participative style of management that allows people to report problems and suggestions for improvement without fear or intimidation or enquiries aimed at apportioning blame. These must then be addressed with statistical thinking by all members of the organization.

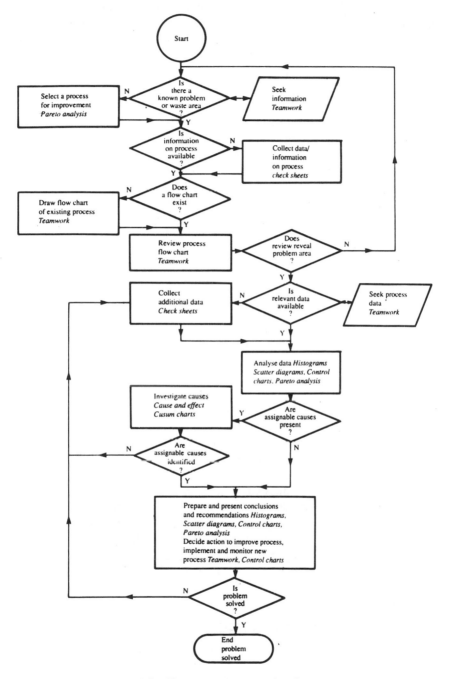

Figure 13.2 *The systematic approach to improvement*

Activities to improve quality must include the assignment of various people in the organization to work on random causes and assignable causes. The appropriate people to identify assignable or special causes are usually different from those needed to identify random or common causes. The same is true of those needed to remove causes. Removal of random causes is the responsibility of management, often with the aid of experts in the process such as engineers, chemists and systems analysts. Assignable causes can frequently be handled at a local level by those working in the process such as supervisors and operators. Without some knowledge of the likely origins of random and assignable causes it is difficult to efficiently allocate human resources to improve quality.

Most of the improvements in quality will require action by management, and in almost all cases the removal of assignable causes will make a fundamental change in the way processes are operated. For example, an assignable cause of variation in a production process may result when there is a change from one supplier's material to another. To prevent this assignable cause from occurring in the particular production processes, a change in the way the organization chooses and works with suppliers may be needed. Improvements in conformance are often linked to a policy of single sourcing.

Another area in which the knowledge of random and assignable causes of variation is vital is in the supervision of people. A mistake which is often made is the assignment of faults in the process (random causes) to those working on the process, e.g. operators and staff, rather than to those in charge of the process: management. Clearly, it is important for a supervisor to know whether problems, mistakes, or rejected material are a result of random causes, assignable causes related to the system, or assignable causes related to the people under his or her supervision. Again the use of the systematic approach and the appropriate techniques will help the supervisor to accomplish this.

Management must demonstrate commitment to this by providing leadership and the necessary resources. These resources will include training on the job, time to effect the improvements, improvement techniques and a commitment to institute changes for ongoing improvement. This will move the organisation from having a reactive management system to having one of prevention. This all requires time and effort by everyone, every day.

Process control charts and improvements

The emphasis which must be placed on never-ending improvement has important implications for the way in which process control charts are applied. They should not be used purely for control, but as an aid in the

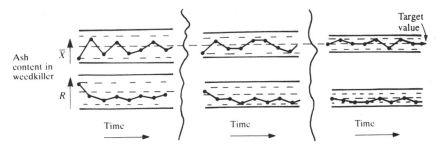

Figure 13.3 *Continuous process improvement – reduction in variability*

reduction of variability by those at the point of operation capable of observing and removing assignable causes of variation. They can be used effectively in the identification and gradual elimination of random causes of variation. In this way the process of continuous improvement may be charted, and adjustments made to the control charts in use to reflect the improvements.

This is shown in Figure 13.3 where progressive reductions in the variability of ash content in a weedkiller have led to decreasing sample ranges. If the control limits on the mean and range charts are recalculated periodically or after a step change, their positions will indicate the improvements which have been made over a period of time, and *ensure*

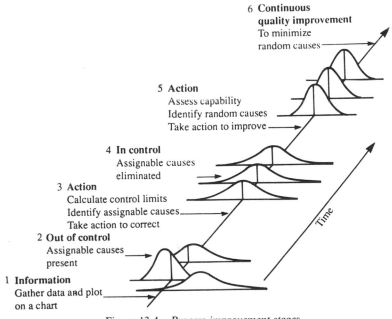

Figure 13.4 *Process improvement stages*

that the new level of process capability is maintained. **Further** improvements can then take place (Figure 13.4). Similarly, attribute or cusum charts may be used to show a decreasing level of number of errors, or proportion of defects and to indicate improvements in capability.

Often in process control situations, action signals are given when the assignable cause results in a desirable event, such as the reduction of an impurity level, a decrease in error rate, or an increase in order intake. Clearly, assignable or special causes which result in deterioration of the process must be investigated and eliminated, but those that result in improvements must also be sought out and managed so that they become part of the process operation. Significant variation between batches of material, operators, or differences between suppliers are frequent causes of action signals on control charts. The continuous improvement philosophy demands that these are all investigated and the results used to take another step on the long ladder to perfection. Action signals and assignable or special causes of variation should stimulate enthusiasm for solving a problem or understanding an improvement, rather than gloom and despondency.

The never-ending improvement cycle

Prevention of failure is the primary objective of the quality improvement process and is caused by a management team that is focused on improvement. The journey to cause prevention will require a defined

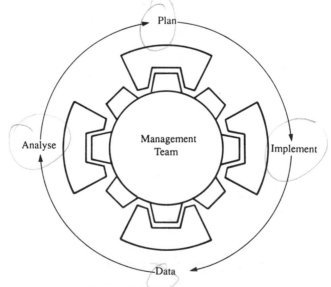

Figure 13.5   *The Deming cycle*

process for improvement. The system which will allow ongoing improvement is the so-called Deming cycle (Figure 13.5). This will provide the strategy in which the SPC tools will be most useful, and will identify the steps for improvement.

Plan

The first phase of the system – plan – helps to focus the effort of the improvement team on the process. The following questions should be addressed by the quality improvement team.

1 What are the requirements of the output from the process?
2 Who are the customers of the output? Both internal and external customers should be included.
3 What are the objectives of the improvement effort? The objectives of the improvement effort may include one or all of the following:

- Improve customer satisfaction
- Eliminate internal difficulties
- Eliminate unnecessary work
- Eliminate failure costs
- Eliminate non-conforming output

Every process has many opportunities for improvement, and resources should be directed to assure that all efforts will have a positive impact on these objectives. When the objectives of the improvement effort are established, output identified, and the customers noted, then the team is ready for the implementation stage.

Implement

The implementation effort will have the purpose of:

1 Defining the processes that will be improved.
2 Identifying and selecting opportunities for improvement.

The quality improvement team should accomplish the following steps during implementation:

1 Define the scope of the system to be improved and flowchart the processes within this system.
2 Identify the key processes which will contribute to the objectives identified in the planning stage.
3 Identify the customer/supplier relationships for the key processes.

These steps can be completed by the quality improvement team through their present knowledge of the system. This knowledge will be advanced throughout the improvement effort, and with each cycle the flowchart and cause and effect diagrams should be updated. The improvement team should launch each effort for improvement with selected theories on causes of variation. The following stages will help the team make improvements on the selected process:

1 Identify and select the process in the system that will offer the greatest opportunities for improvement. (The team may find that a completed process flowchart will facilitate and communicate understanding of the selected process to all team members.)
2 Document the steps and actions that are necessary to make improvements. (It is often useful to consider what the flowchart would look like if every job was done right the first time, often called 'imagineering'.)
3 Define the cause and effect relationships in the process using a cause and effect diagram.
4 Identify the important sources of data concerning the process. (The team should develop a data collection plan.)
5 Identify the measurements which will be used for the various parts of the process.
6 Identify the largest contributors to variation in the process. The team should use their collective experience and brainstorm the possible causes of variation.

During the next phase of the improvement effort, the team will apply the knowledge and understanding gained from these efforts and gain additional knowledge about the process.

Data
The data collection phase has the following objectives:

1 To collect data from the process as determined in the planning and implementation phase.
2 To determine the stability of the process using the appropriate control chart method(s).
3 If the process is stable, to determine the capability of the process.
4 To prove or disprove any theories established in the earlier phases.
5 If the team observed any unplanned events during data collection, to determine the impact this will have on the improvement effort.
6 To update the flowcharts and cause and effect diagrams, so the data collection adds to current knowledge.

Analyse
The purpose of this phase is to analyse the findings of the prior phases and help plan for the next effort of improvement. During this phase of process improvement, the following should be accomplished:

1 Determine the action on the process which will be required. It will identify the inputs or combinations of inputs that will need to be improved. These should be noted on an updated flowchart of the process.
2 Develop greater understanding of the causes and effects.
3 Ensure that the agreed changes have the anticipated impact on the specified objectives.
4 Identify the departments and organizations which will be involved in analysis, implementation and management of the recommended changes.
5 Determine the objectives for the next round of improvement. Problems and opportunities discovered in this stage should be considered as objectives for these efforts. Pareto charts should be consulted from the earlier work and revised to assist in this process.

13.4 Taguchi methods

Genichi Taguchi has defined a number of methods to simultaneously reduce costs and improve quality management. The current popularity of his approach is a fitting testimony to the merits of this work. The following discussion seeks to emphasize the correlation between his methods and the narrower range of topics covered in this book.

The Taguchi methods may be considered under four main headings:

- Total loss function
- Design of products, processes and production
- Reduction in variation
- Statistically planned experiments

Total loss function

The essence of Taguchi's definition of total loss function is that the smaller the loss generated by a product or service from the time it is transferred to the customer, the more desirable it is. Any variation about a target value for a product or service will result in some loss to the customer and such losses should be minimized. It is clearly reasonable to spend on quality improvements provided that they result in larger savings for either the producer or the customer. Earlier chapters have illustrated ways in which non-conforming products, when assessed and controlled by variables, can be reduced to events which will occur at probabilities of the order of 1 in 100,000 – such reductions will have a large potential impact on the customer's losses.

Taguchi's loss function is developed by using a statistical method which need not concern us here – but the concept of loss by the customer as a measure of quality performance is clearly a useful one.

Design of products, process and production

For any product or service we may identify three stages of design – the product (or service) design, the process (or method) design and the production (or operation) design. Each of these overlapping stages has many steps, the outputs of which are often the inputs to other steps. For all the steps, the matching of the outputs to the requirements of the inputs of the next step clearly affects the quality and cost of the resultant final product or service. Taguchi's clear classification of these three stages may be used to direct management's effort not only to the three stages but also the separate steps and their various interfaces. Following this model, management is moved to select for study 'narrowed down' subjects, to achieve 'focused' activity, to increase the depth of understanding, and to greatly improve the probability of success towards higher quality levels.

Design must include consideration of the potential problems which will arise as a consequence of the operating and environmental conditions under which the product or service will be both produced and used. Equally, the costs incurred during production will be determined by the actual manufacturing process. Controls, including SPC techniques, will always cost money but the amount expended can be reduced by careful consideration of control during the initial design of the process. In these, and many other ways, there is a large interplay between the three stages of development.

In this context, Taguchi distinguishes between 'on-line' and 'off-line' quality management. On-line methods are technical aids used for the control of a process or the control of quality during the production of products and services – broadly the subject of this book. Off-line methods use technical aids in the design of products and processes. Too often the off-line methods are based on the evaluation of products and processes rather than their improvement. Effort is directed towards assessing reliability rather than to reviewing the design of both product and process with a view to removing potential imperfections by design. Off-line methods are directed towards improving the capability of design. A variety of techniques are possible in this quality planning activity and include structured teamwork, the use of formal quality management systems, the auditing of control procedures, the review of control procedures and failure mode and effect analysis applied on a company-wide basis.

Reduction in variation

Reducing the variation of key processes, and hence, product parameters about their target values is the primary objective of a quality improvement programme. The widespread practice of stating specifications in terms of simple upper and lower limits conveys the idea that the customer is equally satisfied with all the values within the specification limits and is suddenly not satisfied when a value slips outside the specification band. The practice of stating a tolerance band may lead to manufacturers aiming to produce and despatch products whose parameters are just inside the specification band. In any assembly, whether mechanical, electrical, chemical, processed food, processed data as in banking, civil construction, etc., there will be a multiplicity of activities and hence a multiplicity of sources of variation which all combine to give the total variation.

For variables, the mid-specification or some other target value should be stated along with a specified variability about this value. For those performance characteristics which cannot be measured on a continuous scale it is better to employ a scale such as: excellent, very good, good, fair, unsatisfactory, very poor; rather than a simple pass or fail, good or bad.

Taguchi introduces a three-step approach to assigning nominal values and tolerances for product and process parameters:

Design system

The application of scientific, engineering and technical knowledge to produce a basic functional prototype design requires a fundamental understanding of both the needs of customers and the production possibilities. One is not seeking trade-offs at this stage, but a clear definition of the customer's real requirements, possibly classified as critical, important and desirable, and an equally clear definition of the supplier's known capabilities to respond to these requirements, possibly distinguishing between the use of existing technology and the development of new techniques.

Parameter design

This entails a study of the whole process system design aimed at achieving the most robust operational settings – those which will react least to variations of inputs.

Process developments tend to move through cycles. The most revolutionary developments tend to start life as either totally unexpected results (fortunately observed and understood) or success in achieving expected results, but often only after considerable, and sometimes frustrating, effort. Development moves on through further cycles of attempting to increase the reproducibility of the transformation of inputs to outputs, and includes the optimization of the process conditions to those which are more robust to variations in all the inputs. An ideal process would accommodate wide variations in the inputs with relatively small impacts on the variations in the outputs. Some processes, and the environments in which they are carried out, are less prone to multiple variations than others. Types of cereal and domestic animals have been bred to produce cross-breeds which can tolerate wide variations in climate, handling, soil, feeding, etc. Machines have been designed to allow for a wide range of the physical dimensions of the operators (cars, for example). Industrial techniques for the processing of food will accommodate wide variations in the raw materials with the least influence on the taste of the final product. The textile industry constantly handles, at one end, the wide variations which exist among natural and man-made fibres and, at the other end, garment designs which allow a limited range of sizes to be acceptable to the highly variable geometry of the human form. Specifying the conditions under which such robustness can be achieved is the object of parameter design.

Tolerance design
A knowledge of the nominal settings advanced by parameter design enables tolerance design to begin. This requires a trade-off between the costs of production or operation and the losses acceptable to the customer arising from performance variation. It is at this stage that the tolerance design of cars or clothes ceases to allow for all versions of the human form, and that either blandness or artificial flavours may begin to dominate the taste of processed food.

These three steps pass from the original concept of the potential for a process or product, through the development of the most robust conditions of operation, to the compromise involved when setting 'commercial' tolerances – and focus on the need to consider actual or potential variations at all stages. When considering variations within an existing process it is clearly beneficial to similarly examine their contributions from the three points of view.

Statistically planned experiments
Experimentation is necessary under various circumstances and in particular in order to establish the optimum conditions which give the most robust process – to assess the parameter design. 'Accuracy' and 'precision', as defined in Chapter 5, may now be regarded as 'nominal settings' (target or optimum values of the various parameters of both processes and products) and 'noise' (both the random variation and the 'room' for adjustment around the nominal setting). If there is a problem it

will not normally be an unachievable nominal setting but unacceptable noise. Noise is recognized as the combination of the random variations and the ability to detect and adjust for drifts of the nominal setting. Experimentation should therefore be directed towards maintaining the nominal setting and assessing the associated noise under various experimental conditions. Some of the steps in such research will already be familiar to the reader. These include grouping data together, in order to reduce the effect on the observations of the random component of the noise and expose more readily the effectiveness of the control mechanism, the identification of assignable causes, the search for their origins and the evaluation of individual components of some of the sources of random variation.

Noise is divided into three classes: outer, inner and between. Outer noise includes those variations whose sources lie outside the management's controls, such as variations in the environment which influence the process (for example ambient temperature fluctuations). Inner noise arises from sources which are within managements' control but not the subject of the normal routine for process control, such as the condition or age of a machine. Between noise is that tolerated as a part of the control techniques in use – this is the 'room' needed to detect change and correct for it. Trade-off between these different types of noise is sometimes necessary. Taguchi quotes the case of a tile manufacturer who had invested in a large and expensive kiln for baking tiles, and in which the heat transfer through the oven and the resultant temperature cycle variation gave rise to an unacceptable degree of product variation. While a redesign of the oven was not impossible, both cost and time made this solution unavailable – the kiln gave rise to 'outer' noise. Effort had, therefore, to be directed towards finding other sources of variation, either 'inner' or 'between', and, by reducing the noise they contributed, bringing the total noise to an acceptable level. It is only at some much later date, when specifying the requirements of a new kiln, that the problem of the outer noise becomes available and can be addressed.

In many processes, the number of variables which can be the subject of experimentation is vast, and each variable will be the subject of a number of sources of noise within each of the three classes. So the possible combinations for experimentation are seemingly endless. The 'statistically planned experiment' is a system directed towards minimizing the amount of experimentation to yield the maximum of results and in doing this to take account of both accuracy and precision – nominal settings and noise. Taguchi recognizes that in any ongoing industrial process the list of the major sources of variation and the critical parameters which are affected by 'noise' are already known. So the combination of useful experiments may be reduced to a manageable number by making use of this inherent knowledge. Experimentation can be used to identify:

1   The design parameters which have a large impact on the product's parameters and/or performance.
2   The design parameters which have no influence on the product or process performance characteristics.
3   The setting of design parameters at levels which minimize the noise within the performance characteristics.
4   The setting of design parameters which will reduce variation without adversely affecting cost.

As with nearly all the techniques and facets of SPC, the 'design of experiments' is not new; Tippet used these techniques in the textile industry more than fifty years ago. Along with the other quality gurus (Juran, Deming, Crosby and Ishikawa), Taguchi has enlarged the world's view of the applications of established techniques. His major contributions are in emphasizing the cost of quality by use of the total loss function and the sub-division of complex 'problem solving' into manageable component parts. The authors hope that this book will make a similar, modest, contribution towards the understanding and adoption of underutilized quality management technology.

Summarizing improvement

Improving products or service quality is achieved through improvements in the processes that produce the product or operate the service. Each activity and each job are part of a process which can be improved. Improvement is derived from people learning and the approaches presented above provide a 'road map' for progress to be made. The main thrust of the approach is a team with common objectives – using the improvement cycle, defining current knowledge, building on that knowledge, and making changes in the process. Integrated into the cycle are methods and tools that will enhance the learning process.

When this strategy is employed, the quality of products and services is improved, job satisfaction is enhanced, communications are strengthened, productivity is increased, costs are lowered, market share rises, new jobs are provided, and additional profits flow. In other words, quality improvement as a business strategy provides rewards to everyone involved: customers receive value for their money, employees gain job security, and owners or shareholders are rewarded with a healthy organization capable of paying real dividends. This strategy will be the common thread of all companies who will remain competitive in world markets for the remainder of the twentieth century and well into the twenty-first century.

Chapter highlights

- For successful SPC there must be management commitment to quality, a quality policy, and a documented quality management system.
- The main objective of the system is to cause improvements through reduction in variation in processes. The system should apply to, and interact with, all activities of the organization.
- The role of the quality system is to define and control processes, procedures and methods. The system audit and review will ensure the procedures are followed or changed.
- Measurement is an essential part of the quality and SPC systems. The activities, materials, equipment, etc., to be measured must be identified precisely. The measurements must be accepted by the people involved and, in their use, the emphasis must be on providing assistance to solve problems.
- Control charts may be designed on the basis of economics, rather than on purely statistical grounds, but some of these methods contradict current SPC thinking. Parameters such as sample size (n), however, are influenced by the process capability, the chart performance, and the need to distinguish between process accuracy and precision.
- Teamwork plays a vital role in continuous improvement. In most organizations it means moving from 'independence' to 'interdependence'. Inputs from all relevant processes are required to make changes to complex systems. Good communication mechanisms are essential for successful SPC teamwork and meetings must be managed.
- A quality improvement team (QIT) is a group brought together by management to tackle a particular problem. Flowcharts, cause and effect diagrams and brainstorming are useful in building the QIT around the process, both in manufacturing and service organizations. Problem solving groups will eventually give way to problem prevention teams.
- All processes deteriorate with time. Process improvement requires an understanding of who is responsible, what resources are required, and which SPC tools will be used. This requires action by management.
- Control charts should not be used only for control, but as an aid in reducing variability. The progressive identification and elimination of causes of variation may be charted and the limits adjusted accordingly to reflect the improvements.

- Never-ending improvement takes place in the Deming cycle of plan, implement, record data, analyse.
- The Japanese engineer Taguchi has defined a number of methods to reduce costs and improve quality. His methods appear under four headings: the total loss function; design of products, processes and production; reduction in variation; and statistically planned experiments. Taguchi's main contribution is to enlarge people's views of the applications of some established techniques.
- Improvements, based on teamwork and the techniques of SPC, will lead to quality products and services, lower costs, better communications and job satisfaction, increased productivity, market share and profits, and higher employment.

14 SPC in non-manufacturing

14.1 The process and the data

Organizations which embrace the concepts of total quality should recognize the value of SPC techniques in areas such as sales, purchasing, invoicing, finance, distribution, training, etc., which are outside production or operations – the traditional area for SPC use. A Pareto analysis, a histogram, a flowchart, or a control chart is a vehicle for communication. Data is data and, whether the numbers represent defects or invoice errors, the information relates to machine settings, process variables, or prices, quantities, discounts, customers, or supply points is irrelevant, the techniques can always be used.

Some of the most exciting applications of SPC have emerged from organizations and departments which, when first introduced to the methods, could see little relevance to their own activities. Following appropriate training, however, they have learned how to, for example:

- *Pareto analyse* sales turnover by product and injury data.
- *Brainstorm* and *cause and effect analyse* reasons for late payment and poor purchase invoice matching.
- *Histogram* absenteeism and arrival times of trucks during the day.
- *Control chart* the movement in currency and weekly demand of a product.

Distribution staff have used p charts to monitor the proportion of deliveries which are late and Pareto analysis to look at complaints involving the distribution system. Word processor operators have used cause and effect analysis and histograms to represent errors in output from their service. Moving average and cusum charts have immense potential for improving forecasting in all areas including: marketing, demand, output, currency values and commodity prices.

Those organizations which have made most progress in implementing a company-wide approach to quality have recognized at an early stage that

SPC is for the whole organization. Restricting it to traditional manufacturing or operations activities means that a window of opportunity for improvement has been closed. Applying the methods and techniques outside manufacturing will make it easier, not harder, to gain maximum benefit from an SPC programme.

Sales and Marketing is one area which often resists training in SPC on the basis that it is difficult to apply. Personnel in this vital function need to be educated in SPC methods for two reasons:

1 They need to understand the way the manufacturing and/or service producing processes in their organizations work. This enables them to have more meaningful and involved dialogues with customers about the whole product/service system capability and control. It will also enable them to influence customers' thinking about specifications and create and maintain a competitive advantage from improved process capabilities.

2 They need to identify and improve the marketing processes and activities. A significant part of the sales and marketing effort is clearly associated with building relationships, which are best built on facts (data) and not opinions. There are also opportunities to use SPC techniques directly in such areas as forecasting, demand levels, market requirements, monitoring market penetration, marketing control, and product development, all of which must be viewed as processes.

SPC has considerable applications for non-manufacturing organizations, in both the public and private sectors. Data and information on patients in hospitals, students in universities, polytechnics, colleges and schools, people who pay (and do not pay) tax, draw social security benefit, shop at Sainsbury's or Macy's, are available in abundance. If they were to be used in a systematic way, and all operations treated as processes, far better decisions could be made concerning the past, present, and future performances of non-manufacturing operations.

The remainder of this chapter is devoted to a number of case studies in which SPC has been applied in a non-manufacturing or service environment.

14.2 Process capability analysis in a bank

A project team in a small bank was studying the productivity of the cashier operations. Work during the implementation of SPC had identified variation in transaction (deposit/withdrawal) times as a potential area for improvement. The cashiers agreed to collect data on transaction times to study the process.

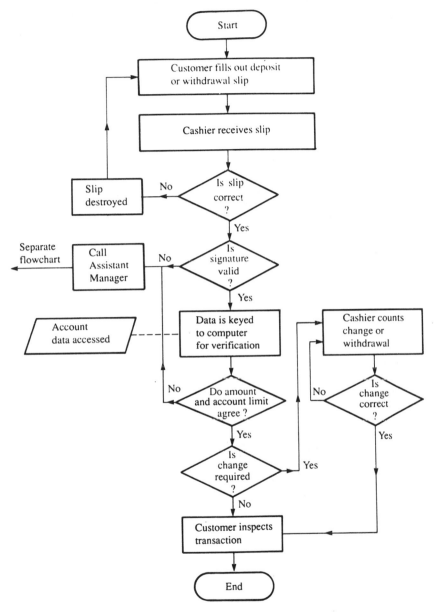

Figure 14.1 *Flowchart for bank transactions*

Once an hour, each cashier recorded in time the seconds required to complete the next seven transactions. After three days, the operators developed control charts for this data. All the cashiers calculated control limits for their own data. The totals of the $\overline{X}$s and Rs for twenty-four subgroups (three days times eight hours per day) for one cashier were: $\Sigma\overline{X}$ = 5640 seconds, ΣR = 1900 seconds. Control limits for this cashier's $\overline{X}$ and R chart were calculated and the process was shown to be stable.

An 'efficiency standard' had been laid down that transactions should average three minutes (180 seconds), with a maximum of five minutes (300 seconds) for any one transaction. The process capability was calculated as follows:

$$\overline{\overline{X}} = \frac{\Sigma\overline{X}}{k} = \frac{5640}{24} = 235 \text{ seconds}$$

$$\overline{R} = \frac{\Sigma R}{k} = \frac{1900}{24} = 79.2 \text{ seconds}$$

$$\sigma = \overline{R}/d_n, \text{ for n} = 7, \sigma = 79.2/2.704 = 29.3 \text{ seconds}$$

$$Cpk = \frac{\text{USL} - \overline{\overline{X}}}{3\sigma} = \frac{300 - 235}{3 \times 29.3} = 0.74,$$

i.e. not capable, and not centred.

As the process was not capable of meeting the requirements, management led an effort to improve transaction efficiency. This began with a flowcharting of the process as shown in Figure 14.1. In addition, a brainstorm session involving the cashiers was used to generate the cause and effect diagram of Figure 14.2. A quality improvement team was formed, further data collected, and the 'vital' areas of incompletely documented procedures and cashier training were tackled. This resulted over a period of six months, in a reduction in average transaction time to 190 seconds, with standard deviation of 15 seconds. (Cpk = 2.44).

14.3   Profits on sales

An airline company had been keeping track of the sales and the percent of the turnover as profit. The sales for the last twenty-five months had remained relatively constant due to the large percentage of charter business. During the previous few months profits, as a percentage of turnover, had been below average, and the information in Table 14.1 had been collected.

After attending an SPC course at Bradford Management Centre, the company accountant decided to analyse the data using a cusum chart. He calculated the average profit over the period to be 8.0 per cent (assuming

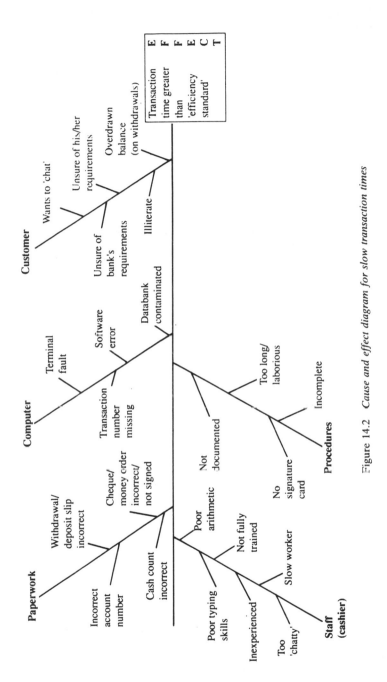

Figure 14.2   *Cause and effect diagram for slow transaction times*

Table 14.1   *Profit, as percentage of turnover, for each twenty-five months*

Year 1		Year 2	
Month	Profit %	Month	Profit %
January	7.8	January	9.2
February	8.4	February	9.6
March	7.9	March	9.0
April	7.6	April	9.9
May	8.2	May	9.4
June	7.0	June	8.0
July	6.9	July	6.9
August	7.2	August	7.0
September	8.0	September	7.3
October	8.8	October	6.7
November	8.8	November	6.9
December	8.7	December	7.2
		January Year 3	7.6

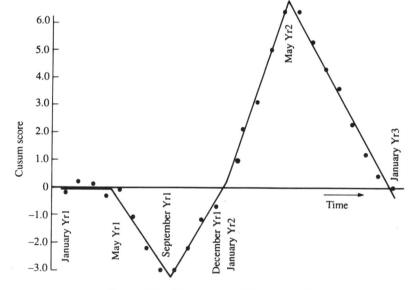

Figure 14.3   *Cusum chart of data on profits*

constant sales), and subtracted this value from each month's profit figure. He then cumulated the differences and plotted them as in Figure 14.3.

The dramatic changes which took place in approximately May and September in Year 1, and in May in Year 2 were investigated and found to be associated with the following assignable causes:

May   Year 1 Introduction of 'efficiency' bonus payment scheme.

September Year 1 Introduction of quality improvement teams.

May   Year 2 Revision of efficiency bonus payment scheme.

The motivational (or otherwise) impact of managerial decisions and actions often manifests itself in business performance results in this way. The cusum technique is useful in highlighting the change points so that causes may be assigned.

14.4 Forecasting income

The three divisions of a service company were required to forecast income on an annual basis and update the forecasts each month. These forecasts were critical to staffing and prioritizing resources in the organization.

Forecasts were normally made one year in advance. The one month forecast was thought to be reasonably reliable. If the three month forecast had been reliable, the material scheduling could have been done more efficiently. Table 14.2 shows the three month forecasts made by the

Table 14.2 *Three month income forecast (unit × 1000) and actual (unit × 1000)*

| | Division A | | Division B | | Division C | |
Month	Forecast	Actual	Forecast	Actual	Forecast	Actual
1	200	210	250	240	350	330
2	220	205	300	300	420	430
3	230	215	130	120	310	300
4	190	200	210	200	340	345
5	200	200	220	215	320	345
6	210	200	210	190	240	245
7	210	205	230	215	200	210
8	190	200	240	215	300	320
9	210	220	160	150	310	330
10	200	195	340	355	320	340
11	180	185	250	245	320	350
12	180	200	340	320	400	385
13	180	240	220	215	400	405
14	220	225	230	235	410	405
15	220	215	320	310	430	440
16	220	220	320	315	330	320
17	210	200	230	215	310	315
18	190	195	160	145	240	240
19	190	185	240	230	210	205
20	200	205	130	120	330	320

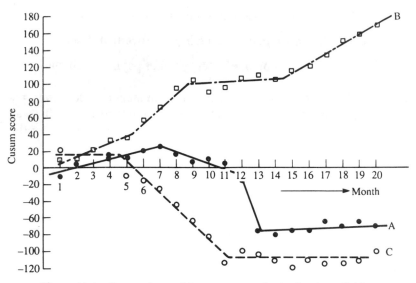

Figure 14.4   *Cusum charts of forecast v. actual sales for three divisions*

three divisions for twenty consecutive months. The actual income for each month is also shown.

Again the cusum chart was used to examine the data. This time, the actual sales were subtracted from the forecast and the differences cumulated. The resulting cusum graphs are shown in Figure 14.4. Clearly there is a vast difference in forecasting performance of the three divisions. Overall, division B is underforecasting resulting in a constantly rising cusum. A and C were generally overforecasting during months 7 to 12 but, during the latter months of the period, their forecasting improved resulting in a stable, almost horizontal line cusum plot. Periods of improved performance such as this may be useful in identifying the causes of the earlier overforecasting and the generally poor performance of division B's forecasting system. The points of change in slope may also be useful indicators of assignable causes, if the management system can provide the necessary information.

Other techniques useful in forecasting include the moving mean and moving range charts and exponential smoothing (see Chapter 8).

14.5   Ranking in managing product range

Some figures were produced by a small chemical company concerning the company's products, their total volume ($), and direct costs. These are given in Table 14.3. The products were ranked in order of income and contribution for the purpose of Pareto analysis, and the results are given

Table 14.3 *Some products and their total volume, direct costs and contribution*

Code number	Description	Total volume ($)	Total direct costs ($)	Total contribution ($)
001	Captine	1,040	1,066	26
002	BHD-DDB	16,240	5,075	11,165
003	DDB-Sulphur	16,000	224	15,776
004	Nicotine-Phos.	42,500	19,550	22,950
005	Fensone	8,800	4,800	4,000
006	Aldrone	106,821	45,642	61,179
007	DDB	2,600	1,456	1,144
008	Dimox	6,400	904	5,496
009	DNT	288,900	123,264	165,636
010	Parathone	113,400	95,410	17,990
011	HETB	11,700	6,200	5,500
012	Mepofox	12,000	2,580	9,420
013	Derros-Pyrethene	20,800	20,800	0
014	Dinosab	37,500	9,500	28,000
015	Maleic Hydrazone	11,300	2,486	8,814
016	Thirene-BHD	63,945	44,406	19,539
017	Dinosin	38,800	25,463	13,337
018	2,4-P	23,650	4,300	19,350
019	Phosphone	13,467	6,030	7,437
020	Chloropicrene	14,400	7,200	7,200

in Table 14.4. To consider either income or contribution in the absence of the other could lead to incorrect conclusions; for example, Product 013 which is ranked ninth in income actually makes zero contribution.

One way of handling this type of ranked data is to plot an income-contribution rank chart. In this the abscissae are the income ranks, and the ordinates are the contribution ranks. Thus, product 010 has an income rank of 2 and a contribution rank of 7. Hence, product 010 is represented by the point (2,7) in Figure 14.5, on which all the points have been plotted in this way.

Ideally volume and contribution ranks should correspond, giving a straight line at 45° to either axis. Products/services lying above this line should be examined by asking:

• Can costs be reduced?
• Can prices be increased?

The question which should be asked about the products/services lying below the line should be:

• Can the sales volume be increased?

Table 14.4   *Income rank/contribution rank table*

Code number	Description	Income rank	Contribution rank
001	Captine	20	20
002	BHB-DDD	10	10
003	DDB-Sulphur	11	8
004	Nicotine-Phos.	5	4
005	Fensone	17	17
006	Aldrone	3	2
007	DDB	19	18
008	Dimox	18	16
009	DNT	1	1
010	Parathone	2	7
011	HETB	15	15
012	Mepofox	14	11
013	Derros-Pyrethene	9	19
014	Dinosab	7	3
015	Maleic Hydrazone	16	12
016	Thirene-BHD	4	5
017	Dinosin	6	9
018	2,4-P	8	6
019	Phosphone	13	13
020	Chloropicrene	12	14

The answers to these questions alone cannot determine the action to be taken, but they can provide excellent guidance.

One company reduced its product range from 1500 to 220 using this approach and:

1   Eliminating all products whose sales were less than 0.01 per cent of total turnover.
2   Examination of all products whose gross profit was below a certain level.
3   Detailed consideration of the sales trends of each remaining product.

To assist their customers, whose goodwill might have been jeopardized by the abandoning of certain products, they negotiated the transfer of orders for discontinued lines to other firms in the same field which were still making these lines. This drastic reduction of range was accompanied by a reduction in the total number of orders received, but the size of the orders received increased and there was, in fact, an expansion in total turnover.

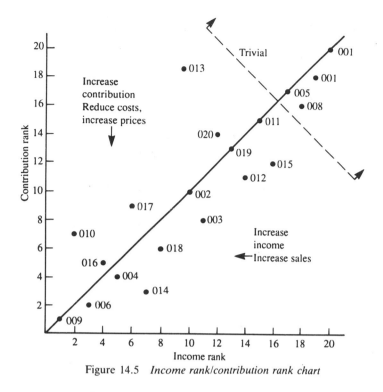

Figure 14.5   *Income rank/contribution rank chart*

14.6   Activity sampling

Activity or work sampling is, as the name implies, a sampling technique based on the binomial theory (see Chapter 11). It is used to obtain a realistic picture of productive time, or time spent on particular activities both by human and technological resources.

    An exercise should begin with discussions with the staff involved, explaining to them the observation process, and the reasons for the study. This would be followed by an examination of the processes, establishing the activities to be identified. A preliminary study is normally carried out to confirm that the set of activities identified is complete, familiarize people with the method and reduce the intrusive nature of work measurement, and to generate some preliminary results in order to establish the number of observations required in the full study. The preliminary study would normally cover 50–100 observations, made at random points during a representative period of time, and may include the design of a check sheet on which to record the data. After the study it should be possible to determine the number of observations required in the full study using the formula:

$$N = \frac{4P(100-P)}{L^2} \text{ (for 95 per cent confidence)}$$

where N = Number of observations

P = Percentage occurrence of any one activity

L = Required precision in the estimate of P

If the first study indicated that 45 per cent of the time is spent on productive work, and it is felt that a precision of 2 per cent is desirable for the full study (that is, we want to be reasonably confident that the actual value lies between 43 per cent and 47 per cent assuming the study confirms the value of 45 per cent), then the formula tells us we should make

$$\frac{4 \times 45 \times (100 - 45)}{2 \times 2} = 2475 \text{ observations}$$

If the work centre concerned has five operators, this implies 495 tours of the centre in the full study. It is now possible to plan the main study with the 495 tours covering a representative period of time.

Having carried out the full study, it is possible to use the same formula, suitably arranged, to establish the actual precision in the percentage occurrence of each activity:

$$L = \sqrt{\frac{4P(100-P)}{N}}$$

The technique of activity sampling, although quite simple, is very powerful. It can be used in a variety of ways, in a variety of environments, both manufacturing and non-manufacturing. While it can be used to indicate areas which are worthy of further analysis, using for example method study techniques, it can also be used to establish time standards themselves. While time study requires training and experience before a practitioner can employ it effectively, activity sampling can be used to advantage with much less training. This and other techniques of industrial engineering or work study are described fully in *Production and Operations Management* by Lockyer, Muhlemann and Oakland (Pitman, 1988).

14.7   Absenteeism

Figure 14.6 is a simple demonstration of how analysis of attribute data may be helpful in a non-manufacturing environment. A manager joined the Personnel Department of a gas supply company at the time shown

Date	UAL	11.5	UWL		9.5	Mean		4.83	LWL		0.5	LAL		Specification						
Time/sample no.	1	2	3	4	5	6	7	8	9	10	11	12	13	14	15	16	17	18	19	20	21
Total inspected. *n*		C		O		N		S		T		A		N		T					
Total absent, *np*	6	4	3	2	7	8	5	6	1	3	5	2	8	4	3	5	8	7	5	4	5

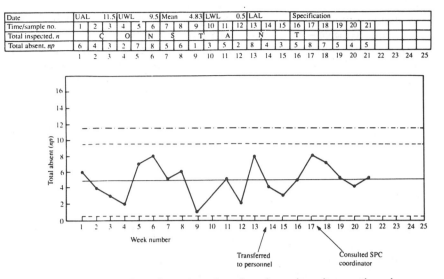

Figure 14.6 *Attribute chart of number of employee-days absent each week*

by the plot for week 14 on the 'employees absent in one week chart'. She attended an SPC course two weeks later (week 16), but at this time control charts were not being used in the Personnel Department. She started plotting the absenteeism data from week 15 onwards. When she plotted the dreadful result for week 17, she decided to ask the SPC coordinator for his opinion of the action to be taken, and to set up a meeting to discuss the alarming increase in absenteeism. The SPC coordinator examined the history of absenteeism and established the average value as well as the warning and action lines, both of which he added to the plot. Based on this he persuaded her to take no action and to cancel the proposed meeting since there was no significant event to discuss.

Did the results settle down to a more acceptable level after this? No, the results continued to be randomly scattered about the average – there had been no assignable cause for the observation in week 17 and hence no requirement for a solution. In many organizations the meeting would not only have taken place, but the management would have congratulated themselves on their 'evident' success in reducing absenteeism. Over the whole period there were no significant changes in the 'process' and absenteeism was running at an average of approximately five per week, with random or common variation about that value. No assignable or special causes had occurred. If there was an item for the agenda of a meeting about absenteeism, it should have been to discuss the way in which the average could be reduced and the discussion would be helped by looking at the general causes which give rise to this average – we go back to Pareto and cause and effect analyses.

In both manufacturing and non-manufacturing, and when using both attributes and variables, the temptation to take action when a 'change' is assumed to have occurred is high, and reacting to changes which are not significant is a frequent cause of adding variation to otherwise stable processes. This is sometimes known as management interference; it may be recognized as stable running of a process during the night shift, or at weekends, when the managers are at home!

14.8   Errors on invoices

The accountant of the Bruddersford Chemical Company was asked how many invoices were the subject of errors. He had no immediate measure but thought that double entry book-keeping principles and the use of computerized accounting systems would ensure that the number of erroneous invoices would be very low. When pressed he suggested 1 in 1000 or 0.1 per cent.

Table 14.5   *Data on invoices and credit notes*
Annual turnover £50 mn – total value of 1988 credit notes
£2.5 mn

	Date	Invoices issued in month	Credit notes in month
Year 1			
	January	502	25
	February	442	23
	March	483	21
	April	525	24
	May	503	23
	June	530	21
	July	475	23
	August	413	24
	September	533	26
	October	556	26
	November	607	22
	December	438	24
Year 2			
	January	464	20
	February	483	25
	March	537	20
	April	492	23

He was not sure of the number of invoices produced in a year nor of the number of credit notes, but agreed that no credit note could be raised without there having been an error on the invoice – an error being defined as a recognized and agreed difference between the statement of the transaction on the invoice and the customer's actual requirements. The company were at pains to ensure that all such errors were strictly controlled and the procedure for authorizing and reporting the issue of credit notes was strict and rigorously applied. Confusion was avoided by a house rule which required that each credit note exactly cancelled its corresponding invoice and, when appropriate, a second modified invoice was issued.

Invoices and credit notes were numbered chronologically and, with comparative ease, the data in Table 14.5 were obtained.

The number of invoices issued seems to vary about some central tendency, as do the number of credit notes. If the number of invoices issued per month was randomly distributed about a mean value, use could perhaps be made of a p chart, but plotting the invoices issued as a c chart will indicate the degree of randomness. Figure 14.7 shows the c chart of the number of invoices issued per month ($\bar{c} = 498$, UAL = 564.5, UWL = 542.5, LWL = 453.5, and LAL = 431.5). This shows trends and action signals around August (low sales) and December (catch-up with invoices for year-end). There are explanations for these assignable causes and at other times the number of invoices issued is reasonably random. It follows that the number of credit notes should be reasonably random.

Figure 14.8 is a c chart for number of credit notes per month, with $c = 23$ (i.e. 4.6 per cent of all invoices), UAL = 37.5, UWL = 32.5, LWL = 12.5 and LAL = 9.5. The c chart for credit notes may be random but the scatter is unbelievably low. If the credit notes come

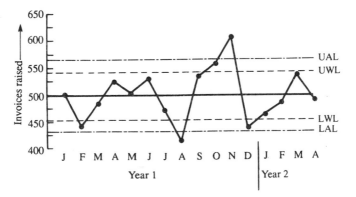

Figure 14.7   c chart – invoices/month

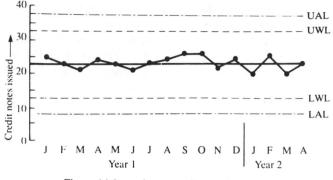

Figure 14.8 c *chart – credit notes/month*

from a population which obeys the binomial distribution and credit notes arise randomly, their distribution should also obey the binomial distribution. In this case it means that the scatter should be larger than indicated by the c chart. So some assignable explanation should be sought. It could be that the credit notes are being smoothed out to avoid having to explain difficult peaks and exposing over encouraging troughs. In the case from which this data was collected, the accountant admitted that he had massaged the results. He and the Chief Executive subsequently agreed that this would not be done in future, and that the CE would also not scream unless there was evidence of a *real* trend or change in the number of credit notes.

The average value of an invoice was £8,326, whereas the average value of a credit note was £8,865. This might infer that customers do not check small value invoices as rigorously as they do large ones and, hence, credit notes have a higher average value. Customers will probably not complain when the error is in their favour, so a more reasonable estimate of the number of errors would be double the 4.6 per cent noted above.

14.9 Injury data

In an effort to improve safety in their plant, a company decided to chart the number of injuries that required first aid, each month. Since approximately the same amount of hours were worked each month, a c chart was utilized. Table 14.6 contains the data collected over a two-year period.

From this data, the average number of injuries per month ($\bar{c}$) was calculated:

$$\bar{c} = \frac{\Sigma c}{k} = \frac{133}{24} = 5.54 \text{ (Centreline)}$$

Table 14.6   *Injury data*

Year 1 Month	Number of injuries (c)	Year 2 Month	Number of injuries (c)
January	6	January	10
February	2	February	5
March	4	March	9
April	8	April	4
May	5	May	3
June	4	June	2
July	23	July	2
August	7	August	1
September	3	September	3
October	5	October	4
November	12	November	3
December	7	December	1
			$\Sigma c = 133$

The control limits were as follows:

$$\text{UAL/LAL} = \bar{c} \pm 3\sqrt{\bar{c}} = 5.54 \pm 3\sqrt{5.54}$$

Upper Action Line = 12.6 injuries (there is no lower action line)

$$\text{UWL/LWL} = \bar{c} \pm 2\sqrt{\bar{c}} = 5.54 \pm 2\sqrt{5.54}$$
$$= 10.3 \text{ and } 0.8$$

Figure 14.9 shows the control chart. In July Year 1, the reporting of twenty-three injuries resulted in a point above the upper control limit.

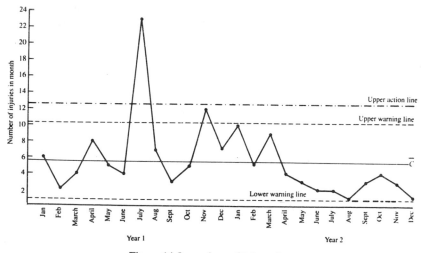

Figure 14.9   c *chart of injury data*

The assignable cause was a large amount of holiday leave taken during that month. Untrained people and excessive overtime were used to achieve the normal number of hours worked for a month. There was also a run of nine points in a row below the centreline starting in April Year 2. This indicated that the average number of reported first aid cases per month had been reduced. This reduction was attributed to a switch from wire to plastic baskets for the carrying and storing of parts and tools which greatly reduced the number of injuries due to cuts. If this trend has continued, the control limits should have been recalculated when sufficient data was available.

14.10   Summarizing SPC in non-manufacturing

It is clear from the examples presented in this chapter that the principles and techniques of SPC may be applied to any human activity, provided that it is regarded as a process. The only way to control process outputs, whether they be artefacts, paperwork, services, or communications, is to manage the inputs systematically. Data from the outputs, the process itself, or the inputs, in the form of numbers or information, may then be used to modify and improve the operation.

Presenting data in an efficient and easy to understand manner is as vital in the office as it is on the factory floor and, as we have seen in this chapter, the basic tools of SPC have a great deal to offer in all areas of management. Data obtained from processes must be analysed quickly so that continual reduction in the variety of ways of doing things will lead to never-ending improvement.

In many non-manufacturing operations there is an 'energy barrier' to be surmounted in convincing people that the SPC approach and techniques have a part to play. Everyone must be educated so that they understand and look for potential SPC applications. Training in the basic approach of:

- No process without data collection
- No data collection without analysis
- No analysis without action

will ensure that every possible opportunity is given to use these powerful methods to greatest effect.

Chapter highlights

- SPC may be used, as part of total quality management (TQM), in all areas of an organization, including sales, marketing, purchasing, finance, etc. Indeed, some of the best examples of use of flowcharts, histograms, Pareto analysis, cause and effect analysis, and other tools have been in these areas.

- SPC has wide application in the so-called 'service' industries, in both private and public sectors. It provides a systematic approach to data analysis and continuous improvement.
- Examples of the use of SPC in 'non-manufacturing' include:
  - Process capability studies in banks
  - Tracking sales and profits in airlines
  - Forecasting income
  - Managing a product range through use of Pareto rankings for income and contribution
  - Activity sampling
  - Monitoring absenteeism
  - Checking errors on invoices
  - Examining injury data.
- The principles and techniques of SPC can be applied to any human activity – *everything* is a process. The efficient presentation and use of data are important in every walk of life.
- The 'energy barriers' to SPC which may exist in the non-manufacturing areas must be overcome to ensure that, for all processes, there is: data collection, data analysis and determined action (or no action) following the analysis.

15 The implementation of statistical process control

15.1 Introduction

The authors and their colleagues have carried out structured investigations into the applications of statistical methods of process control in a wide range of manufacturing and service industries. Many of the various techniques have been available to industry since the 1920s, when Shewhart first published his book on control charts. There is now a vast and ever-increasing academic literature on the subject and most post-1980 production and operations management textbooks discuss quality management in terms other than simple verification and measurement techniques. Research work within the European Centre for Total Quality Management at the Management Centre of the University of Bradford, has demonstrated that there was little application of SPC techniques during the early 1980s and that the rate at which applications have developed in more recent years has been high.

Sometimes the introduction of SPC has been suspect, resulting from a lack of understanding and acceptance of the fundamental hypotheses on which the techniques are based. In many cases, the simple collection and recording of data has resulted in marked advances in process control. In other parts of industry, the walls of manufacturing and non-manufacturing units have been plastered with charts which, to the informed eye, clearly demonstrate the capability of the processes in use. But the charts are often filled in only to comply with a directive, which was not adequately explained and the consequences of which may not have been understood by the person or persons giving it.

The authors recall the case of a plastic injection moulding component manufacturer who had introduced the use of 'np' charts, run by the process operators. They were not only delighted to see that the defective rate had fallen considerably but were convinced that the 'magic' of the statistically-based chart had been responsible. The mechanism at work in this case was simply an improved operator performance resulting from the combination of making operators responsible for assessing their own performance and for recording their findings. The sceptic would argue that the operators probably cheated and started to record lower results. The fact is that people only cheat when they have an incentive to do so and when the chances of being found out are low. We also recall the case of a garment manufacturer who had introduced the charts widely, found that the time taken to examine the whole sample was somewhat long, reduced the sample to half the agreed size and had then plotted *twice* the number of defectives found. There was no record of the type of non-conformance and the control charts were filed away without being used, either for product classification or process evaluation. In this case, the operators not only cheated but were happy to talk about it – it was clear to them that this was another 'MFU' (management foul up). So SPC techniques can be both used and abused.

SPC techniques are frequently adopted at the insistence of the customer. Major customers can clearly use commercial clout to get what they want. Without entering into the more general subject of management and motivation, one should recognize that external threat is a powerful motivator.

Once tried out, companies either quickly reject SPC techniques or go on using them to earn greater reputation within the market, an increase in market share, greater job satisfaction, higher productivity – in general, better management. The difference between these two possible outcomes is explained by the degree of understanding of SPC. A major launch of what was then called SQC (statistical quality control) took place in the American automotive industry in about 1960. It was not understood and, contrary to the original expectations, resulted in the introduction of more inspection, the use of 'acceptance sampling' schemes, and little prevention. At the same time the techniques were given added emphasis in the Japanese automotive industry. The ultimate impact that this has had on the market share of Japanese owned and managed companies is a well-recorded event of recent business history. The Japanese made sure that their managers and operators were involved in the whole implementation process and understood SPC before it was applied.

At the moment of first implementing SPC, it is absolutely imperative to ensure that there is a very clear understanding of what is involved. This does not require the employment of statisticians, but the acceptance of:

- The fact that all work is always carried out by a *process*.
- The necessity to share the *responsibility* for the management of quality.
- The inevitable presence of *variation* and the need to manage it.

- The obligation to do nothing to a process unless and until there is clear *evidence* that change is required.
- The recognition that *stability* and variation are partners in all processes.
- The requirement for *prevention* and not detection.
- The belief that *progress* is made by both ceasing to do what is known to be undesirable and by ensuring that it is possible to do *continuously* what is known to be desirable.
- The acceptance of failure as an event requiring *investigation* and *remedial action* and not the apportionment of blame.
- The achievement of common *training* as a route to common *understanding* for, without common understanding, barriers within the hierarchy will be generated and these will prevent proper management and control.

15.2 Successful users of SPC and the benefits derived

In-depth examination of companies which use SPC incorrectly, invariably shows that no real commitment, understanding, or encouragement from senior managers exists. It is apparent in some firms that lack of knowledge, and even positive deceit, can lead to unjustifiable claims being made to either employees or to customers. No system of quality management will survive the lack of *full* commitment to the achievement of never-ending improvement. The failure to understand this leads to the steady loss of control of quality and the very costly consequences.

Truly successful users of SPC can only remain so when the senior management is both aware of, and committed to, the continued use and development of the techniques. A commonly occurring influence contributing to the use and development of SPC is that exerted by an enthusiastic employee. Typically, this person works within either the manufacturing, operations, or quality function and coordinates all the activities associated with SPC. (S)he does not attempt to control quality but manages the quality system.

Other themes which recur in successful user companies are:

- One problem is tackled at a time and dealt with thoroughly before moving on to others.
- All those involved in the use of the techniques understand what they are being asked to do and how this will help them.
- Clear, written and agreed instructions exist for all procedures.
- There is evidence of a long-term commitment to quality improvement at all levels.

- Training is a continuous and structured process.

The benefits which follow from the application of statistical methods to the control of processes are many and varied. A major spin-off is the continuously improving reputation of the company for high-quality products and services. This leads to a steady or expanding, always healthy, share of the market. The improved process and product uniformity causes a direct reduction in the external failure costs – warranty claims, customer complaints, and the intractable 'loss of goodwill'. The corresponding reductions in the cost of internal failure – scrap, rework, blending, repair, second class or low value product, etc. generate a bonus in the form of increased productivity, by reducing the size of the 'hidden plant', which is devoted to producing non-conforming products.

The benefits are not confined to the lowering of the total quality related costs, since additional information leads to more efficient management. For example, precisely determined vendor evaluations lead to better purchasing through closer and more honest associations with suppliers. Similarly, good process control and very low non-conformance rates result in very robust forecasting of production or operations outputs and their input requirements. This is one of the precursors to the reduced stocks and work-in-progress, which are currently associated with just-in-time (JIT) programmes.

Two major requirements appear necessary for the successful implementation of SPC, and are present in all companies which continue to use the techniques successfully and derive the consequential benefits:

1   A real commitment from senior management.
2   A dedicated and well informed quality-related employee.

To this may be added the need for some external advice and influence. The authors have observed the effect of the intervention of a 'third party', such as a consultant or external trainer. This is seldom the same as a visit from the man from the head office! The 'third party' can contribute and influence in a number of ways which include:

- Specific commitments for action, made in the presence of a third party, have a much higher probability of being executed.
- The wider knowledge of an external expert can often contribute to immediate solutions to old, 'chestnut', problems.
- The external expert can always afford to be dismissed from further participation and should feel freer to comment, criticize and question.

One has to recognize a tendency for all companies to be self-congratulatory and introvert. This is often counterbalanced, in part, by a direct and sometimes forceful input of advice from customers or, more generally, the market. But feedback from this source is not always objective, not always listened to, and not always given – 'voting with their feet' frequently occurs. In all the other liberal professions, such as medicine, law, accounting, the need for external expertise is recognized. The requirement exists equally for management, in general, and quality and process management in particular.

15.3   The initial barriers to the implementation of SPC

The selection of a specific project with which to launch the implementation of SPC should take account of the knowledge available and the control of the process being:

- Highly desirable.
- Possible within a reasonable time period.
- Possible by the use of techniques requiring simple training for their introduction.

Many companies are simply not aware of the measured capabilities of their processes and often have exaggerated ideas about them. Consequently, the control of them is nebulous.

The first barrier which usually has to be overcome is that companies pay too little attention to the need for training outside the technological requirements imposed on them by the nature of their business. With a few notable exceptions, they are unsympathetic to the devotion of anything beyond minimal effort and time for training in the techniques of management and control. Where time is made available, it is sometimes wasted by the use of 'cheap' self-developed techniques which achieve little more than improvements in communications. These often seem like talk shops, not an environment in which the relevance of new techniques to old procedures can be readily embraced. Sometimes, operator training is totally absent, except when new processes or new legislation force it upon the organization. The lack of quality management and process control training reflects the insipid support and commitment offered by senior management. Lame excuses such as 'the operators will never understand it', 'why do we need to be bothered with the detail', 'it seems like a lot of extra work', 'we lack the necessary facilities', or 'can't a computer do it', abound and should not be tolerated. The knowledgeable enthusiast can play an important role in overcoming this apathy, provided that senior management allow and encourage it.

The impact of a third party here can be remarkable. The third party's views will seldom differ significantly from those of some of the more informed management, but the third party's presence has a dramatic effect on the acceptability of such views, and the consequential need for an agreement to action (A prophet is not without honour, save in his own country – Matthew Chapter 13 verse 57).

15.4 A proposed methodology for SPC implementation

The task of implementing SPC can appear daunting and the person charged with this responsibility will find little comfort or relevant advice in much of the available literature. Numerous attempts have been made to devise a simple plan of action for the implementation of SPC but, frequently, such plans are frankly ridiculous when applied outside the specific industry to which they relate and from which they were drawn. Implementation depends on a vast number of factors including the source of the motivation, the availability of recorded data, the frequency of such data, the nature of current product quality or process control problems, the activities of the competition, the previous experience of the employees, the ability of the organization to demonstrate its long-term commitment to any policy or plan, etc.

The conclusions of the authors' and their colleagues' work, over more than a decade and involving several hundred separate projects, cannot be distilled into a simple formula. A summary can be attempted, however, by detailing some of the 'steps' and indicating roughly the order in which they should be taken. The relevance of each step and the speed with which it may be introduced depend on the procedures present within an organization and the objectives to which its senior management is commited.

The 'quality status' of a company has no bearing on the possibility of SPC techniques being of value – the company may or may not be aware of its quality problems, in any event it will always benefit from a review of its process control procedures. The first formal objective should be the preparation of a set of 'terms of reference' in which the objectives, the programme of work, the anticipated timing (both total man-days' work and the period over which they are likely to be spread), the leader, the other participants, the reporting mechanism and the proposed method of measuring progress are specified.

Quality understanding, commitment and policy

The foundation on which the implementation of any management technique rests is the understanding of, and commitment to, a defined policy. It is a nonsense to be committed to the use of any technique without a proper understanding of its 'philosophy', hypotheses, theory and applications. Commitment without understanding is an act of faith, and management or control solely by faith cannot be recommended. Once there is understanding and commitment, it needs to be translated into a statement of policy so that the commitment can begin to percolate through the organization.

Numerous organizations have fallen into the trap of following the latest advice from management 'gurus' without understanding, and have developed a practice which is rightly perceived by employees as management's 'latest flavour of the month'. The constant search for action by senior managers is understandable, but no action should be taken without justification, and all organizations need stability as well as change. Organizations whose employees await the next 'flavour of the month' or 'magic ointment' will experience considerable difficulties in the implementation of SPC. An organization may need to address itself to the question of how to persuade the totality of its employees that it is capable of making and living with a long-term commitment. The arrival of a new chief executive and his new 'broom' is one of the opportunities presented to such organizations; the reality of external threat is another.

It is the authors' experience that within organizations everyone learns quickly to follow the example set by the senior management. So there are few problems in obtaining commitment to sensible initiatives at the process operator level, provided that the commitment is demonstrated convincingly at the senior management level. If the senior management are seen to be constantly changing their minds then the rest of the organization will either ignore them or act as undiscerning 'yes-men'.

Of course, the requirement to implement SPC is not always initiated, with or without understanding, at the senior levels of an organization. Where the lower level enthusiast is the person pushing for implementation, (s)he has to recognize that SPC will be of only short-term benefit to the organization, unless and until (s)he has achieved both understanding and commitment at the senior level. The enthusiast is capable at all times of acting as a *focus* for the commitment and understanding.

The long-term nature of a commitment to quality may be made evident by the customers' clearly stated requirements for such a commitment. This should be seen as a reactive response to an external threat. A proactive demonstration of the company's commitment is always to be preferred. A major strength of registration to quality systems standards, such as the International Standard ISO 9000 series, is the commitment by the organization to constant evaluation of its procedures, by an independent third party. A detailed commitment to the use of SPC techniques, included in the quality manual of such a registered company, would be a clear statement of policy and a powerful demonstration of the long-term nature of its commitment.

Quality organization, monitoring, planning and design

While everyone is responsible for the quality of products and services supplied to customers, someone has to carry an explicit responsibility for the coordination of quality management and process control. The organization must show that such a person has been nominated and this person's behaviour must demonstrate competence in carrying out the task. An alarming tendency still exists to appoint as quality managers people who have no qualification for the post, except that they have 'warm' feelings about quality, not always associated with the need for process control.

Process control is an activity which needs to be monitored. This is only possible when data is recorded. The first simple step of recording data and making records available has been shown to produce dramatic improvements in process control particularly in those industries where the part played by the operators' personal skills is large. A recent example within a textile making-up factory showed that assessing operator performance by regular sampling, and displaying the results publicly, resulted in a tenfold reduction in the defect rate within a period of a few weeks.

Monitoring process control involves a review of daily results and their examination for trends, longer-term reviews, the reassessment on a regular basis of the capability of processes, the classification of types of defect for Pareto analysis, ensuring that once an improved procedure has been adopted it remains in constant use, etc.

Plans are needed to drive continuing improvement forward. These plans need to cover the identification of problems and the certainty that solutions will be sought, the recognition of solutions and their implementation, the changes of procedures and the consequential need for informing, training, etc. The elimination of waste is only achieved by prioritizing and proper planning.

Design is essentially concerned with the translation of requirements into a form suitable for operation, production or use. One of the first activities in design is to check the specification – does it exist, is it reasonable? While it is clearly not possible to offer either products or services without a specification, written specifications are often absent, out of date, or totally unachievable. Perhaps even worse, these are often widely known and unchallenged facts. Design includes redesign and hence the preparation of plans for experimentation to optimize process conditions, to develop measurement techniques, to improve existing processes, to act and evaluate ideas. Design also extends to research or the search for novel techniques, ideas, information and systems. Market research is included here.

Quality systems

As the implementation of SPC techniques proceeds it influences an ever-increasing area of the total activities of the organization, its customers and suppliers. The subject of quality systems has already been covered in Chapter 13 and need not be repeated here. Its importance to the total and permanent implementation of SPC must be emphasized, however.

Process capability and control

Process capability must be assessed and not assumed. The capability of all processes can be measured. This is true when the output is assessed by either measured variables or counted attributes. Once the capability has been assessed it can be compared with the specification. The comparison will show that the process is:

- Incapable, when the delivery of unidentified non-conforming output is inevitable.
- Barely capable, when the possibility of delivering defective product is small but real.
- Highly capable, when the producer and the customer can both have a high degree of confidence in the delivery of output always being within specification.

Where the capability of the process is adequate, control charts can be set up to monitor the process parameters and outputs, with a view to altering the process only when an adjustment is clearly required. The data thus collected and recorded can be used to classify the outputs, particularly during periods of erratic process behaviour.

Assessing the capability of processes is possible when a bank of existing data is available. The analysis of such data is a good way to begin examining existing practices. The analysis may reveal that processes are not in statistical control, in which case the assignable causes need to be investigated. When using historic data, there are sometimes problems in identifying the assignable causes, even though the control charts indicate when they occurred. If this is so, extending the analysis into the present and future process operation, without any change to the established procedures for control, will enable the continued presence of the assignable causes of variation to be ascertained and their causes to be investigated in real time. The identification of assignable causes is a major step towards their elimination or control, and results in improved process capability as well as improved control. Where assignable causes are known to exist, the introduction of control charts becomes both necessary and certain to yield process and product quality improvements. As the data builds up, the use of cause and effect, Pareto analysis and other simple statistically-based techniques described in Chapters 3 and 4 can assist in concentrating effort and identifying remedial action.

Quality teamwork and training

Both teamwork and training have been discussed in Chapter 13. They are included here so that the list of the components which make up our proposed methodology for the implementation of SPC is complete.

15.5   How to start the implementation of SPC

'Would you tell me, please, which way I ought to go from here?'
'That depends a good deal on where you want to get to', said the Cat.
'I don't much care where - - -', said Alice.
'Then it doesn't matter which way you go', said the Cat.
                    (*Alice's Adventures in Wonderland* – Lewis Carroll)

Only very infrequently is one presented with a 'green-field' situation in which there are no previously established procedures and practices. In such a situation, the opportunity must not be missed to start process control using SPC techniques. No site is ever truly 'green-field' in that the organization will already possess, or will have acquired, the know-how of the processes required to transform inputs to outputs – the personnel will all have some experience of work in other environments, and the equipment will probably not be untested and untried. The absence of firmly established previous practices and the inevitable requirement to train the process operators presents a unique opportunity to study in advance the capability of processes operated elsewhere, to establish the use of charting for control, and to train operators in the various SPC techniques which will be used during the initial operation of the processes and refined as improvement to control is achieved.

Much more frequently, the equipment, personnel, inputs, methods of control, etc. are all established and accepted, often with built-in views about the capability of the processes and the suitability of the methods of control and quality achievement. Under these circumstances the 'start' point in implementation is that in which the organization finds itself. It might be easier if it were not so, but basically you start from here, the point at which you are. The direction in which one proceeds has then to be determined by establishing an objective. Even knowing both the starting point and the destination does not then determine what initial step to take since that is also a function of the existing facilities, the motivation and support likely to be made available, the speed with which the destination should be reached, the distance of the destination, etc.

To begin to answer some of these questions it is a good idea to start by putting together a few flowcharts, starting with those which reflect the present processes and ending with objectives which a little brainstorming brings to mind. It may be necessary to separate the long-term objectives from the medium and short-term objectives. For example, the chief executive of an organization, assuming that (s)he has a clear understanding of SPC, may take as long-term objectives the improvements of products and services so as to become 'number one' in the market and maintain that position. The medium-term objectives might then include the assessment of the capability of all current processes, bringing them into statistical control and monitoring them by charting. The short-term objective could be to convince the quality manager that SPC techniques are what is required and then to start motivating him/her to become the enthusiast that the organization will require to maintain momentum throughout the period of implementation. But an enthusiastic quality manager who has either recently joined the organization or acquired enthusiasm for SPC techniques, will have different objectives. In the short term, these may need to include the requirement to demonstrate the relevance of the techniques to the organization's processes and the benefits which would accrue from their use. The enthusiastic quality manager will not be able to engage the enthusiasm of others with the facility, given to the chief executive, of setting an example which it is difficult to totally ignore.

While having clearly in mind the long-term objectives, a good initial policy is to look for a problem capable of clarification within the short term, and where any progress is likely to have a significant impact on current thinking and control procedures. A host of possible questions may arise, for example:

- How are SPC techniques relevant to our processes?
- What is our measured capability and how does it compare with our competitors?
- Is the market interested in process capability and how can it be made to respond to a knowledge of capable processes?
- What current problems are recognized in quality and process control?
- What is the history of customer complaints and comments?
- Can a specific problem be nominated and is it suitable for an initial pilot scheme?
- To whom must the validity of the techniques be demonstrated?

The flowchart shown in Figure 13.2 gives a systematic approach to SPC implementation and may assist in the preparation of a flowchart relevant to the implementation within an explicit environment.

The reader will see that the flowchart in Figure 13.2 starts by a series of steps aimed at identifying a problem which merits action, proceeds to the collection, analysis and interpretation of data and moves onto the presentation of the results of the analysis before concluding that the problem is solved. It may be argued that in an organization which has adopted the philosophy of never-ending improvement, no problem is ever solved but merely advanced one more step towards perfection, so that it is now ready for further study in order to make the next step. Progress is often made in a series of small steps rather than in gigantic leaps. Progress is of little value unless it is recognized. Publication and presentation of the results are vital in maintaining momentum within any organization. A major advantage of SPC presentations is that they have to be supported by both the facts and a valid analysis. Unjustified claims, no matter how loudly they are shouted, can only be accepted as hypotheses which need to be justified. The round of hypothesizing, finding the relevant data, analysing its statistical significance and presenting the conclusions as a justified claim then begins.

A final comment

A good, documented quality management system provides a foundation for the successful application of SPC techniques. It is not possible to 'graft' statistical methods on to a poor quality system. SPC will lie dormant without adequate written procedures for process verification, inspection or test, the recording and analysis of data, and the presentation and use of the results.

All manufacturing and service industries benefit from the implementation of statistically-based methods of analysis for process control and decision making. No matter what the process, the risk of drawing erroneous conclusions and taking erroneous action will be reduced by the use of SPC. The systematic structured approach to the implementation, which is recommended here, provides a powerful spearhead with which to improve process control and management decisions at all levels of an organization. Conformance, or consistency of manufacture and operations, is the common goal which will lead to customer satisfaction, increased market share, and ultimately to the 'number one' position in the market. Increased knowledge of process capability leads to better marketing decisions, product design and product reliability.

The 1980s saw major changes in world trade with the emergence of strength in previously weak nations and the decline in the previous strength of others. The 1990s will hopefully not see an increase in protectionism but an increasingly active competition in which the interests of the customer are paramount. These interests will be served by buying at the 'best value for money' from suppliers on whom one may count with very high degrees of confidence. Suppliers will only achieve such reputations by the use of SPC in all their processes.

Chapter highlights

- Implementing SPC is not an easy process. There is no unique methodology which meets the requirements of all organizations.
- The requirement to adopt SPC techniques is growing rapidly in all industries. Failure to follow this evolution puts competitiveness at large, and possibly fatal, risk.
- The simple recording and publication of process performance data can have a significant impact on understanding, enthusiasm for quality and subsequent performance.
- Customers are increasingly pressing for knowledge of process capability and external threat is a very effective source of motivation.
- The implementation of SPC techniques without a proper understanding of their meaning is normally disastrous. Understanding includes:
  - All work is carried out by a process.
  - The responsibility for the management of quality must be shared.
  - The inevitable presence of variation and the need to manage it.
  - The obligation to do nothing to a process without evidence that change is required.
  - Stability and variation are partners in all processes.
  - Prevention and not detection.
  - Progress is made by either ceasing to do the undesirable or making possible the desirable.
  - Failure requires investigation and remedial action, not apportionment of blame.
  - Common training is a route to common understanding.
- Without understanding, commitment and support from senior management, no management system, including SPC, can succeed.
- An SPC enthusiast with special training is a vital part of any implementation programme.
- The role that can be played by a third party during implementation and development of the use of the techniques is considerable.

- The benefits of using SPC are numerous and include reduced costs of external failure, internal failure, and waste; higher productivity; increased self-esteem and reputation; and, privileged positions within the market.
- A major barrier to the effective introduction of SPC is often an unwillingness to devote adequate resources to proper training at all levels of the organization.
- Progress in implementing SPC starts by recognizing the processes and control techniques currently in use along with their limitations, and establishing a provisional list of long-term, medium-term and short-term objectives.
- Further steps include:
  - *Understanding, commitment and policy* – commitment without understanding is an act of faith – policy reflects the commitment and understanding.
  - *Organization, monitoring, planning and design* – process control and quality management need to feature in the organization, planning and running of an operation.
  - *Systems* for the management of quality.
  - *Process capability and control* – capability must be assessed and not assumed.
  - *Teamwork and training* – progress will be ineffective without engaging the interest, enthusiasm and understanding of others, teams need both a common language and a common understanding. Both can be provided by training as a continuous process.
- Start the implementation of SPC now or run the risk of being left behind as manufacturing, services industries and non-profit making organizations all move to increased levels of achievement and performance.

Appendices

Appendix A   The normal distribution and non-normality

The mathematical equation for the normal curve (alternatively known as the Gaussian distribution) is:

$$y = \frac{1}{\sigma\sqrt{2\pi}} \; e^{-(x-\bar{x})^2/2\sigma^2}$$

where y = height of curve at any point x along the scale of the variable
       σ = standard deviation of the population
       $\bar{x}$ = average value of the variable for the distribution
       π = ratio of circumference of a circle to its diameter ($\pi = 3.1416$).
If $z = (x - \bar{x})/\sigma$, then the equation becomes:

$$y = \frac{1}{\sigma\sqrt{2\pi}} \; e^{-z^2/2}$$

The constant $1/\sqrt{2\pi}$ has been chosen to ensure that the area under this curve is equal to unity, or probability 1.0. This allows the area under the curve between any two values of z to represent the probability that any item chosen at random will fall between the two values of z. The values given in Table A.1 show the proportion of process output beyond a single specification limit that is z standard deviation units away from the process average. It must be remembered, of course, that the process must be in statistical control and the variable must be normally distributed (see Chapters 5 and 6).

Normal probability paper

A convenient way to examine variables data is to plot it in a cumulative form on probability paper. This enables the proportion of items outside a given limit to be read directly from the diagram. It also allows the data to

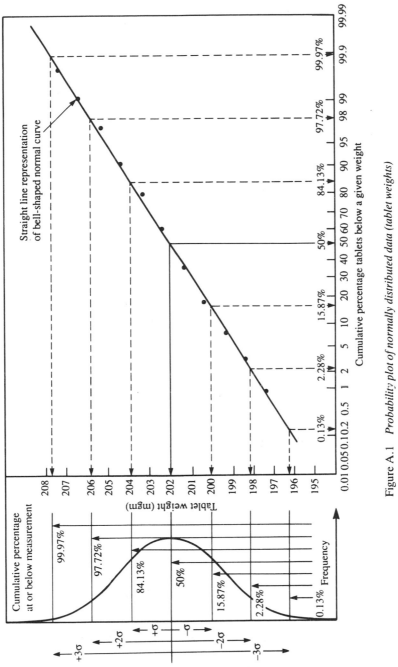

Figure A.1   *Probability plot of normally distributed data (tablet weights)*

Table A.1 *Proportions under the tail of the normal distribution*

$Z = \dfrac{(x - \mu)}{\sigma}$	.00	.01	.02	.03	.04	.05	.06	.07	.08	.09
0.0	.5000	.4960	.4920	.4880	.4840	.4801	.4761	.4721	.4681	.4641
0.1	.4602	.4562	.4522	.4483	.4443	.4404	.4364	.4325	.4286	.4247
0.2	.4207	.4168	.4129	.4090	.4052	.4013	.3974	.3936	.3897	.3859
0.3	.3821	.3783	.3745	.3707	.3669	.3632	.3594	.3557	.3520	.3483
0.4	.3446	.3409	.3372	.3336	.3300	.3264	.3228	.3192	.3156	.3121
0.5	.3085	.3050	.3015	.2981	.2946	.2912	.2877	.2843	.2810	.2776
0.6	.2743	.2709	.2676	.2643	.2611	.2578	.2546	.2514	.2483	.2451
0.7	.2420	.2389	.2358	.2327	.2296	.2266	.2236	.2206	.2177	.2148
0.8	.2119	.2090	.2061	.2033	.2005	.1977	.1949	.1922	.1894	.1867
0.9	.1841	.1814	.1788	.1762	.1736	.1711	.1685	.1660	.1635	.1611
1.0	.1587	.1562	.1539	.1515	.1492	.1469	.1446	.1423	.1401	.1379
1.1	.1357	.1335	.1314	.1292	.1271	.1251	.1230	.1210	.1190	.1170
1.2	.1151	.1131	.1112	.1093	.1075	.1056	.1038	.1020	.1003	.0985
1.3	.0968	.0951	.0934	.0918	.0901	.0885	.0869	.0853	.0838	.0823
1.4	.0808	.0793	.0778	.0764	.0749	.0735	.0721	.0708	.0694	.0681
1.5	.0668	.0655	.0643	.0630	.0618	.0606	.0594	.0582	.0571	.0559
1.6	.0548	.0537	.0526	.0516	.0505	.0495	.0485	.0475	.0465	.0455
1.7	.0446	.0436	.0427	.0418	.0409	.0401	.0392	.0384	.0375	.0367
1.8	.0359	.0351	.0344	.0336	.0329	.0322	.0314	.0307	.0301	.0294
1.9	.0287	.0281	.0274	.0268	.0262	.0256	.0250	.0244	.0239	.0233

z	.00	.01	.02	.03	.04	.05	.06	.07	.08	.09
2.0	.0228	.0222	.0216	.0211	.0206	.0201	.0197	.0192	.0187	.0183
2.1	.0179	.0174	.0170	.0165	.0161	.0157	.0153	.0150	.0146	.0142
2.2	.0139	.0135	.0132	.0128	.0125	.0122	.0119	.0116	.0113	.0110
2.3	.0107	.0104	.0101	.0099	.0096	.0093	.0091	.0088	.0086	.0084
2.4	.0082	.0079	.0077	.0075	.0073	.0071	.0069	.0067	.0065	.0063
2.5	.0062	.0060	.0058	.0057	.0055	.0053	.0052	.0050	.0049	.0048
2.6	.0046	.0045	.0044	.0042	.0041	.0040	.0039	.0037	.0036	.0035
2.7	.0034	.0033	.0032	.0031	.0030	.0029	.0028	.0028	.0027	.0026
2.8	.0025	.0024	.0024	.0023	.0022	.0021	.0021	.0020	.0019	.0019
2.9	.0018	.0018	.0017	.0016	.0016	.0015	.0015	.0014	.0014	.0013
3.0	.0013									
3.1	.0009									
3.2	.0006									
3.3	.0004									
3.4	.0003									
3.5	.00025									
3.6	.00015									
3.7	.00010									
3.8	.00007									
3.9	.00005									
4.0	.00003									

be tested for normality – if it is normal the cumulative frequency plot will be a straight line.

The type of graph paper shown in Figure A.1 is readily obtainable. The variable is marked along the linear vertical scale, while the horizontal scale shows the percentage of items with variables below that value. The method of using probability paper depends upon the number of values available.

Large sample size

Columns 1 and 2 in Table A.2 give a frequency table for weights of tablets. The cumulative total of tablets with the corresponding weights are given in column 3. The cumulative totals are expressed as percentages of $(n + 1)$ in column 4, where n is the total number of tablets. These percentages are plotted against the upper boundaries of the class intervals on probability paper in Figure A.1. The points fall approximately on a straight line indicating that the distribution is normal. From the graph we can read, for example, that about 2 per cent of the tablets in the population weigh 198.0 mg or less. This may be useful information if that weight represents a specification tolerance. We can also read off the median value as 202.0 mg – a value below which half (50 per cent) of the tablet weights will lie. If the distribution is normal, the median is also the mean weight.

It is possible to estimate the standard deviation of the data, using Figure A.1. We know that 68.3 per cent of the data from a normal

Table A.2   *Tablet weights*

Column 1 tablet weights (mgm)	Column 2 frequency (f)	Column 3 cumulative (i)	Column 4 percentage: $\left(\dfrac{i}{n+1}\right) \times 100$
196.5–197.4	3	3	0.82
197.5–198.4	8	11	3.01
198.5–199.4	18	29	7.92
199.5–200.4	35	64	17.49
200.5–201.4	66	130	35.52
201.5–202.4	89	219	59.84
202.5–203.4	68	287	78.42
203.5–204.4	44	331	90.44
204.5–205.4	24	355	96.99
205.5–206.4	7	362	98.91
206.5–207.4	3	365(n)	99.73

distribution will lie between the values μ ± σ. Consequently if we read off the tablet weights corresponding to 15.87 per cent and 84.13 per cent of the population, the difference between the two values will be equal to twice the standard deviation (σ).

Hence, from Figure A.1:

Weight at 84.13% = 203.85 mg

Weight at 15.87% = 200.15 mg

2σ = 3.70 mg

σ = 1.85 mg

Small sample size

The procedure for sample sizes of less than twenty is very similar. A sample of ten light bulbs have lives as shown in Table A.3. Once again the cumulative number failed by a given life is computed (second column) and expressed as a percentage of $(n + 1)$ where n is the number of bulbs examined (third column). The results have been plotted on probability paper in Figure A.2. Estimates of mean and standard deviation may be made as before.

Table A.3   *Lives of light bulbs*

Bulb life in hours (ranked in ascending order)	Cumulative number of bulbs failed by a given life (i)	Percentage: $\dfrac{i}{n + 1} \times 100$
460	1	9.1
520	2	18.2
550	3	27.3
580	4	36.4
620	5	45.5
640	6	54.5
660	7	63.6
700	8	72.7
740	9	81.8
800	10(n)	90.9

Non-normality

There are situations in which the data are not normally distributed. Non-normal distributions are indicated on linear probability paper by non-straight lines. The reasons for this type of data include:

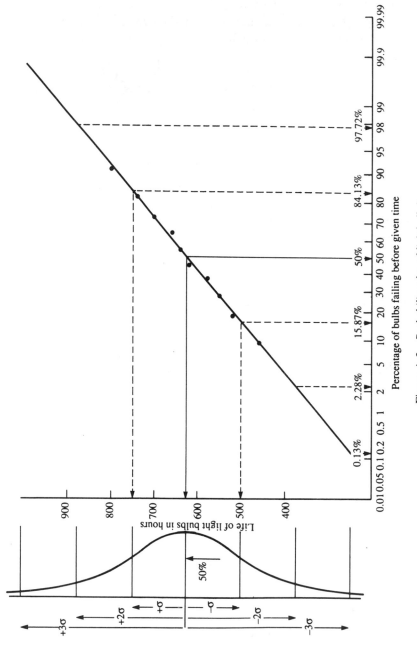

Figure A.2   *Probability plot of lightbulb lives*

1 The underlying distribution fits a standard statistical model other than normal. Ovality, impurity, flatness and other characteristics bounded by zero often have skew, which can be measured. Kurtosis is another measure of the shape of the distribution being the degree of 'flattening' or 'peaking'.

2 The underlying distribution is complex and does not fit a standard model. Self-adjusting processes, such as those controlled by computer, often exhibit a non-normal pattern. The combination of outputs from several similar processes may not be normally distributed, even if the individual process outputs give normal patterns. Movement of the process mean due to gradual changes, such as tool wear, may also cause non-normality.

3 The underlying distribution is normal, but assignable causes of variation are present causing non-normal patterns. A change in material, operator interference, or damaged equipment are a few of the many examples which may cause this type of behaviour.

The standard probability paper may serve as a diagnostic tool to detect divergences from normality and to help decide future actions:

1 If there is a scattering of the points and no distinct pattern emerges, a technological investigation of the process is called for.

2 If the points make up a particular pattern, various interpretations of the behaviour of the characteristic are possible. Examples are given in Figure A.3. In A.3(a), selection of output has taken place to screen out that which is outside the specification. A.3(b) shows selection to one specification limit or a drifting process. A.3(c) shows a case where two distinct distribution patterns have been mixed. Two separate analyses should be performed by stratifying the data. If the points make up a smooth curve, as in A.3(d), this indicates a distribution other than normal. Interpretation of the pattern may suggest the use of an alternative probability paper.

In some cases, if the data is plotted on logarithmic probability paper, a straight line is obtained. This indicates that the data is taken from a log-normal distribution, which may then be used to estimate the appropriate descriptive parameters. Another type of probability paper which may be used is Weibull. Points should be plotted on these papers against the appropriate measurement and cumulative percentage frequency values, in the same way as for normal data. The paper giving the best straight line fit should then be selected. When a satisfactory distribution fit has been achieved, capability indices (see Chapter 7) may be estimated by reading off the values at the points where the best fit line intercepts the 0.13 per cent and 99.87 per cent lines. These values are then

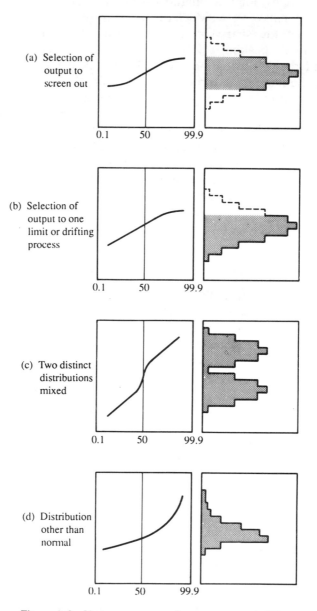

(a) Selection of
    output to
    screen out

0.1   50   99.9

(b) Selection of
    output to one
    limit or drifting
    process

0.1   50   99.9

(c) Two distinct
    distributions
    mixed

0.1   50   99.9

(d) Distribution
    other than
    normal

0.1   50   99.9

Figure A.3   *Various non-normal patterns on probability paper*

used in the formulae:

$$Cp = \frac{USL - LSL}{99.87 \text{ percentile} - 0.13 \text{ percentile}}$$

$$Cpk = \text{minimum of } \frac{USL - \overline{\overline{X}}}{99.87 \text{ percentile} - \overline{\overline{X}}} \text{ or } \frac{\overline{\overline{X}} - LSL}{\overline{\overline{X}} - 0.13 \text{ percentile}}$$

Computer methods

Some standard SPC packages, such as 'Quality Analyst'* have routine procedures for testing for normality. These will carry out a probability plot and calculate indices for both skewness and kurtosis. As with all indices, these are only meaningful to those who understand them.

* North West Analytical; Distributors in UK: Adept Ltd.

Appendix B   Constants used in the design of control charts for mean

Sample size (n)	Hartley's Constant (d_n or d_2)	Constants for mean charts using						
		Sample standard deviation		Sample range		Average sample standard deviation	
		A_1	$2/3\ A_1$	A_2	$2/3 A_2$	A_3	$2/3\ A_3$
2	1.128	2.12	1.41	1.88	1.25	2.66	1.77
3	1.693	1.73	1.15	1.02	0.68	1.95	1.30
4	2.059	1.50	1.00	0.73	0.49	1.63	1.09
5	2.326	1.34	0.89	0.58	0.39	1.43	0.95
6	2.534	1.20	0.82	0.48	0.32	1.29	0.86
7	2.704	1.13	0.76	0.42	0.28	1.18	0.79
8	2.847	1.06	0.71	0.37	0.25	1.10	0.73
9	2.970	1.00	0.67	0.34	0.20	1.03	0.69
10	3.078	0.95	0.63	0.31	0.21	0.98	0.65
11	3.173	0.90	0.60	0.29	0.19	0.93	0.62
12	3.258	0.87	0.58	0.27	0.18	0.89	0.59

Formulae $\sigma = \dfrac{\bar{R}}{d_n}$ or $\dfrac{\bar{R}}{d_2}$

Mean charts

Action lines $= \bar{\bar{X}} \pm A_1\ \sigma$ Warning lines $= \bar{\bar{X}} \pm 2/3\ A_1\ \sigma$

$\qquad\qquad = \bar{\bar{X}} \pm A_2\ \bar{R}$ $\qquad\qquad\quad = \bar{\bar{X}} \pm 2/3\ A_2\ \bar{R}$

$\qquad\qquad = \bar{\bar{X}} \pm A_3\ \bar{s}$ $\qquad\qquad\quad = \bar{\bar{X}} \pm 2/3\ A_3\ \bar{s}$

Process capability

$$Cp = \frac{USL - LSL}{\sigma}$$

$$Cpk = \text{minimum of} \frac{USL - \bar{\bar{X}}}{3\sigma} \text{ or } \frac{\bar{\bar{X}} - LSL}{3\sigma}$$

Appendix C Constants used in the design of control charts for range

Sample size (n)	Constants for use with mean range ($\bar{R}$)						Constants for use with standard deviation (σ)		Constants for use in USA range charts based on $\bar{R}$	
	$D'_{0.099}$	$D'_{0.001}$	$D'_{0.975}$	$D'_{0.025}$	$D_{0.999}$	$D_{0.001}$	$D_{0.975}$	$D_{0.025}$	D_2	D_4
2	0.00	4.12	0.04	2.81	0.00	4.65	0.04	3.17	0	3.27
3	0.04	2.98	0.18	2.17	0.06	5.05	0.30	3.68	0	2.57
4	0.10	2.57	0.29	1.93	0.20	5.30	0.59	3.98	0	2.28
5	0.16	2.34	0.37	1.81	0.37	5.45	0.85	4.20	0	2.11
6	0.21	2.21	0.42	1.72	0.54	5.60	1.06	4.36	0	2.00
7	0.26	2.11	0.46	1.66	0.69	5.70	1.25	4.49	0.08	1.92
8	0.29	2.04	0.50	1.62	0.83	5.80	1.41	4.61	0.14	1.86
9	0.32	1.99	0.52	1.58	0.96	5.90	1.55	4.70	0.18	1.82
10	0.35	1.93	0.54	1.56	1.08	5.95	1.67	4.79	0.22	1.78
11	0.38	1.91	0.56	1.53	1.20	6.05	1.78	4.86	0.26	1.74
12	0.40	1.87	0.58	1.51	1.30	6.10	1.88	4.92	0.28	1.72

Formulae

Action lines: Upper $= D'_{0.001} \bar{R}$ Lower $= D'_{0.999} \bar{R}$

               or $D_{0.001}\sigma$ or $D_{0.999}\sigma$

Warning lines: Upper $= D'_{0.025} \bar{R}$ Lower $= D'_{0.975} \bar{R}$

               or $D_{0.025}\sigma$ or $D_{0.975}\sigma$

Control limits (USA):   Upper $= D_4\bar{R}$ Lower $= D_2\bar{R}$

Appendix D   Constants used in the design of control charts for median and range

Sample size (n)	Constants for median charts A_4	Constants for median charts $2/3\ A_4$	Constants for range charts $D^m_{.001}$	Constants for range charts $D^m_{.025}$
2	2.22	1.48	3.98	2.53
3	1.27	0.84	2.83	1.79
4	0.83	0.55	2.45	1.55
5	0.71	0.47	2.24	1.42
6	0.56	0.37	2.12	1.34
7	0.52	0.35	2.03	1.29
8	0.44	0.29	1.96	1.24
9	0.42	0.28	1.91	1.21
10	0.37	0.25	1.88	1.18

Formulae

Median chart	Action Lines	$= \tilde{\tilde{X}} \pm A_4 \tilde{R}$
	Warning lines	$= \tilde{\tilde{X}} \pm 2/3\ A_4 \tilde{R}$
Range chart	Upper action line	$= D^m_{.001}\ \tilde{R}$
	Upper warning line	$= D^m_{.025}\ \tilde{R}$

Appendix E Constants used in the design of control charts for standard deviation

Sample size (n)	c_n	Constants used with $\bar{s}$				Constants used with σ			
		$B'_{.001}$	$B'_{.025}$	$B'_{.975}$	$B'_{.999}$	$B_{.001}$	$B_{.025}$	$B_{.975}$	$B_{.999}$
2	1.253	4.12	2.80	0.04	0.02	3.29	2.24	0.03	0.01
3	1.128	2.96	2.17	0.18	0.04	2.63	1.92	0.16	0.03
4	1.085	2.52	1.91	0.29	0.10	2.32	1.76	0.27	0.09
5	1.064	2.28	1.78	0.37	0.16	2.15	1.67	0.35	0.15
6	1.051	2.13	1.69	0.43	0.22	2.03	1.61	0.41	0.21
7	1.042	2.01	1.61	0.47	0.26	1.92	1.55	0.45	0.25
8	1.036	1.93	1.57	0.51	0.30	1.86	1.51	0.49	0.29
9	1.032	1.87	1.53	0.54	0.34	1.81	1.48	0.52	0.33
10	1.028	1.81	1.49	0.56	0.37	1.76	1.45	0.55	0.36
11	1.025	1.78	1.49	0.58	0.39	1.73	1.45	0.57	0.38
12	1.023	1.73	1.44	0.60	0.42	1.69	1.41	0.59	0.41
13	1.021	1.69	1.42	0.62	0.44	1.66	1.39	0.61	0.43
14	1.019	1.67	1.41	0.63	0.46	1.64	1.38	0.62	0.45
15	1.018	1.64	1.40	0.65	0.47	1.61	1.37	0.63	0.47
16	1.017	1.63	1.38	0.66	0.49	1.60	1.35	0.65	0.48
17	1.016	1.61	1.36	0.67	0.50	1.58	1.34	0.66	0.50
18	1.015	1.59	1.35	0.68	0.52	1.56	1.33	0.67	0.51
19	1.014	1.57	1.34	0.69	0.53	1.55	1.32	0.68	0.52
20	1.013	1.54	1.34	0.69	0.54	1.52	1.32	0.68	0.53
21	1.013	1.52	1.33	0.70	0.55	1.50	1.31	0.69	0.54
22	1.012	1.51	1.32	0.71	0.56	1.49	1.30	0.70	0.56
23	1.011	1.50	1.31	0.72	0.57	1.48	1.30	0.71	0.56
24	1.011	1.49	1.30	0.72	0.58	1.47	1.29	0.71	0.57
25	1.011	1.48	1.30	0.73	0.59	1.46	1.28	0.72	0.58

Formulae $\sigma = \bar{s}c_n$

Standard deviation chart
- Upper action line $= B'_{.001}\bar{s}$ or $B_{.001}\sigma$
- Upper warning line $= B'_{.025}\bar{s}$ or $B_{.025}\sigma$
- Lower warning line $= B'_{.975}\bar{s}$ or $B_{.975}\sigma$
- Lower action line $= B'_{.999}\bar{s}$ or $B_{.999}\sigma$

Appendix F   Cumulative Poisson probability tables

The table gives the probability that x or more defects (or defectives) will be found when the average number of defects (or defectives) is $\bar{c}$.

$\bar{c} =$	0.1	0.2	0.3	0.4	0.5	0.6	0.7	0.8	0.9	1.0
$x = 0$	1.0000	1.0000	1.0000	1.0000	1.0000	1.0000	1.0000	1.0000	1.0000	1.0000
1	.0952	.1813	.2592	.3297	.3935	.4512	.5034	.5507	.5934	.6321
2	.0047	.0175	.0369	.0616	.0902	.1219	.1558	.1912	.2275	.2642
3	.0002	.0011	.0036	.0079	.0144	.0231	.0341	.0474	.0629	.0803
4		.0001	.0003	.0008	.0018	.0034	.0058	.0091	.0135	.0190
5				.0001	.0002	.0004	.0008	.0014	.0023	.0037
6							.0001	.0002	.0003	.0006
7										.0001

$\bar{c} =$	1.1	1.2	1.3	1.4	1.5	1.6	1.7	1.8	1.9	2.0
$x = 0$	1.0000	1.0000	1.0000	1.0000	1.0000	1.0000	1.0000	1.0000	1.0000	1.0000
1	.6671	.6988	.7275	.7534	.7769	.7981	.8173	.8347	.8504	.8647
2	.3010	.3374	.3732	.4082	.4422	.4751	.5068	.5372	.5663	.5940
3	.0996	.1205	.1429	.1665	.1912	.2166	.2428	.2694	.2963	.3233
4	.0257	.0338	.0431	.0537	.0656	.0788	.0932	.1087	.1253	.1429
5	.0054	.0077	.0107	.0143	.0186	.0237	.0296	.0364	.0441	.0527
6	.0010	.0015	.0022	.0032	.0045	.0060	.0080	.0104	.0132	.0166
7	.0001	.0003	.0004	.0006	.0009	.0013	.0019	.0026	.0034	.0045
8			.0001	.0001	.0002	.0003	.0004	.0006	.0008	.0011
9							.0001	.0001	.0002	.0002

$\bar{c} =$	2.1	2.2	2.3	2.4	2.5	2.6	2.7	2.8	2.9	3.0
$x = 0$	1.0000	1.0000	1.0000	1.0000	1.0000	1.0000	1.000	1.0000	1.0000	1.0000
1	.8775	.8892	.8997	.9093	.9179	.9257	.9328	.9392	.9450	.9502
2	.6204	.6454	.6691	.6916	.7127	.7326	.7513	.7689	.7854	.8009
3	.3504	.3773	.4040	.4303	.4562	.4816	.5064	.5305	.5540	.5768
4	.1614	.1806	.2007	.2213	.2424	.2640	.2859	.3081	.3304	.3528
5	.0621	.0725	.0838	.0959	.1088	.1226	.1371	.1523	.1682	.1847
6	.0204	.0249	.0300	.0357	.0420	.0490	.0567	.0651	.0742	.0839
7	.0059	.0075	.0094	.0116	.0142	.0172	.0206	.0244	.0287	.0335
8	.0015	.0020	.0026	.0033	.0042	.0053	.0066	.0081	.0099	.0119
9	.0003	.0005	.0006	.0009	.0011	.0015	.0019	.0024	.0031	.0038
10	.0001	.0001	.0001	.0002	.0003	.0004	.0005	.0007	.0009	.0011
11					.0001	.0001	.0001	.0002	.0002	.0003
12									.0001	.0001

$\bar{c} =$	3.1	3.2	3.3	3.4	3.5	3.6	3.7	3.8	3.9	4.0
$x = 0$	1.0000	1.0000	1.0000	1.0000	1.0000	1.0000	1.0000	1.0000	1.0000	1.0000
1	.9550	.9592	.9631	.9666	.9698	.9727	.9753	.9776	.9798	.9817
2	.8153	.8288	.8414	.8532	.8641	.8743	.8838	.8926	.9008	.9084
3	.5988	.6201	.6406	.6603	.6792	.6973	.7146	.7311	.7469	.7619
4	.3752	.3975	.4197	.4416	.4634	.4848	.5058	.5265	.5468	.5665
5	.2018	.2194	.2374	.2558	.2746	.2936	.3128	.3322	.3516	.3712
6	.0943	.1054	.1171	.1295	.1424	.1559	.1699	.1844	.1994	.2149
7	.0388	.0446	.0510	.0579	.0653	.0733	.0818	.0909	.1005	.1107
8	.0142	.0168	.0198	.0231	.0267	.0308	.0352	.0401	.0454	.0511
9	.0047	.0057	.0069	.0083	.0099	.0117	.0137	.0160	.0185	.0214

x										
10	.0014	.0018	.0022	.0027	.0033	.0040	.0048	.0058	.0069	.0081
11	.0004	.0005	.0006	.0008	.0010	.0013	.0016	.0019	.0023	.0028
12	.0001	.0001	.0002	.0002	.0003	.0004	.0005	.0006	.0007	.0009
13					.0001	.0001	.0001	.0002	.0002	.0003
14									.0001	.0001

$\bar{c}$ =	4.1	4.2	4.3	4.4	4.5	4.6	4.7	4.8	4.9	5.0
x = 0	1.0000	1.0000	1.0000	1.0000	1.0000	1.0000	1.0000	1.0000	1.0000	1.0000
1	.9834	.9850	.9864	.9877	.9889	.9899	.9909	.9918	.9926	.9933
2	.9155	.9220	.9281	.9337	.9389	.9437	.9482	.9523	.9561	.9596
3	.7762	.7898	.8026	.8149	.8264	.8374	.8477	.8575	.8667	.8753
4	.5858	.6046	.6228	.6406	.6577	.6743	.6903	.7058	.7207	.7350
5	.3907	.4102	.4296	.4488	.4679	.4868	.5054	.5237	.5418	.5595
6	.2307	.2469	.2633	.2801	.2971	.3142	.3316	.3490	.3665	.3840
7	.1214	.1325	.1442	.1564	.1689	.1820	.1954	.2092	.2233	.2378
8	.0573	.0639	.0710	.0786	.0866	.0951	.1040	.1133	.1231	.1334
9	.0245	.0279	.0317	.0358	.0403	.0451	.0503	.0558	.0618	.0681
10	.0095	.0111	.0129	.0149	.0171	.0195	.0222	.0251	.0283	.0318
11	.0034	.0041	.0048	.0057	.0067	.0078	.0090	.0104	.0120	.0137
12	.0011	.0014	.0017	.0020	.0024	.0029	.0034	.0040	.0047	.0055
13	.0003	.0004	.0005	.0007	.0008	.0010	.0012	.0014	.0017	.0020
14	.0001	.0001	.0002	.0002	.0003	.0003	.0004	.0005	.0006	.0007
15				.0001	.0001	.0001	.0001	.0001	.0002	.0002
16									.0001	.0001

$\bar{c} =$	5.2	5.4	5.6	5.8	6.0	6.2	6.4	6.6	6.8	7.0
$x = 0$	1.0000	1.0000	1.0000	1.0000	1.0000	1.0000	1.0000	1.0000	1.0000	1.0000
1	.9945	.9955	.9963	.9970	.9975	.9980	.9983	.9986	.9989	.9991
2	.9658	.9711	.9756	.9794	.9826	.9854	.9877	.9897	.9913	.9927
3	.8912	.9052	.9176	.9285	.9380	.9464	.9537	.9600	.9656	.9704
4	.7619	.7867	.8094	.8300	.8488	.8658	.8811	.8948	.9072	.9182
5	.5939	.6267	.6579	.6873	.7149	.7408	.7649	.7873	.8080	.8270
6	.4191	.4539	.4881	.5217	.5543	.5859	.6163	.6453	.6730	.6993
7	.2676	.2983	.3297	.3616	.3937	.4258	.4577	.4892	.5201	.5503
8	.1551	.1783	.2030	.2290	.2560	.2840	.3127	.3419	.3715	.4013
9	.0819	.0974	.1143	.1328	.1528	.1741	.1967	.2204	.2452	.2709
10	.0397	.0488	.0591	.0708	.0839	.0984	.1142	.1314	.1498	.1695
11	.0177	.0225	.0282	.0349	.0426	.0514	.0614	.0726	.0849	.0985
12	.0073	.0096	.0125	.0160	.0201	.0250	.0307	.0373	.0448	.0534
13	.0028	.0038	.0051	.0068	.0088	.0113	.0143	.0179	.0221	.0270
14	.0010	.0014	.0020	.0027	.0036	.0048	.0063	.0080	.0102	.0128
15	.0003	.0005	.0007	.0010	.0014	.0019	.0026	.0034	.0044	.0057
16	.0001	.0002	.0002	.0004	.0005	.0007	.0010	.0014	.0018	.0024
17		.0001	.0001	.0001	.0002	.0003	.0004	.0005	.0007	.0010
18					.0001	.0001	.0001	.0002	.0003	.0004
19								.0001	.0001	.0001

$\bar{c} =$	7.2	7.4	7.6	7.8	8.0	8.2	8.4	8.6	8.8	9.0
$x = $ 0	1.0000	1.0000	1.0000	1.0000	1.0000	1.0000	1.0000	1.0000	1.0000	1.0000
1	.9993	.9994	.9995	.9996	.9997	.9997	.9998	.9998	.9998	.9999
2	.9939	.9949	.9957	.9964	.9970	.9975	.9979	.9982	.9985	.9988
3	.9745	.9781	.9812	.9839	.9862	.9882	.9900	.9914	.9927	.9938
4	.9281	.9368	.9446	.9515	.9576	.9630	.9677	.9719	.9756	.9788
5	.8445	.8605	.8751	.8883	.9004	.9113	.9211	.9299	.9379	.9450
6	.7241	.7474	.7693	.7897	.8088	.8264	.8427	.8578	.8716	.8843
7	.5796	.6080	.6354	.6616	.6866	.7104	.7330	.7543	.7744	.7932
8	.4311	.4607	.4900	.5188	.5470	.5746	.6013	.6272	.6522	.6761
9	.2973	.3243	.3518	.3796	.4075	.4353	.4631	.4906	.5177	.5443
10	.1904	.2123	.2351	.2589	.2834	.3085	.3341	.3600	.3863	.4126
11	.1133	.1293	.1465	.1648	.1841	.2045	.2257	.2478	.2706	.2940
12	.0629	.0735	.0852	.0980	.1119	.1269	.1429	.1600	.1780	.1970
13	.0327	.0391	.0464	.0546	.0638	.0739	.0850	.0971	.1102	.1242
14	.0159	.0195	.0238	.0286	.0342	.0405	.0476	.0555	.0642	.0739
15	.0073	.0092	.0114	.0141	.0173	.0209	.0251	.0299	.0353	.0415
16	.0031	.0041	.0052	.0066	.0082	.0102	.0125	.0152	.0184	.0220
17	.0013	.0017	.0022	.0029	.0037	.0047	.0059	.0074	.0091	.0111
18	.0005	.0007	.0009	.0012	.0016	.0021	.0027	.0034	.0043	.0053
19	.0002	.0003	.0004	.0005	.0006	.0009	.0011	.0015	.0019	.0024
20	.0001	.0001	.0001	.0002	.0003	.0003	.0005	.0006	.0008	.0011
21				.0001	.0001	.0001	.0002	.0002	.0003	.0004
22							.0001	.0001	.0001	.0002
23										.0001

$\bar{c}=$	9.2	9.4	9.6	9.8	10.0	11.0	12.0	13.0	14.0	15.0
$x=$ 0	1.0000	1.0000	1.0000	1.0000	1.0000	1.0000	1.0000	1.0000	1.0000	1.0000
1	.9999	.9999	.9999	.9999	1.0000	1.0000	1.0000	1.0000	1.0000	1.0000
2	.9990	.9991	.9993	.9994	.9995	.9998	.9999	1.0000	1.0000	1.0000
3	.9947	.9955	.9962	.9967	.9972	.9988	.9995	.9998	.9999	1.0000
4	.9816	.9840	.9862	.9880	.9897	.9951	.9977	.9990	.9995	.9998
5	.9514	.9571	.9622	.9667	.9707	.9849	.9924	.9963	.9982	.9991
6	.8959	.9065	.9162	.9250	.9329	.9625	.9797	.9893	.9945	.9972
7	.8108	.8273	.8426	.8567	.8699	.9214	.9542	.9741	.9858	.9924
8	.6990	.7208	.7416	.7612	.7798	.8568	.9105	.9460	.9684	.9820
9	.5704	.5958	.6204	.6442	.6672	.7680	.8450	.9002	.9379	.9626
10	.4389	.4651	.4911	.5168	.5421	.6595	.7576	.8342	.8906	.9301
11	.3180	.3424	.3671	.3920	.4170	.5401	.6528	.7483	.8243	.8815
12	.2168	.2374	.2588	.2807	.3032	.4207	.5384	.6468	.7400	.8152
13	.1393	.1552	.1721	.1899	.2084	.3113	.4240	.5369	.6415	.7324
14	.0844	.0958	.1081	.1214	.1355	.2187	.3185	.4270	.5356	.6368
15	.0483	.0559	.0643	.0735	.0835	.1460	.2280	.3249	.4296	.5343
16	.0262	.0309	.0362	.0421	.0487	.0926	.1556	.2364	.3306	.4319
17	.0135	.0162	.0194	.0230	.0270	.0559	.1013	.1645	.2441	.3359
18	.0066	.0081	.0098	.0119	.0143	.0322	.0630	.1095	.1728	.2511
19	.0031	.0038	.0048	.0059	.0072	.0177	.0374	.0698	.1174	.1805
20	.0014	.0017	.0022	.0028	.0035	.0093	.0213	.0427	.0765	.1248
21	.0006	.0008	.0010	.0012	.0016	.0047	.0116	.0250	.0479	.0830
22	.0002	.0003	.0004	.0005	.0007	.0023	.0061	.0141	.0288	.0531
23	.0001	.0001	.0002	.0002	.0003	.0010	.0030	.0076	.0167	.0327
24			.0001	.0001	.0001	.0005	.0015	.0040	.0093	.0195

$\bar{c} =$	16.0	17.0	18.0	19.0	20.0	21.0	22.0	23.0	24.0	25.0
$x = 0$	1.0000	1.0000	1.0000	1.0000	1.0000	1.0000	1.0000	1.0000	1.0000	1.0000
1	1.0000	1.0000	1.0000	1.0000	1.0000	1.0000	1.0000	1.0000	1.0000	1.0000
2	1.0000	1.0000	1.0000	1.0000	1.0000	1.0000	1.0000	1.0000	1.0000	1.0000
3	1.0000	1.0000	1.0000	1.0000	1.0000	1.0000	1.0000	1.0000	1.0000	1.0000
4	.9999	1.0000	1.0000	1.0000	1.0000	1.0000	1.0000	1.0000	1.0000	1.0000
5	.9996	.9998	.9999	1.0000	1.0000	1.0000	1.0000	1.0000	1.0000	1.0000
6	.9986	.9993	.9997	.9998	.9999	1.0000	1.0000	1.0000	1.0000	1.0000
7	.9960	.9979	.9990	.9995	.9997	.9999	.9999	1.0000	1.0000	1.0000
8	.9900	.9946	.9971	.9985	.9992	.9996	.9998	.9999	1.0000	1.0000
9	.9780	.9874	.9929	.9961	.9979	.9989	.9994	.9997	.9998	.9999
10	.9567	.9739	.9846	.9911	.9950	.9972	.9985	.9992	.9996	.9998
11	.9226	.9509	.9696	.9817	.9892	.9937	.9965	.9980	.9989	.9994
12	.8730	.9153	.9451	.9653	.9786	.9871	.9924	.9956	.9975	.9986
13	.8069	.8650	.9083	.9394	.9610	.9755	.9849	.9909	.9946	.9969
14	.7255	.7991	.8574	.9016	.9339	.9566	.9722	.9826	.9893	.9935
15	.6325	.7192	.7919	.8503	.8951	.9284	.9523	.9689	.9802	.9876
16	.5333	.6285	.7133	.7852	.8435	.8889	.9231	.9480	.9656	.9777
17	.4340	.5323	.6249	.7080	.7789	.8371	.8830	.9179	.9437	.9623
18	.3407	.4360	.5314	.6216	.7030	.7730	.8310	.8772	.9129	.9395
19	.2577	.3450	.4378	.5305	.6186	.6983	.7675	.8252	.8717	.9080
25						.0002	.0007	.0020	.0050	.0112
26						.0001	.0003	.0010	.0026	.0062
27							.0001	.0005	.0013	.0033
28							.0001	.0002	.0006	.0017
29								.0001	.0003	.0009
30									.0001	.0004
31									.0001	.0002
32										.0001

20	.8664	.8197	.7623	.6940	.6157	.5297	.4394	.3491	.2637	.1878
21	.8145	.7574	.6899	.6131	.5290	.4409	.3528	.2693	.1945	.1318
22	.7527	.6861	.6106	.5284	.4423	.3563	.2745	.2009	.1385	.0892
23	.6825	.6083	.5277	.4436	.3595	.2794	.2069	.1449	.0953	.0582
24	.6061	.5272	.4449	.3626	.2840	.2125	.1510	.1011	.0633	.0367
25	.5266	.4460	.3654	.2883	.2178	.1568	.1067	.0683	.0406	.0223
26	.4471	.3681	.2923	.2229	.1623	.1122	.0731	.0446	.0252	.0131
27	.3706	.2962	.2277	.1676	.1174	.0779	.0486	.0282	.0152	.0075
28	.2998	.2323	.1726	.1225	.0825	.0525	.0313	.0173	.0088	.0041
29	.2366	.1775	.1274	.0871	.0564	.0343	.0195	.0103	.0050	.0022
30	.1821	.1321	.0915	.0602	.0374	.0218	.0118	.0059	.0027	.0011
31	.1367	.0958	.0640	.0405	.0242	.0135	.0070	.0033	.0014	.0006
32	.1001	.0678	.0436	.0265	.0152	.0081	.0040	.0018	.0007	.0003
33	.0715	.0467	.0289	.0169	.0093	.0047	.0022	.0010	.0004	.0001
34	.0498	.0314	.0187	.0105	.0055	.0027	.0012	.0005	.0002	.0001
35	.0338	.0206	.0118	.0064	.0032	.0015	.0006	.0002	.0001	.0001
36	.0225	.0132	.0073	.0038	.0018	.0008	.0003	.0001		
37	.0146	.0082	.0044	.0022	.0010	.0004	.0002	.0001		
38	.0092	.0050	.0026	.0012	.0005	.0002	.0001			
39	.0057	.0030	.0015	.0007	.0003	.0001				
40	.0034	.0017	.0008	.0004	.0001	.0001				
41	.0020	.0010	.0004	.0002	.0001					
42	.0012	.0005	.0002	.0001						
43	.0007	.0003	.0001							
44	.0004	.0002	.0001							
45	.0002	.0001								
46	.0001									

$\bar{c} =$	40.0	38.0	36.0	34.0	32.0	30.0	29.0	28.0	27.0	26.0
$x = 9$	1.0000	1.0000	1.0000	1.0000	1.0000	1.0000	1.0000	1.0000	1.0000	1.0000
10	1.0000	1.0000	1.0000	1.0000	1.0000	1.0000	1.0000	1.0000	.9999	.9999
11	1.0000	1.0000	1.0000	1.0000	1.0000	1.0000	1.0000	.9999	.9998	.9997
12	1.0000	1.0000	1.0000	1.0000	1.0000	.9999	.9999	.9998	.9996	.9992
13	1.0000	1.0000	1.0000	1.0000	1.0000	.9998	.9997	.9994	.9990	.9982
14	1.0000	1.0000	1.0000	1.0000	.9999	.9996	.9993	.9987	.9978	.9962
15	1.0000	1.0000	1.0000	.9999	.9997	.9991	.9984	.9973	.9954	.9924
16	1.0000	1.0000	.9999	.9998	.9993	.9981	.9967	.9946	.9912	.9858
17	1.0000	1.0000	.9998	.9995	.9986	.9961	.9937	.9899	.9840	.9752
18	.9999	.9999	.9997	.9990	.9972	.9927	.9885	.9821	.9726	.9580
19	.9999	.9998	.9993	.9980	.9948	.9871	.9801	.9700	.9555	.9354
20	.9998	.9995	.9986	.9963	.9907	.9781	.9674	.9522	.9313	.9032
21	.9996	.9990	.9973	.9932	.9841	.9647	.9489	.9273	.8985	.8613
22	.9993	.9981	.9951	.9884	.9740	.9456	.9233	.8940	.8564	.8095
23	.9986	.9965	.9915	.9809	.9594	.9194	.8896	.8517	.8048	.7483
24	.9974	.9938	.9859	.9698	.9390	.8854	.8471	.8002	.7441	.6791
25	.9955	.9897	.9776	.9540	.9119	.8428	.7958	.7401	.6758	.6041
26	.9924	.9834	.9655	.9326	.8772	.7916	.7363	.6728	.6021	.5261
27	.9877	.9741	.9487	.9047	.8344	.7327	.6699	.6003	.5256	.4481
28	.9807	.9611	.9264	.8694	.7838	.6671	.5986	.5251	.4491	.3730
29	.9706	.9435	.8977	.8267	.7259	.5969	.5247	.4500	.3753	.3033
30	.9568	.9204	.8621	.7765	.6620	.5243	.4508	.3774	.3065	.2407
31	.9383	.8911	.8194	.7196	.5939	.4516	.3794	.3097	.2447	.1866
32	.9145	.8552	.7697	.6573	.5235	.3814	.3126	.2485	.1908	.1411
33	.8847	.8125	.7139	.5911	.4532	.3155	.2521	.1949	.1454	.1042
34	.8486	.7635	.6530	.5228	.3850	.2556	.1989	.1495	.1082	.0751
35	.8061	.7086	.5885	.4546	.3208	.2027	.1535	.1121	.0787	.0528
36	.7576	.6490	.5222	.3883	.2621	.1574	.1159	.0822	.0559	.0363
37	.7037	.5862	.4558	.3256	.2099	.1196	.0856	.0589	.0388	.0244
38	.6453	.5216	.3913	.2681	.1648	.0890	.0619	.0413	.0263	.0160
39	.5840	.4570	.3301	.2166	.1268	.0648	.0438	.0283	.0175	.0103

40	.5210	.3941	.2737	.1717	.0956	.0463	.0303	.0190	.0113	.0064
41	.4581	.3343	.2229	.1336	.0707	.0323	.0205	.0125	.0072	.0039
42	.3967	.2789	.1783	.1019	.0512	.0221	.0136	.0080	.0045	.0024
43	.3382	.2288	.1401	.0763	.0364	.0148	.0089	.0050	.0027	.0014
44	.2838	.1845	.1081	.0561	.0253	.0097	.0056	.0031	.0016	.0008
45	.2343	.1462	.0819	.0404	.0173	.0063	.0035	.0019	.0009	.0004
46	.1903	.1139	.0609	.0286	.0116	.0040	.0022	.0011	.0005	.0002
47	.1521	.0872	.0445	.0199	.0076	.0025	.0013	.0006	.0003	.0001
48	.1196	.0657	.0320	.0136	.0049	.0015	.0008	.0004	.0002	.0001
49	.0925	.0486	.0225	.0091	.0031	.0009	.0004	.0002	.0001	
50	.0703	.0353	.0156	.0060	.0019	.0005	.0002	.0001		
51	.0526	.0253	.0106	.0039	.0012	.0003	.0001	.0001		
52	.0387	.0178	.0071	.0024	.0007	.0002	.0001			
53	.0281	.0123	.0047	.0015	.0004	.0001				
54	.0200	.0084	.0030	.0009	.0002	.0001				
55	.0140	.0056	.0019	.0006	.0001					
56	.0097	.0037	.0012	.0003	.0001					
57	.0066	.0024	.0007	.0002						
58	.0044	.0015	.0005	.0001						
59	.0029	.0010	.0003	.0001						
60	.0019	.0006	.0002							
61	.0012	.0004	.0001							
62	.0008	.0002	.0001							
63	.0005	.0001								
64	.0003	.0001								
65	.0002									
66	.0001									
67	.0001									

For values of $\bar{c}$ greater than 40, use the table of areas under the normal curve (Appendix A) to obtain approximate Poisson probabilities, putting $\mu = \bar{c}$ and $\sigma = \sqrt{\bar{c}}$.

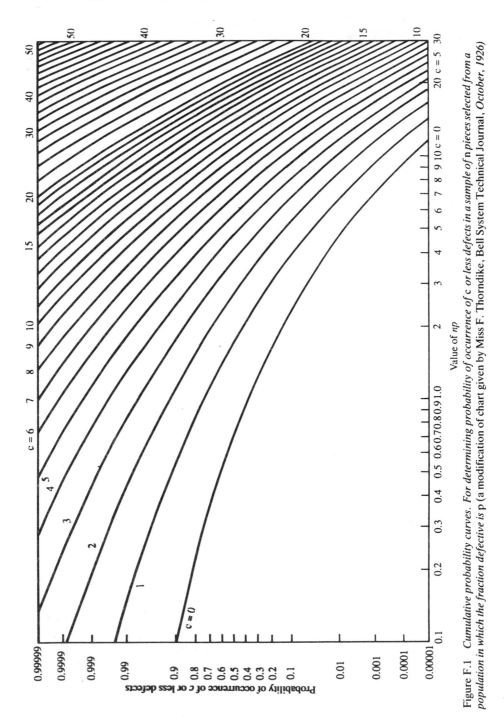

Figure F.1   *Cumulative probability curves. For determining probability of occurrence of c or less defects in a sample of n pieces selected from a population in which the fraction defective is* p *(a modification of chart given by Miss F. Thorndike, Bell System Technical Journal, October, 1926)*

Appendix G Confidence limits and tests of significance

Confidence limits

When an estimate of the mean of a parameter has been made it is desirable to know, not only the estimated mean value, which should be the most likely value, but also how precise the estimate is.

If, for example, eighty results on weights of tablets give a mean, $\overline{X} = 1\ 250.5$ mg and standard deviation, $\sigma = 4.5$ mg, have these values come from a process with mean, $\mu = 250.0$ mg? If the process has a mean $\mu = 250.0$, 99.7 per cent of all sample means $(\overline{X})$ should have a value between:

$$\mu \pm 3\sigma/\sqrt{n}$$
$$\text{i.e. } \mu - 3\sigma/\sqrt{n} < \overline{X} < \mu + 3\sigma \sqrt{n}$$

therefore:

$$\overline{X} - 3\sigma/\sqrt{n} < \mu < \overline{X} + 3\sigma \sqrt{n}$$

i.e. μ will lie between:

$$\overline{X} \pm 3\sigma/\sqrt{n}$$

this is the *confidence interval* at the *confidence coefficient* of 99.7 per cent.

Hence, for the tablet example, the 99.7 per cent interval for μ is:

$$250.5 \pm (3 \times 4.5/\sqrt{80}) \text{ mg}$$
$$\text{i.e. } 249.0 \text{ to } 252.0 \text{ mg}$$

which says that we may be 99.7 per cent confident that the true mean of the process lies between 249 mg and 252 mg, provided that the process was in statistical control at the time of the data collection. A 95 per cent confidence interval may be calculated in a similar way, using the range $\pm 2\sigma/\sqrt{n}$. This is, of course, the basis of the control chart for means.

Difference between two mean values

A problem that frequently arises is to assess the magnitude of the differences between two mean values. The difference between the two observed means is calculated: $\overline{X}_1 - \overline{X}_2$, together with the standard error of the difference. These values are then used to calculate confidence limits for the true difference, $\mu_1 - \mu_2$. If the upper limit is less than zero, μ_2 is greater than μ_1; if the lower limit is greater than zero, μ_1 is greater than μ_2. If the limits are too wide to lead to reliable conclusions, more observations are required.

If we have for sample size n_1, $\overline{X}_1$ and σ_1, and for sample size n_2, $\overline{X}_2$ and σ_2, the standard error of $\overline{X}_1 - \overline{X}_2$,

$$SE = \sqrt{\frac{\sigma_1^2}{n_1} + \frac{\sigma_2^2}{n_2}}$$

when σ_1 and σ_2 are more or less equal:

$$SE = \sigma \sqrt{\frac{1}{n_1} + \frac{1}{n_2}}$$

The 99.7 per cent confidence limits are, therefore:

$$(\overline{X}_1 - \overline{X}_2) \pm 3\sigma \sqrt{\frac{1}{n_1} + \frac{1}{n_2}}$$

Tests of significance

A common procedure to aid interpretation of data analysis is to carry out a 'test of significance'. When applying such a test, we calculate the probability P that a certain result would occur if a 'null hypothesis' were true, i.e. that the result does not differ from a particular value. If this probability is equal to or less than a given value, α, the result is said to be significant at the α level. When $P = 0.05$, the result is usually referred to as 'significant' and when $P = 0.01$ as 'highly significant'.

The *t* test for means

There are two types of tests for means, the normal test given above and the 'students' *t*-test. The normal test applies when the standard deviation σ is known or is based on a large sample, and the *t* test is used when σ must be estimated from the data and the sample size is small ($n<30$). The *t* test is applied to the difference between two means, μ_1 and μ_2 and two examples are given below to illustrate the *t* test method.

1   In the first case μ_1 is known and μ_2 is estimated as $\overline{X}$. The first step is to calculate the *t* statistic:

$$t = (\overline{X} - \mu_1)/s/\sqrt{n}$$

where s is the $(n-1)$ estimate of σ. We then refer to Table G.1 to determine the significance. The following results were obtained for the percentage iron in ten samples of furnace slag material: 15.3, 15.6,

Table G.1   *Probability points of the* t-*distribution (single-sided)*

Degrees of freedom ($n - 1$)	P				
	0.1	0.05	0.025	0.01	0.005
1	3.08	6.31	12.70	31.80	63.70
2	1.89	2.92	4.30	6.96	9.92
3	1.64	2.35	3.18	4.54	5.84
4	1.53	2.13	2.78	3.75	4.60
5	1.48	2.01	2.57	3.36	4.03
6	1.44	1.94	2.45	3.14	3.71
7	1.42	1.89	2.36	3.00	3.50
8	1.40	1.86	2.31	2.90	3.36
9	1.38	1.83	2.26	2.82	3.25
10	1.37	1.81	2.23	2.76	3.17
11	1.36	1.80	2.20	2.72	3.11
12	1.36	1.78	2.18	2.68	3.05
13	1.35	1.77	2.16	2.65	3.01
14	1.34	1.76	2.14	2.62	2.98
15	1.34	1.75	2.13	2.60	2.95
16	1.34	1.75	2.12	2.58	2.92
17	1.33	1.74	2.11	2.57	2.90
18	1.33	1.73	2.10	2.55	2.88
19	1.33	1.73	2.09	2.54	2.86
20	1.32	1.72	2.09	2.53	2.85
21	1.32	1.72	2.08	2.52	2.83
22	1.32	1.72	2.07	2.51	2.82
23	1.32	1.71	2.07	2.50	2.81
24	1.32	1.71	2.06	2.49	2.80
25	1.32	1.71	2.06	2.48	2.79
26	1.32	1.71	2.06	2.48	2.78
27	1.31	1.70	2.05	2.47	2.77
28	1.31	1.70	2.05	2.47	2.76
29	1.31	1.70	2.05	2.46	2.76
30	1.31	1.70	2.04	2.46	2.75
40	1.30	1.68	2.02	2.42	2.70
60	1.30	1.67	2.00	2.39	2.66
120	1.29	1.66	1.98	2.36	2.62
∞	1.28	1.64	1.96	2.33	2.58

16.0, 15.4, 16.4, 15.8, 15.7, 15.9, 16.1, 15.7. Do the analyses indicate that the material is significantly different from the declared specification of 16.0 per cent?

$$\overline{X} = \frac{\Sigma X}{n} = \frac{157.9}{10} = 15.79\%$$

$$s_{(n-1)} = \sqrt{\frac{\Sigma(X_i - \overline{X})^2}{n - 1}} = 0.328\%$$

$$t_{calc} = \frac{\mu_1 - \overline{X}}{s/\sqrt{n}} = \frac{16.0 - 15.79}{0.328/\sqrt{10}}$$

$$= \frac{0.21}{0.1037} = 2.025$$

Consultation of Table G.1 for $(n-1) = 9$ (i.e. the 'number of degrees of freedom') gives a tabulated value for $t_{0.05}$ of 1.83, i.e. at the 5 per cent level of significance. Hence, there is only a 5 per cent chance that the calculated value of t will exceed 1.83, if there is no significant difference between the mean of the analyses and the specification. So we may conclude that the mean analysis differs significantly (at 5 per cent level) from the specification. Note, the result is not highly significant, since the tabulated value of $t_{0.01}$, i.e. at the 1 per cent level, is 2.82 and this has not been exceeded.

2   In the second case, results from two sources are being compared. This situation requires the calculation of the t statistic from the mean of the differences in values and the standard error of the differences. The example should illustrate the method. To check on the analysis of percentage impurity present in a certain product, a manufacturer took twelve samples, halved each of them and had one half tested in his own laboratory (A) and the other half tested by an independent laboratory (B). The results obtained were:

Sample No.	1	2	3	4	5	6
Laboratory A	0.74	0.52	0.32	0.67	0.47	0.77
Laboratory B	0.79	0.50	0.43	0.77	0.67	0.68
Difference, d = A−B	−0.05	+0.02	−0.11	−0.10	−0.20	+0.09

Sample No.	7	8	9	10	11	12
Laboratory A	0.72	0.80	0.70	0.69	0.94	0.87
Laboratory B	0.91	0.80	0.98	0.67	0.93	0.82
Difference, d = A−B	−0.19	0	−0.28	+0.02	+0.01	+0.05

Is there any significant difference between the test results from the two laboratories?

$$\text{Total difference} |\Sigma d| = 0.74$$

$$\text{Mean difference } |\bar{d}| = \frac{|\Sigma d|}{n} = \frac{0.74}{12} = 0.062$$

Standard deviation estimate,

$$s_{(n-1)} = \sqrt{\frac{\Sigma(\bar{d} - d_i)^2}{n-1}} = 0.115$$

$$t_{calc} = \frac{|\bar{d}|}{s/\sqrt{n}} = \frac{0.062}{0.115/\sqrt{12}} = 1.868$$

From Table G.1 and for $(n-1) = 11$ degrees of freedom, the tabulated value of t is obtained. As we are looking for a difference in means, irrespective of which is greater, the test is said to be *double sided*, and it is necesary to double the probabilities in Table G.1 for the critical values of t. From Table G.1 then:

$$t_{0.025}(11) = 2.20$$
$$\text{since } 1.868 < 2.20$$
$$\text{i.e. } t_{calc} < t_{0.025}(11)$$

and there is insufficient evidence, at the 5 per cent level, to suggest that the two laboratories differ.

The *F* test for variances

The *F* test is used for comparing two variances. If it is required to compare the values of two variances σ_1^2 and σ_2^2 from estimates s_1^2 and s_2^2, based on (n_1-1) and (n_2-1) degrees of freedom respectively, and the alternative to the Null Hypothesis $(\sigma_1^2 = \sigma_1^2)$ is $\sigma_1^2 > \sigma_2^2$, we calculate the ratio $F = s_1^2/s_2^2$ and refer to Table G.2 for the critical values of F, with (n_1-1) and (n_2-1) degrees of freedom, where s_1^2 is always the highest variance and n_1 is the corresponding sample size. The levels tabulated in G.2 refer to the single upper tail area of the F distribution. If the alternative to the Null Hypothesis is σ_1^2 not equal to σ_2^2, the test is double sided, and we calculate the ratio of the larger estimate to the smaller one and the probabilities in Table G.2 are doubled to give the critical values for this ratio. In each case the calculated values of F must be greater than the tabulated critical values, for significant differences at the appropriate level shown in the probability point column.

For example, in the filling of cans of beans, it is suspected that the variability in the morning is greater than that in the afternoon. From collected data:

Morning $n_1 = 40$, $\overline{X}_1 = 451.78$, $s_1 = 1.76$

Afternoon $n_2 = 40$, $\overline{X}_2 = 450.71$, $s_2 = 1.55$

Degrees of freedom $(n_1-1) = (n_2-1) = 39$

$$F = \frac{s_1^2}{s_2^2} = \frac{1.76^2}{1.55^2} = \frac{3.098}{2.403} = 1.29$$

(note if $s_1^2 < s_2^2$ the test statistic would have been $F = \frac{s_2^2}{s_1^2}$)

If there is a good reason for the variability in the morning to be greater than in the afternoon (e.g. equipment and people 'settling down') then the test will be a one-tail test. For $\alpha = 0.05$, from Table G.2, the critical value for the ratio is $F_{0.05} \simeq 1.70$ by interpolation. Hence, the sample value of s_1^2/s_2^2 is not above $F_{0.05}$, and we accept the Null Hypothesis that $\sigma_1 = \sigma_2$, and the variances are the same in the morning and afternoon.

For confidence limits for the variance ratio, we require both the upper *and lower* tail areas of the distribution. The lower tail area is given by the reciprocal of the corresponding F value in the upper tail. Hence, to obtain the 95 per cent confidence limits for the variance ratio, we require the values of $F_{0.975}$ and $F_{0.025}$. For example, if $(n_1 - 1) = 9$ and $(n_2 - 1) = 15$ then:

$$F_{0.975}(9,15) = \frac{1}{F_{0.025}(15,9)} = \frac{1}{3.77} = 0.27$$

and $F_{0.025}(9,15) = 3.12$.

If s_1^2/s_2^2 exceeds 3.12 or falls short of 0.27, we shall reject the hypothesis that $\sigma_1 = \sigma_2$.

Table G.2   *Critical values of F for variances*

Proba-bility point	Degree of Free-dom n_2-1	Degrees of Freedom n_1-1 *(corresponding to greater variance)*																		
		1	*2*	*3*	*4*	*5*	*6*	*7*	*8*	*9*	*10*	*12*	*15*	*20*	*24*	*30*	*40*	*60*	*120*	*∞*
0.100	1	39.9	49.5	53.6	55.8	57.2	58.2	58.9	59.4	59.9	60.2	60.7	61.2	61.7	62.0	62.3	62.5	62.8	63.1	63.3
0.050		161	199	216	225	230	234	237	239	241	242	244	246	248	249	250	251	252	253	254
0.025		648	800	864	900	922	937	948	957	963	969	977	985	993	997	1001	1006	1010	1014	1018
0.010		4052	4999	5403	5625	5764	5859	5928	5982	6022	6056	6106	6157	6209	6235	6261	6287	6313	6339	6366
0.100	2	8.53	9.00	9.16	9.24	9.29	9.33	9.35	9.37	9.38	9.39	9.41	9.42	9.44	9.45	9.46	9.47	9.47	9.48	9.49
0.050		18.5	19.0	19.2	19.2	19.3	19.3	19.4	19.4	19.4	19.4	19.4	19.4	19.4	19.5	19.5	19.5	19.5	19.5	19.5
0.025		38.5	39.0	39.2	39.2	39.3	39.3	39.4	39.4	39.4	39.4	39.4	39.4	39.4	39.5	39.5	39.5	39.5	39.5	39.5
0.010		98.5	99.0	99.2	99.2	99.3	99.3	99.4	99.4	99.4	99.4	99.4	99.4	99.4	99.5	99.5	99.5	99.5	99.5	99.5
0.100	3	5.54	5.46	5.39	5.34	5.31	5.28	5.27	5.25	5.24	5.23	5.22	5.20	5.18	5.18	5.17	5.16	5.15	5.14	5.13
0.050		10.1	9.55	9.28	9.12	9.01	8.94	8.89	8.85	8.81	8.79	8.74	8.70	8.66	8.64	8.62	8.59	8.57	8.55	8.53
0.025		17.4	16.0	15.4	15.1	14.9	14.7	14.6	14.5	14.5	14.4	14.3	14.3	14.2	14.1	14.1	14.0	14.0	13.9	13.9
0.010		34.1	30.8	29.5	28.7	28.2	27.9	27.7	27.5	27.3	27.2	27.1	26.9	26.7	26.6	26.5	26.4	26.3	26.2	26.1
0.100	4	4.54	4.32	4.19	4.11	4.05	4.01	3.98	3.95	3.94	3.92	3.90	3.87	3.84	3.83	3.82	3.80	3.79	3.78	3.76
0.050		7.71	6.94	6.59	6.39	6.26	6.16	6.09	6.04	6.00	5.96	5.91	5.86	5.80	5.77	5.75	5.72	5.69	5.66	5.63
0.025		12.2	10.6	10.0	9.60	9.36	9.20	9.07	8.98	8.90	8.84	8.75	8.66	8.56	8.51	8.46	8.41	8.36	8.31	8.26
0.010		21.2	18.0	16.7	16.0	15.5	15.2	15.0	14.8	14.7	14.5	14.4	14.2	14.0	13.9	13.8	13.7	13.7	13.6	13.5
0.100	5	4.06	3.78	3.62	3.52	3.45	3.40	3.37	3.34	3.32	3.30	3.27	3.24	3.21	3.19	3.17	3.16	3.14	3.12	3.10
0.050		6.61	5.79	5.41	5.19	5.05	4.95	4.88	4.82	4.77	4.74	4.68	4.62	4.56	4.53	4.50	4.46	4.43	4.40	4.36
0.025		10.0	8.43	7.76	7.39	7.15	6.98	6.85	6.76	6.68	6.62	6.52	6.43	6.33	6.28	6.23	6.18	6.12	6.07	6.02
0.010		16.3	13.3	12.1	11.4	11.0	10.7	10.5	10.3	10.2	10.1	9.89	9.72	9.55	9.47	9.38	9.29	9.20	9.11	9.02

Degrees of Freedom n_1-1 (corresponding to greater variance)

Degree of Freedom n_2-1	Probability point	1	2	3	4	5	6	7	8	9	10	12	15	20	24	30	40	60	120	∞
6	0.100	3.78	3.46	3.29	3.18	3.11	3.05	3.01	2.98	2.96	2.94	2.90	2.87	2.84	2.82	2.80	2.78	2.76	2.74	2.72
	0.050	5.99	5.14	4.76	4.53	4.39	4.28	4.21	4.15	4.10	4.06	4.00	3.94	3.87	3.84	3.81	3.77	3.74	3.70	3.67
	0.025	8.81	7.26	6.60	6.23	5.99	5.82	5.70	5.60	5.52	5.46	5.37	5.27	5.17	5.12	5.07	5.01	4.96	4.90	4.85
	0.010	13.7	10.9	9.78	9.15	8.75	8.47	8.26	8.10	7.98	7.87	7.72	7.56	7.40	7.31	7.23	7.14	7.06	6.97	6.88
7	0.100	3.59	3.26	3.07	2.96	2.88	2.83	2.78	2.75	2.72	2.70	2.67	2.63	2.59	2.58	2.56	2.54	2.51	2.49	2.47
	0.050	5.59	4.74	4.35	4.12	3.97	3.87	3.79	3.73	3.68	3.64	3.57	3.51	3.44	3.41	3.38	3.34	3.30	3.27	3.23
	0.025	8.07	6.54	5.89	5.52	5.29	5.12	4.99	4.90	4.82	4.76	4.67	4.57	4.47	4.42	4.36	4.31	4.25	4.20	4.14
	0.010	12.2	9.55	8.45	7.85	7.46	7.19	6.99	6.84	6.72	6.62	6.47	6.31	6.16	6.07	5.99	5.91	5.82	5.74	5.65
8	0.100	3.46	3.11	2.92	2.81	2.73	2.67	2.62	2.59	2.56	2.54	2.50	2.46	2.42	2.40	2.38	2.36	2.34	2.32	2.29
	0.050	5.32	4.46	4.07	3.84	3.69	3.58	3.50	3.44	3.39	3.35	3.28	3.22	3.15	3.12	3.08	3.04	3.01	2.97	2.93
	0.025	7.57	6.06	5.42	5.05	4.82	4.65	4.53	4.43	4.36	4.30	4.20	4.10	4.00	3.95	3.89	3.84	3.78	3.73	3.67
	0.010	11.3	8.65	7.59	7.01	6.63	6.37	6.18	6.03	5.91	5.81	5.67	5.52	5.36	5.28	5.20	5.12	5.03	4.95	4.86
9	0.100	3.36	3.01	2.81	2.69	2.61	2.55	2.51	2.47	2.44	2.42	2.38	2.34	2.30	2.28	2.25	2.23	2.21	2.18	2.16
	0.050	5.12	4.26	3.86	3.63	3.48	3.37	3.29	3.23	3.18	3.14	3.07	3.01	2.94	2.90	2.86	2.83	2.79	2.75	2.71
	0.025	7.12	5.71	5.08	4.72	4.48	4.32	4.20	4.10	4.03	3.96	3.87	3.77	3.67	3.61	3.56	3.51	3.45	3.39	3.33
	0.010	10.6	8.02	6.99	6.42	6.06	5.80	5.61	5.47	5.35	5.26	5.11	4.96	4.81	4.73	4.65	4.57	4.48	4.40	4.31
10	0.100	3.28	2.92	2.73	2.61	2.52	2.46	2.41	2.38	2.35	2.32	2.28	2.24	2.20	2.18	2.16	2.13	2.11	2.08	2.06
	0.050	4.96	4.10	3.71	3.48	3.33	3.22	3.14	3.07	3.02	2.98	2.91	2.84	2.77	2.74	2.70	2.66	2.62	2.58	2.54
	0.025	6.94	5.46	4.83	4.47	4.24	4.07	3.95	3.85	3.78	3.72	3.62	3.52	3.42	3.37	3.31	3.26	3.20	3.14	3.08
	0.010	10.0	7.56	6.55	5.99	5.64	5.39	5.20	5.06	4.94	4.85	4.71	4.56	4.41	4.33	4.25	4.17	4.08	4.00	3.91

12	0.100	3.18	2.81	2.61	2.48	2.39	2.33	2.28	2.24	2.21	2.19	2.15	2.10	2.06	2.04	2.01	1.99	1.96	1.93	1.90
	0.050	4.75	3.89	3.49	3.26	3.11	3.00	2.91	2.85	2.80	2.75	2.69	2.62	2.54	2.51	2.47	2.43	2.38	2.34	2.30
	0.025	6.55	5.10	4.47	4.12	3.89	3.73	3.61	3.51	3.44	3.37	3.28	3.18	3.07	3.02	2.96	2.91	2.85	2.79	2.72
	0.010	9.33	6.93	5.95	5.41	5.06	4.82	4.64	4.50	4.39	4.30	4.16	4.01	3.86	3.78	3.70	3.62	3.54	3.45	3.36
15	0.100	3.07	2.70	2.49	2.36	2.27	2.21	2.16	2.12	2.09	2.06	2.02	1.97	1.92	1.90	1.87	1.85	1.82	1.79	1.76
	0.050	4.54	3.68	3.29	3.06	2.90	2.79	2.71	2.64	2.59	2.54	2.48	2.40	2.33	2.29	2.25	2.20	2.16	2.11	2.07
	0.025	6.20	4.77	4.15	3.80	3.58	3.41	3.29	3.20	3.12	3.06	2.96	2.86	2.76	2.70	2.64	2.59	2.52	2.46	2.40
	0.010	8.68	6.36	5.42	4.89	4.56	4.32	4.14	4.00	3.89	3.80	3.67	3.52	3.37	3.29	3.21	3.13	3.05	2.96	2.87
20	0.100	2.97	2.59	2.38	2.25	2.16	2.09	2.04	2.00	1.96	1.94	1.89	1.84	1.79	1.77	1.74	1.71	1.68	1.64	1.61
	0.050	4.35	3.49	3.10	2.87	2.71	2.60	2.51	2.45	2.39	2.35	2.28	2.20	2.12	2.08	2.04	1.99	1.95	1.90	1.84
	0.025	5.87	4.46	3.86	3.51	3.29	3.13	3.01	2.91	2.84	2.77	2.68	2.57	2.46	2.41	2.35	2.29	2.22	2.16	2.09
	0.010	8.10	5.85	4.94	4.43	4.10	3.87	3.70	3.56	3.46	3.37	3.23	3.09	2.94	2.86	2.78	2.69	2.61	2.52	2.42
24	0.100	2.93	2.54	2.33	2.19	2.10	2.04	1.98	1.94	1.91	1.88	1.83	1.78	1.73	1.70	1.67	1.64	1.61	1.57	1.53
	0.050	4.26	3.40	3.01	2.78	2.62	2.51	2.42	2.36	2.30	2.25	2.18	2.11	2.03	1.98	1.94	1.89	1.84	1.79	1.73
	0.025	5.72	4.32	3.72	3.38	3.15	2.99	2.87	2.78	2.70	2.64	2.54	2.44	2.33	2.27	2.21	2.15	2.08	2.01	1.94
	0.010	7.82	5.61	4.72	4.22	3.90	3.67	3.50	3.36	3.26	3.17	3.03	2.89	2.74	2.66	2.58	2.49	2.40	2.31	2.21
30	0.100	2.88	2.49	2.28	2.14	2.05	1.98	1.93	1.88	1.85	1.82	1.77	1.72	1.67	1.64	1.61	1.57	1.54	1.50	1.46
	0.050	4.17	3.32	2.92	2.69	2.53	2.42	2.33	2.27	2.21	2.16	2.09	2.01	1.93	1.89	1.84	1.79	1.74	1.68	1.62
	0.025	5.57	4.18	3.59	3.25	3.03	2.87	2.75	2.65	2.57	2.51	2.41	2.31	2.20	2.14	2.07	2.01	1.94	1.87	1.79
	0.010	7.56	5.39	4.51	4.02	3.70	3.47	3.30	3.17	3.07	2.98	2.84	2.70	2.55	2.47	2.39	2.30	2.21	2.11	2.01

Degrees of Freedom $n_1 - 1$ (corresponding to greater variance)

Probability point / n_2-1	1	2	3	4	5	6	7	8	9	10	12	15	20	24	30	40	60	120	∞
40																			
0.100	2.84	2.44	2.23	2.09	2.00	1.93	1.87	1.83	1.79	1.76	1.71	1.66	1.61	1.57	1.54	1.51	1.47	1.42	1.38
0.050	4.08	3.23	2.84	2.61	2.45	2.34	2.25	2.18	2.12	2.08	2.00	1.92	1.84	1.79	1.74	1.69	1.64	1.58	1.51
0.025	5.42	4.05	3.46	3.13	2.90	2.74	2.62	2.53	2.45	2.39	2.29	2.18	2.07	2.01	1.94	1.88	1.80	1.72	1.64
0.010	7.31	5.18	4.31	3.83	3.51	3.29	3.12	2.99	2.89	2.80	2.66	2.52	2.37	2.29	2.20	2.11	2.02	1.92	1.80
60																			
0.100	2.79	2.39	2.18	2.04	1.95	1.87	1.82	1.77	1.74	1.71	1.66	1.60	1.54	1.51	1.48	1.44	1.40	1.35	1.29
0.050	4.00	3.15	2.76	2.53	2.37	2.25	2.17	2.10	2.04	1.99	1.92	1.84	1.75	1.70	1.65	1.59	1.53	1.47	1.39
0.025	5.29	3.93	3.34	3.01	2.79	2.63	2.51	2.41	2.33	2.27	2.17	2.06	1.94	1.88	1.82	1.74	1.67	1.58	1.48
0.010	7.08	4.98	4.13	3.65	3.34	3.12	2.95	2.82	2.72	2.63	2.50	2.35	2.20	2.12	2.03	1.94	1.84	1.73	1.60
120																			
0.100	2.75	2.35	2.13	1.99	1.90	1.82	1.77	1.72	1.68	1.65	1.60	1.54	1.48	1.45	1.41	1.37	1.32	1.26	1.19
0.050	3.92	3.07	2.68	2.45	2.29	2.18	2.09	2.02	1.96	1.91	1.83	1.75	1.66	1.61	1.55	1.50	1.43	1.35	1.25
0.025	5.15	3.80	3.23	2.89	2.67	2.52	2.39	2.30	2.22	2.16	2.05	1.94	1.82	1.76	1.69	1.61	1.53	1.43	1.31
0.010	6.85	4.79	3.95	3.48	3.17	2.96	2.79	2.66	2.56	2.47	2.34	2.19	2.03	1.95	1.86	1.76	1.66	1.53	1.38
∞																			
0.100	2.71	2.30	2.08	1.94	1.85	1.77	1.72	1.67	1.63	1.60	1.55	1.49	1.42	1.38	1.34	1.30	1.24	1.17	1.00
0.050	3.84	3.00	2.60	2.37	2.21	2.10	2.01	1.94	1.88	1.83	1.75	1.67	1.57	1.52	1.46	1.39	1.32	1.22	1.00
0.025	5.02	3.69	3.12	2.79	2.57	2.41	2.29	2.19	2.11	2.05	1.94	1.83	1.71	1.64	1.57	1.48	1.39	1.27	1.00
0.010	6.63	4.61	3.78	3.32	3.02	2.80	2.64	2.51	2.41	2.32	2.18	2.04	1.88	1.79	1.70	1.59	1.47	1.32	1.00

Degree of Freedom n_2-1 (leftmost column); Probability point.

Appendix H   OC curves and ARL curves for $\overline{X}$ and R charts

OC curves for an R chart (based on upper action line only). Figure H.1 shows, for several different sample sizes, a plot of the probability or chance that the first sample point will fall below the upper action line, following a given increase in process standard deviation. The x axis is the ratio of the new standard deviation (after the change) to the old; the ordinate axis is the probability that this shift will not be detected by the first sample.

It is interesting to compare the OC curves for samples of various sizes. For example, when the process standard deviation increases by a factor of 3, the probability of not detecting the shift with the first sample is:

ca 0.62 for $n = 2$

and   *ca* 0.23 for $n = 5$

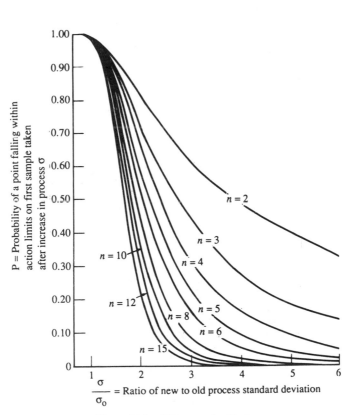

Figure H.1   *OC curves for R chart*

The probabilities of *detecting* the change in the first sample are, therefore:

$$1 - 0.62 = 0.38 \text{ for } n = 2$$
and $$1 - 0.23 = 0.77 \text{ for } n = 5$$

The average run length (ARL) to detection is the *reciprocal of the probability of detection*. In the example of a tripling of the process standard deviation, the ARLs for the two sample sizes will be:

$$\text{for } n = 2, \text{ ARL} = 1/0.38 = 2.6$$
and $$\text{for } n = 5, \text{ ARL} = 1/0.77 = 1.3$$

Clearly the R chart for sample size $n = 5$ has a better 'performance' than the one for $n = 2$, in detecting an increase in process variability.

OC curves for an $\overline{X}$ chart (based on action lines only). If the process standard deviation remains constant, the OC curve for an $\overline{X}$ chart is relatively easy to construct. The probability that a sample will fall within the control limits or action lines can be obtained from the normal distribution table in Appendix A, assuming the sample size $n \geq 4$ or the parent distribution is normal. This is shown in general by Figure H.2, in

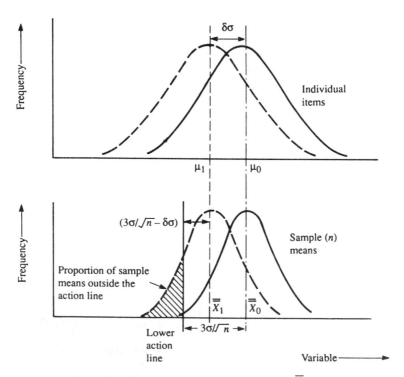

Figure H.2   *Determination of OC curves for an $\overline{X}$ chart*

which action lines for an $\bar{X}$ chart have been set up when the process was stable at mean μ_0, with standard deviation σ. The $\bar{X}$ chart action lines were set at $\bar{\bar{X}}''_0 \pm 3\sigma/\sqrt{n}$.

If the process mean decreases by $\delta\sigma$ to a new mean μ_1, the distribution of sample means will become centred at $\bar{\bar{X}}_1$, and the probability of the first sample mean falling outside the lower action line will be equal to the shaded proportion under the curve. This can be found from the table in Appendix A.

An example should clarify the method. For the steel rod cutting process, described in Chapters 5 and 6, the process mean, $\bar{\bar{X}}_0 = 150.1$ mm and the standard deviation, $\sigma = 5.25$ mm. The lower action line on the mean chart, for a sample size $n = 4$,

$$= \bar{\bar{X}}_0 - 3\,\sigma/\sqrt{n}$$

$$= 150.1 - 3 \times 5.25/\sqrt{4}$$

$$= 142.23 \text{ mm}$$

If the process mean decreases by one σ value (5.25 mm), the distance between the action line and the new mean of the distribution of sample means ($\bar{\bar{X}}_1$) is given by:

$$(3\sigma/\sqrt{n} - \delta\sigma)$$

$$= 3 \times 5.25/\sqrt{4} - 1 \times 5.25 = 2.625 \text{ mm}$$

This distance in terms of number of standard errors of the mean (the standard deviation of the distribution) is:

$$\frac{(3\sigma/\sqrt{n} - \delta\sigma)}{\sigma/\sqrt{n}} \qquad \text{standard errors} \qquad\qquad \text{Formula A}$$

$$\text{or} \quad \frac{2.625}{5.25/\sqrt{4}} = \quad 1 \text{ standard error}$$

The formula A may be further simplified to:

$$(3 - \delta\sqrt{n}) \text{ standard errors}$$

In the example: $3 - 1 \times \sqrt{4} = 1$ standard error, and the shaded proportion under the distribution of sample means is 0.1587 (from Appendix A). Hence, the probability of detecting, with the first sample on the mean chart ($n = 4$), a change in process mean of one standard deviation is 0.1587. The probability of *not detecting* the change is $1 - 0.1587 = 0.8413$ and this value may be used to plot a point on the OC curve. The average run length (ARL) to detection of such a change using this chart, with action lines only, is $1/0.1587 = 6.3$.

Clearly the ARL will depend upon whether or not we incorporate the decision rules based on warning lines, runs, and trends. Figures H.3 and H.4 show how the mean chart operating characteristic and ARL to action signal (point in zone 3), respectively, vary with the sample size, and these curves may be used to decide which sample size is appropriate, when inspection costs and the magnitude of likely changes have been considered. It is important to consider also the frequency of available data and in certain process industries ARLs in time, rather than points plotted, may be more useful. Alternative types of control charts for variables may be more appropriate in these situations (see Chapter 8).

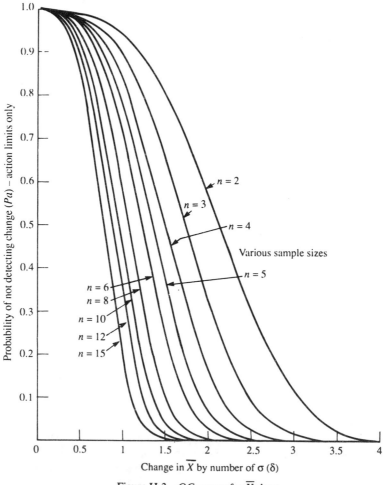

Figure H.3   *OC curves for $\overline{X}$ chart*

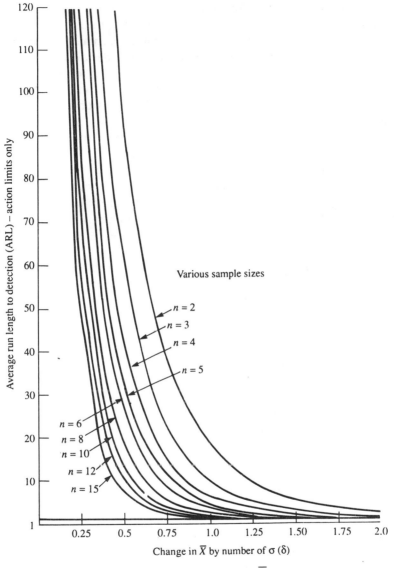

Figure H.4   *ARL curves for X̄ chart*

Appendix I   Autocorrelation

A basic assumption in constructing control charts, such as those for $\overline{X}$, R, moving $\overline{X}$, and moving R, is that the individual data points used are independent of one another. When data are taken in order, there is often a tendency for the observations made close together in time or space to be more alike than those taken further apart. There is often a technological reason for this serial dependence or 'autocorrelation' in the data. For example, physical mixing, residence time or capacitance can produce autocorrelation in continuous processes.

Autocorrelation may be due to shift or day of week effects, or may be due to identifiable causes that are not related to the 'time' order of the data. When groups of batches of material are produced alternatively from two reactors, for example, positive autocorrelation can be explained by the fact that alternate batches are from the same reactor. Trends in data may also produce autocorrelation.

Autocorrelation may be displayed graphically by plotting the data on a scatter diagram, with one axis representing the data in the original order, and the other axis representing the data moved up or down by one or more observations (see Figure I.1).

In most cases, the relationship between the variable and its 'lag' can be summarized by a straight line. The strength of the linear relationship is indicated by the *correlation coefficient*, a number between -1 and 1. The autocorrelation coefficient, often called simply the autocorrelation, is the correlation coefficient of the variable with its lag. Clearly, there is a different autocorrelation for each lag.

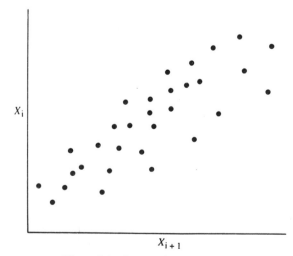

Figure I.1   *Scatter plot of autocorrelated data*

If autocorrelated data is plotted on standard control charts, the process may appear to be out of statistical control for mean, when in fact the data represents a stable process. If action is taken on the process, in an attempt to find the incorrectly identified 'assignable' causes, additional variation will be introduced into the process.

When autocorrelation is encountered, there are four procedures to reduce its impact. These are based on *avoidance* and *correction*:

Avoid
1 Move to 'upstream' measurements to control the process.
2 For continuous processes, sample less often so that the sample interval is longer than the residence time.

Correct
3 For autocorrelation due to special causes, use stratification and rational subgrouping to clarify what is really happening.
4 For intrinsic, stable autocorrelation, use knowledge of the technology to model and 'filter out' the autocorrelation; standard control charts may then be applied to the filtered data.

The mathematics for filtering the data, which can include Laplace transforms, are outside the scope of this book. The reader is referred to the many excellent texts on statistics which deal with these methods.

Appendix J   Approximations to assist in process control of attributes

This Appendix is primarily intended for the reader who does not wish to accept the simple method of calculating control chart limits for sampling of attributes, but would like to set action and warning lines at known levels of probability.

The Poisson approximation

The Poisson distribution is easy to use. The calculation of probabilities is relatively simple and, as a result, concise tables (Appendix F) which cover a range of values of '$\bar{c}$', the defect rate, are readily available. The binomial distribution, on the other hand, is somewhat tedious to handle since it has to cover different values for both 'n', the sample size, and 'p', the proportion defective.

The Poisson distribution can be used to approximate the binomial distribution under certain conditions. Let us examine a particular case and see how the two distributions perform. We are taking samples of size ten from a pottery process which is producing on average 1 per cent defectives. Expansion of the binomial expression, $(0.01 + 0.99)^{10}$ or consultation of the statistical tables will give the following probabilities of finding 0, 1, 2, and 3 defectives:

Number of defectives in sample of 10	Binomial probability of finding that number of defectives
0	0.9044
1	0.0913
2	0.0042
3	0.0001

There is virtually no chance of finding more than three defectives in the sample. The reader may be able to appreciate these figures more easily if we imagine that we have taken 10,000 of these samples of ten. The results should look like this:

Number of defectives in sample of 10	Number of samples out of 10,000 which have that number of defectives
0	9044
1	913
2	42
3	1

We can check the average number of defectives per sample by calculating:

$$\text{Average number of defectives per sample} = \frac{\text{Total number of defectives}}{\text{Total number of samples}}$$

$$n\bar{p} = \frac{913 + (42 \times 2) + (3 \times 1)}{10,000}$$

$$= \frac{1,000}{10,000} = 0.1$$

Now, in the Poisson distribution we must use the average number of defectives, $\bar{c}$ to calculate the probabilities. Hence, in the approximation we let:

$$\bar{c} = n\bar{p} = 0.1$$

so:

$$e^{-\bar{c}} (\bar{c}^x/x!) = e^{-n\bar{p}} ((n\bar{p})^x/x!) = e^{-0.1} (0.1^x/x!)$$

and we find that the probabilities of finding defectives in the sample of ten are:

Number of defectives in sample of 10	Poisson probability of finding that number of defectives	Number of samples out of 10,000 which have that number of defectives
0	0.9048	9048
1	0.0905	905
2	0.0045	45
3	0.0002	2

The reader will observe the similarity of these results to those obtained using the binomial distribution

$$\text{Average number of defectives per sample} = \frac{905 + (45 \times 2) + (2 \times 3)}{10,000}$$

$$n\bar{p} = \frac{1,001}{10,000} = 0.1001$$

We may now compare the calculations for the standard deviation of these results by the two methods:

Binomial $\sigma = \sqrt{n\bar{p}(1-\bar{p})} = \sqrt{10 \times 0.01 \times 0.99} = 0.315$

Poisson $\sigma = \sqrt{\bar{c}} = \sqrt{n\bar{p}} = \sqrt{10 \times 0.01} = 0.316$

The results are very similar because $(1-\bar{p})$ is so close to unity that there is hardly any difference between the formulae for σ. This brings us to the conditions under which the approximation holds. The binomial can be approximated by the Poisson when:

$$p \leqslant 0.10$$

and
$$np \leqslant 5$$

The normal approximation

It is also possible to provide an approximation of the binomial distribution by the normal curve. This applies as the proportion of classified units, p approaches 0.5 (50 per cent), which may not be very often in a quality control situation, but may be very common in an activity sampling application. It is, of course, valid in the case of coin tossing where the chance of obtaining a head in an unbias coin is 1 in 2. The number of heads obtained if twenty coins are tossed have been calculated from the binomial in Table J.1. The results are plotted on a histogram in Figure J.1. The corresponding normal curve has been superimposed on to the histogram. It is clear that, even though the probabilities were derived from a binomial distribution, the results are virtually a normal distribution and that we may use normal tables to calculate probabilities.

Table J.1   *Number of heads obtained from coin tossing*

Number of heads in tossing 20 coins	Probability (binomial $n = 20, p = 0.5$)	Frequency of that number of heads if 20 coins are tossed 10,000 times
2	0.0002	2
3	0.0011	11
4	0.0046	46
5	0.0148	148
6	0.0370	370
7	0.0739	739
8	0.1201	1201
9	0.1602	1602
10	0.1762	1762
11	0.1602	1602
12	0.1201	1201
13	0.0739	739
14	0.0370	370
15	0.0148	148
16	0.0046	46
17	0.0011	11
18	0.0002	2

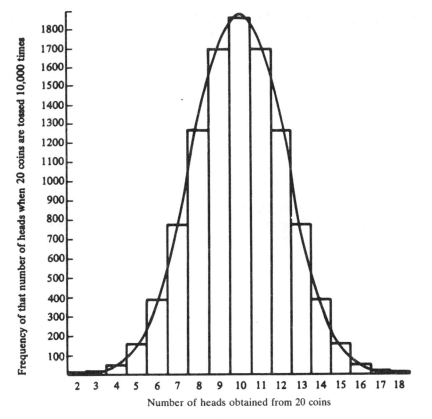

Figure J.1   *Coin tossing – the frequency of obtaining heads when tossing twenty coins*

An example illustrates the usefulness of this method. Suppose we wish to find the probability of obtaining fourteen or more heads when twenty coins are tossed. Using the binomial:

$$P(\geqslant 14) = P(14) + P(15) + P(16) + P(17) + P(18)$$
$$\text{(there is zero probability of finding more than 18)}$$
$$= 0.0370 + 0.0148 + 0.0046 + 0.0011 + 0.0002$$
$$= 0.0577$$

Using the normal tables:

$$\mu = np = 20 \times 0.5 = 10$$
$$\sigma = \sqrt{np(1 - p)} = \sqrt{20 \times 0.5 \times 0.5} = 2.24$$

Since the data must be continuous for the normal curve to operate, the probability of obtaining fourteen or more heads is considered to be from 13.5 upwards.

The general formulae for the z factor is:

$$z = \frac{x - 0.5 - np}{\sigma}$$

Now,

$$z = \frac{14 - 0.5 - 10}{2.24} = 1.563$$

and from the normal tables (Appendix A) the probability of finding fourteen or more heads is 0.058.

The normal curve is an excellent approximation to the binomial when p is close to 0.5 and the sample size n is 10 or more. If n is very large then, even when p is quite small, the binomial distribution becomes quite symmetrical and is well approximated by the normal curve. Obviously the nearer p becomes to 0.5, the smaller n may be for the normal approximation to be applied.

Appendix K   Glossary of terms and symbols

$A_1 - A_4$ Constants used in the calculation of the control lines for mean, moving mean, median and mid-range control chart.

Accuracy   Associated with the nearness of a process to the target value.

Action limit (line)   Line on a control chart beyond which the probability of finding an observation is such that it indicates that a change has occurred to the process and that action should be taken to investigate and/or correct for the change.

Action zone   The zones outside the action limits/lines on a control chart where a result is a clear indication of the need for action.

ARL   The average run length to detection of a change in a process.

Assignable causes   Sources of variation for which an explicit reason exists.

Attribute charts   Control charts used to assess the capability and monitor the performance of parameters assessed as attributes or discrete data.

Attribute data   Qualitative data which can be counted or classified in some meaningful way which does not include measurement.

Average   *See* mean.

B and B'   Constants used in the calculation of control chart lines for standard deviation charts.

Bar   A bar placed above any mathematical symbol indicates that it is the mean value.

Binomial distribution   A probability distribution for samples of attributes which applies when both the number of conforming and non-conforming items is known.

Brainstorming   An activity, normally carried out in groups, in which the participants are encouraged to allow their experience and imagination to run wild, while centred around specific aspects of a problem or effect.

c chart   A control chart used for attributes when the sample size is constant and only the number of non-conformances is known; c is the symbol which represents the number of non-conformances present in samples of a constant size. c-bar ($\bar{c}$) represents the average value of a series of values of c.

Capable   A process which is in statistical control and for which the combination of the degree of random variation and the ability of the control procedure to detect change is consistent with the requirements of the specification.

Cause and effect diagram   A graphic display which illustrates the relationship between an effect and its contributory causes.

Central tendency   The clustering of a population about some preferred value.

Centre line   A line on a control chart at the value of the process mean.

Checklist   A list used to ensure that all steps in a procedure are carried out.

CL   Centre line.

Common causes   *See* random causes.

Conforming   Totally in agreement with the specification.

Continuous data   Quantitative data concerning a parameter in which all values are possible even if limited to a specific range.

Control   The ability or need to observe/monitor a process, record the data observed, interpret the data recorded and adjust the process only when an adjustment can be justified.

Control chart   A graphical method of recording results in order to readily distinguish between random and assignable causes of variation.

Control limits (lines)   Limits or lines set on control charts which separate the zones of stability (no action required), warning (possible problems and the need to seek additional information) and action.

Countable data   A form of discrete data where occurrences or events can only be counted.

Cp   A process capability index based on the ratio of the spread of a frequency distribution to the width of the specification.

Cpk   A process capability index based on both the centring of a frequency distribution and the ratio of the spread of the distribution to the width of the specification. The frequency distribution, and hence the *Cpk*, may reflect the potential for control within the process, the actual performance of the process or the apparent performance after modification by product classification.

Cusum chart   A graphic presentation of the cusum score. The cusum chart is particularly sensitive to the detection of small sustained changes.

Cusum score   The cumulative sum of the differences between a series of observed values and a predetermined target value.

d_n *or* d_2 Symbols which represent Hartley's constant, used to determine the standard deviation from a knowledge of the average range.

D, D_2, D_3, D_4 *and* D' Symbols which represent the constants used to determine the control limits on a range chart.

Data   Facts.

Defect   A fault or flaw which is not permitted by the specification.

Defective   An item which contains one or more defects so that it is judged to be non-conforming.

Detection   The act of discovering.

Deviation   The dispersion between two or more data.

Discrete data   Data not available on a continuous scale.
Dispersion   The spread or scatter about a central tendency.

Frequency   How often something occurs.
Frequency distribution   A table or graph which displays how frequently some values occur by comparison with others. Common distributions include normal, binomial and Poisson.

Grand mean   The mean of either a whole population or the mean of a series of samples taken from the population. The grand mean is an estimate of the true mean – see mu.

Histogram   A diagram which represents the relative frequency of the occurrence of data.

Individual   An isolated result or observation.
Individuals plot   A graph showing a set of individual results.

LAL   Lower action limit or line.
LCL   Lower control limit or line.
LSL   Lower specification limit.
LWL   Lower warning limit or line.

Mean   The average of a set of individual results, calculated by adding together all the individual results and dividing by the number of results. Means are represented by a series of symbols and often carry a bar above the symbol which indicates that it is a mean value.
Mean and range chart   Two charts used to monitor the accuracy and precision of a process being assessed by observing variables.
Mean range   The mean of a series of ranges.
Mean sample size   The average or mean of the sample sizes.
Median   The central value within a population above and below which there are an equal number of members of the population.
Mode   The most frequently occurring value within a population.
Moving mean   A mean value calculated from a series of individual values by moving the sample for calculation of the mean through the series in steps of one individual value and without changing the sample size.
Moving range   A range value calculated from a series of individual values by moving the sample for calculation of the range through the series in steps of one individual value and without changing the sample size.
Mu (μ)   The Greek letter used as the symbol to represent the true mean

of a population as opposed to the various estimates of this value which measurement and calculation make possible.

n   Symbol used to represent the number of individuals within a sample of size *n*. *n*-bar is the average size of a series of samples.

Non-conforming   Not in conformance with any one aspect of the specification.

Normal distribution   Also known as the Gaussian distribution and sometimes referred to as the 'bell-shaped' distribution. The normal distribution has the characteristic that 68.26 per cent of the distribution is contained within ± one standard deviation from the mean value, 95.45 per cent within ± two standard deviations from the mean and 99.73 per cent within ± three standard deviations from the mean.

np chart   A control chart used for attributes when the sample size is constant and the number of conforming and non-conforming items within a sample are both known. *n* is the sample size and *p* the proportion of non-conforming items.

p chart   A control chart used for attributes when the sample sizes are not constant and the number of conforming and non-conforming items are both known. *p* is the proportion of non-conforming items and *p*-bar represents the average of a series of values of *p*.

Pareto analysis   A technique of ranking causes in order to distinguish between the vital few and the trivial many.

Poisson distribution   A probability distribution for samples of attributes which applies when only the number of non-conformances is known.

Population   A large set of data from which samples may be taken.

Precision   Associated with the scatter about a central tendency.

Prevention   The act of seeking to cause something not to occur.

Probability   A measure of the likelihood of an occurrence or incident.

Process   Any activity which converts inputs into outputs.

Process capability   A measure of the capability of a process achieved by assessing the statistical state of control of the process and comparing the amount of random variation present with the tolerance allowed by the specification.

Process capability index   An index of capability, *see Cp and Cpk*.

Process control   The management of a process by observation, analysis, interpretation and action designed to limit variation.

Process mean   The average observed value of an attribute or a variable within a process.

Proportion defective   The ratio of the defectives to the sample size represented by the symbol *p*. *p*-bar represents the mean value of a series of values of *p*.

Quality Meeting the requirements.

R The symbol for the range.
R-bar ($\bar{R}$) The symbol for the mean of a series of value of *R*.
Random causes The contributions to variation which are random in their behaviour.
Randomness A state of disorder in which it is not possible to predict an individual result.
Range The difference between the largest and the smallest result in a sample of individuals – an approximate and easy measure of the degree of scatter. The range is represented by the symbol *R*.
Run A set of results which appears to lie in an ordered series.

Sample A group of individual results, observations or data. A sample is often used for assessment with a view to determining the properties of the whole population or universe from which it is drawn. Samples should be drawn in a random manner.
Sample size The number of individual results included in a sample, or the size of the sample taken. Sample size is represented by the symbol *n*.
Scatter Refers to the dispersion of a distribution.
Scatter diagram The picture which results when simultaneous results for two varying parameters are plotted.
Shewhart chart The control charts for attributes and variables first proposed by Shewhart. These include mean and range, *np*, *p*, *c* and *u* charts.
Sigma (σ) The Greek letter used in SPC to signify the standard deviation of a population.
Skewed distribution A frequency distribution which is not symmetrical about the mean value.
SPC Statistical process control.
Special causes *See* assignable causes.
Specification The requirement against which the acceptability of the inputs or outputs of a process are to be judged.
Spread Refers to the dispersion of a distribution.
SQC Statistical quality control – similar to SPC but with an emphasis on product quality and less emphasis on the need to control processes.
Stable The term used to describe a process when no evidence of assignable causes is present.
Stable zone The central zone between the warning limits on a control chart and within which the results are expected to fall.
Standard deviation A measure of the spread or scatter of a population around its central tendency. Represented by the symbol σ. Estimates of the standard deviation are represented by various symbols such as σ_n, $\sigma_{(n-1)}$, and *s*.

Standard error   The standard deviation of sample mean values – a measure of their spread or scatter around the grand mean. Represented by the symbol SE (and also $\sigma_{\bar{x}}$).

Statistical control   A condition describing a process for which the observed values are scattered about a mean value in such a way as to imply that the origin of the variations is entirely random with no assignable causes of variation and no runs or trends.

Statistical process control   The use of statistically-based techniques for the control of a process of transforming inputs into outputs.

Statistics   The collection and use of data. A method of distilling information from data.

t   The value of a statistic calculated to test significance of the difference between two means.

T   A symbol used to represent a tolerance limit.

Tally chart   A simple tool offered to those who are called upon to record events as they occur or to extract frequencies from existing lists of data.

Target   The objectives to be achieved and against which performance will be assessed, or the mid-point of a specification.

Tolerance   The difference between the lowest and/or the highest value stated in the specification and the mid-point of the specification.

Trend   A series of results which show an upward or downward tendency.

u chart   A control chart used for attributes when the sample size is not constant and only the number of non-conformances is known. *u* is the symbol which represents the number of non-conformances found in a single sample and *u*-bar represents a mean value of *u*.

UAL   Upper action limit or line.

UCL   Upper control limit or line.

Universe   *See* population.

USL   Upper specification limit.

UWL   Upper warning limit or line.

V-mask   A device used in conjunction with a cusum chart to identify trends of known significance.

Variable data   Data which is assessed by measurement.

Variance   A measure of spread equal to the standard deviation squared.

Variation   The inevitable differences between outputs.

Warning limit (line)   Lines on a control chart, on each side of the central line, and within which results are expected to fall, but beyond which the probability of finding an observation is such that it should be regarded as a warning of a possible problem.

Warning zone   The zones on a control chart between the warning and the action limits and within which a result suggests the probability of a change to the process.

x The symbol used to represent an individual value of a variable.
X-bar $(\overline{X})$ The symbol for the mean value of a sample, sometimes the symbol x-bar $(\bar{x})$ is used.
X-bar-bar $(\overline{\overline{X}})$ The symbol for the grand or process mean, sometimes the symbol *X*-bar $(\overline{X})$ is used.

Appendix L   Problems for the reader to solve

Quality management – general – Chapters 1 and 2

1 (a). What is meant by 'quality' and by 'reliability'?
   (b) Differentiate between 'quality of design' and 'quality of conformance'.
   (c) Briefly discuss the three types of cost associated with 'quality of conformance' and discuss how they vary with process capability.

2 'Quality' must be designed and built into a product, it cannot be inspected or advertised in. Discuss.

3 You are the Manufacturing Manager of a small engineering company and have just received the following memo:

Memorandum

To: Manufacturing Manager
From: Sales Manager

Subject: *Order number 2561/3a*

Joe Brown worked very hard to get this order to manufacture an initial 10,000 pistons for PQR Ltd. He now tells me that they are about to return the first batch of 1000 since an initial examination has shown that some of them will not fit into the valve assemblies for which they are intended. I must insist that you give top priority to the examination and rectification of this batch, and that you make sure that this does not recur. As your known PQR are a new customer, and they could put a lot of work our way.

Incidentally, I understand that you are sending operators on a training course in the use of a new semi-automatic measuring gauge for use with that new big machine of yours. I cannot help thinking that you should spend more attention and money on employing more finished product inspection, rather than on training operators to use new fancy equipment.

   (a) Outline how you would investigate the causes of the 'faulty' pistons.
   (b) Discuss the final paragraph of the memo.

Data assembly and presentation – Chapters 3 and 4

Worked examples of many of the following problem types will be found within the text of Chapters 3 and 4.

4 Operators on an assembly line are having difficulties when mounting electronic components onto a printed circuit board. The difficulties include: undersized holes in the board, absence of holes in the board, oversized wires on components, component wires snapping on

bending, components longer than the corresponding hole spacing, wrong components within a batch, and some other less frequent problems. Design a simple tally chart which the operators could be asked to use in order to keep detailed records. How would you make use of such records? How would you engage the interest of the operators in keeping such records?

5   Describe with examples the methods which are available for presentation of information by means of charts, graphs, diagrams, etc. Discuss the advantages of these forms of presentation.

6   The following table shows the recorded thickness of steel plates nominally 3.00 ± 0.10 mm. Plot a frequency distribution histogram of the plate thickness, and comment on the result.

Plate thickness (mm)

2.97	2.92	2.94	3.00	2.94	3.02
2.99	2.97	2.95	2.97	2.92	3.01
3.04	3.00	2.97	2.96	2.97	2.94
2.96	3.04	2.85	2.91	2.99	2.96
2.88	2.95	2.98	2.97	3.09	2.99
3.01	3.13	2.92	2.90	3.03	3.03
3.05	2.90	2.98	3.02	2.98	2.93
3.07	3.01	3.01	3.03	2.91	2.95
3.09	2.99	3.16	2.91	3.00	2.98
2.97	2.92	3.00	3.00	3.01	3.00

7   To establish a manufacturing specification for the thickness of washers, a sample of 200 was taken from the production stream and the thickness of each washer measured. The frequency distribution of the thicknesses is given below. State and explain the conclusions you would draw from this distribution, assuming the following:

(a)   The washers came from one machine.
(b)   The washers came from two machines.

Measured thickness of washers

Thickness (mm)	Number of washers
2.38	2
2.39	13
2.40	32
2.41	29
2.42	18
2.43	21
2.44	20
2.45	22
2.46	22
2.47	13
2.48	3
2.49	0
2.50	1
2.51	1
2.52	0
2.53	1
2.54	0
2.55	2
	200

8   Discuss in detail the application of Pareto analysis as an aid in solving management problems. Give at least three illustrations.

9   You are responsible for a biscuit manufacturing unit, and are concerned about the output from one particular line which makes chocolate-coated wholemeal biscuits. Output is consistently significantly below target. You suspect that this is because the line is frequently stopped, so you initiate an in-depth investigation over a typical two-week period. The table below shows the causes of the stoppages, the number of occasions on which they occurred, and the average amount of output lost on each occasion.

Causes	Number of occurrences	Lost production (00s biscuits)
Wrapping		
Cellophane wrap breakage	1031	3
Cartonner failure	85	100
Enrober		
Chocolate too thin	102	1
Chocolate too thick	92	3
Preparation		
Underweight biscuits	70	25
Overweight biscuits	20	25
Mis-shapen biscuits	58	1
Ovens		
Overcooked biscuits	87	2
Undercooked biscuits	513	1

Use this data and an appropriate technique to indicate where to concentrate remedial action aimed at increasing output.

10 A company manufactures a range of domestic electrical appliances. Particular concern is being expressed about the warranty claims on one product. The customer service department provides the following data relating the claims to the failed component part.

Component part	Number of claims	Average cost of warranty work (per claim)
Drum	110	48.1
Casing	12842	1.2
Worktop	142	2.7
Pump	246	8.9
Electric motor	798	48.9
Heater unit	621	15.6
Door lock mechanism	18442	0.8
Stabilizer	692	2.9
Powder additive unit	7562	1.2
Electric control unit	652	51.9
Switching mechanism	4120	10.2

Discuss what criteria are of importance in identifying those component parts to examine initially. Carry out a full Pareto analysis to identify such components.

11   The principal causes of accidents within a large international organization, their percentage frequency of occurrence, and the estimated annual cost of the lost production arising from each cause is given in the table below.

Accident cause	Percentage frequency	Estimated value of lost production ($ million/annum)
Machinery	16	190
Transport	8	30
Falls from heights > 2 metres	16	100
Tripping	3	10
Collision with fixed objects	9	7
Falling objects	7	20
Handling goods	27	310
Hand tools	7	65
Burns (all types)	5	15
Miscellaneous	2	3

(a)   Using the appropriate data, construct a Pareto curve, and suggest how this may be used most effectively to tackle accident prevention.

(b)   Suggest other ways in which the organization might collect data for alternative accident prevention programmes.

12   A credit card company finds that some 18 per cent of application forms received from customers cannot be processed immediately, owing to the absence of some of the information called for. A sample of 500 incomplete application forms reveals the following data concerning missing information.

Information missing	*Frequency*
Applicant's Age	92
Daytime telephone number	22
Forenames	39
House owner/occupier	6
Home telephone number	1
Income	50
Signature	6
Occupation	15
Bank Account no.	1
Nature of account	10
Postal code	6
Sorting code	85
Credit Limit requested	21
Cards existing	5
Date of application	3
Preferred method of payment	42
Others	46

Using Pareto analysis to determine the major causes of missing information, and suggest ways in which the application form might be redesigned to reduce the incidence of missing information.

13　A company which operates with a four-week accounting period is experiencing difficulties in keeping up with the preparation and issue of sales invoices during the last week of the accounting period. Data collected over two accounting periods is as follows.

Accounting Period 4	Week	1	2	3	4
Number of sales invoices issued		110	272	241	495

Accounting Period 5	Week	1	2	3	4
Number of sales invoices issued		232	207	315	270

Does a simple scatter diagram test suggest a correlation between the week within the period and the demands placed on the invoice department? What action could be proposed?

Handling variable data, process capability and control charts

Chapters 5, 6, 7, 8, 9 and 10 include worked examples of variable data handling.

14   What is meant by the inherent variability of a process?

15   Process control charts may be classified under two broad headings, 'variables' and 'attributes'. Compare these two categories and indicate when each one is most appropriate.

16   State the Central Limits Theorem and explain how it is used in statistical process control.

17   A machine is operated so as to produce ball bearings having a mean diameter of 0.574 inches and with a standard deviation of 0.008 inches. To determine whether the machine is in proper working order a sample of six ball bearings is taken every half-hour and the mean diameter of the six is computed.

    (a)   Design a decision rule whereby one can be fairly certain that the ball bearings constantly meet the requirements.

    (b)   Show how to represent the decision rule graphically.

    (c)   How could even better control of the process be maintained?

18   The following are measures of the impurity, iron, in a fine chemical which is to be used in pharmaceutical products. The data is given in parts per million (ppm).

Sample	X_1	X_2	X_3	X_4	X_5
1	15	11	8	15	6
2	14	16	11	14	7
3	13	6	9	5	10
4	15	15	9	15	7
5	9	12	9	8	8
6	11	14	11	12	5
7	13	12	9	6	10
8	10	15	12	4	6
9	8	12	14	9	10
10	10	10	9	14	14
11	13	16	12	15	18
12	7	10	9	11	16
13	11	7	16	10	14
14	11	7	10	10	7
15	13	9	12	13	17
16	17	10	11	9	8
17	4	14	5	11	11
18	8	9	6	13	9
19	9	10	7	10	13
20	15	10	12	12	16

Set up mean and range charts and comment on the possibility of using them for future control of the iron content.

19 You are a trader in foreign currencies. The spot exchange rates of all currencies are available to you at all times. The following data for one currency was collected at intervals of one minute for a total period of 100 minutes, five consecutive results are shown as one sample.

Sample	Spot exchange rates				
1	1333	1336	1337	1338	1336
2	1335	1335	1332	1337	1335
3	1331	1338	1335	1336	1338
4	1337	1335	1336	1336	1334
5	1334	1335	1336	1336	1337
6	1334	1333	1338	1335	1338
7	1334	1336	1337	1335	1334
8	1336	1337	1335	1332	1331
9	1334	1334	1332	1334	1336
10	1334	1335	1337	1334	1332
11	1334	1334	1335	1336	1332
12	1335	1335	1341	1338	1335
13	1336	1336	1337	1331	1334
14	1335	1335	1332	1332	1339
15	1335	1335	1334	1334	1334
16	1333	1333	1335	1335	1334
17	1334	1340	1336	1338	1342
18	1338	1336	1337	1337	1337
19	1335	1339	1341	1338	1338
20	1339	1340	1342	1339	1339

Use the data to set up mean and range charts; interpret the charts, and discuss the use which could be made of this form of presentation of the data.

20 Using process capability studies, processes may be classified as being in statistical control and capable. Explain the basis and meaning of this classification. Suggest conditions under which control charts may be used, and how they may be adapted to make use of data which is available only infrequently.

21 Conventional control charts are to be used on a process manufacturing small components with a specified length of 60 mm ± 1.5 mm. Two identical machines are involved in making the components and process capability studies carried out on them reveal the following data:

Sample size, n = 5

Sample number	Machine I Mean	Range	Machine II Mean	Range
1	60.10	2.5	60.86	0.5
2	59.92	2.2	59.10	0.4
3	60.37	3.0	60.32	0.6
4	59.91	2.2	60.05	0.2
5	60.01	2.4	58.95	0.3
6	60.18	2.7	59.12	0.7
7	59.67	1.7	58.80	0.5
8	60.57	3.4	59.68	0.4
9	59.68	1.7	60.14	0.6
10	59.55	1.5	60.96	0.3
11	59.98	2.3	61.05	0.2
12	60.22	2.7	60.84	0.2
13	60.54	3.3	61.01	0.5
14	60.68	3.6	60.82	0.4
15	59.24	0.9	59.14	0.6
16	59.48	1.4	59.01	0.5
17	60.20	2.7	59.08	0.1
18	60.27	2.8	59.25	0.2
19	59.57	1.5	61.50	0.3
20	60.49	3.2	61.42	0.4

Calculate the control limits to be used on a mean and range chart for each machine and give the reasons for any differences between them. Compare the results from each machine with the appropriate control chart limits and the specification tolerances. Calculate the relevant process capability indices. Is there any significant correlation between the variations of the two machines? Discuss any implications which should lead to action by the process manager.

22 Distinguish between random and assignable causes of variation. Describe the characteristics of the Normal distribution and construct an example to show how these may be used in answering questions which arise from discussions of specification limits for a product. Define process capability indices and describe how they may be used to monitor the capability of a process, its actual performance and its performance as perceived by a customer.

23 A bottle filling machine is being used to fill 15 ounce bottles of a ketchup. The actual bottles will hold 15.6 ounces. The machine has been set to discharge an average of 15.2 ounces. It is known that the actual amounts discharged follow a normal distribution with a standard deviation of 0.2 ounces.

(a) What proportion of the bottles overflow?
(b) The overflow of bottles causes considerable problems and it has therefore been suggested that the average discharge should be reduced to 15.1 ounces. In order to meet the weights and measures regulations, however, not more than 1 in 40 bottles, on average, must contain less than 14.6 ounces. Will the weights and measures regulations be contravened by the proposed changes?

24 What do you understand by the 'average run length' (ARL) of a control chart? A process, when in statistical control, has a mean of 40 and a standard deviation of 3. Specify action and warning limits for mean and range control charts based on a sample of six. Assuming that the variance remains unchanged, and considering the action limits on the mean chart only, what is the average run length when:

(a) The process mean is 40?
(b) The process mean is 36?

25 Define the average run length (ARL) to detection of a control chart for variables. Based on the action limit only, what is the ARL on the mean chart when the process mean has shifted by one and two standard deviations from the target mean ($n = 4$)? Describe the method of presenting mean values as a cusum chart and the method of constructing a V-mask. Assuming that a V-mask with a decision interval of 5SEs is used, what is the ARL on the cusum chart when the process mean has shifted by one and two standard deviations from the target mean?

26 Explain the principles of Shewhart control charts for sample mean and sample range, and cumulative sum control charts for sample mean and sample range. Compare the performance of these charts

by plotting the following data and interpreting the results. The sample size is four and the specification is 60.0 ± 2.0.

Sample number	Mean	Range	Sample number	Mean	Range
1	60.0	5	26	59.6	3
2	60.0	3	27	60.0	4
3	61.8	4	28	61.2	3
4	59.2	3	29	60.8	5
5	60.4	4	30	60.8	5
6	59.6	4	31	60.6	4
7	60.0	2	32	60.6	3
8	60.2	1	33	63.6	3
9	60.6	2	34	61.2	2
10	59.6	5	35	61.0	7
11	59.0	2	36	61.0	3
12	61.0	1	37	61.4	5
13	60.4	5	38	60.2	4
14	59.8	2	39	60.2	4
15	60.8	2	40	60.0	7
16	60.4	2	41	61.2	4
17	59.6	1	42	60.6	5
18	59.6	5	43	61.4	5
19	59.4	3	44	60.4	5
20	61.8	4	45	62.4	6
21	60.0	4	46	63.2	5
22	60.0	5	47	63.6	7
23	60.4	7	48	63.8	5
24	60.0	5	49	62.0	6
25	61.2	2	50	64.6	4

27 The following table gives the average width in millimetres for each of twenty samples of five panels used in the manufacture of a domestic appliance. The range of each sample is also given.

Sample number	Mean	Range	Sample number	Mean	Range
1	550.8	4.2	11	553.1	3.8
2	552.7	4.2	12	551.7	3.1
3	553.8	6.7	13	561.2	3.5
4	555.8	4.7	14	554.2	3.4
5	553.8	3.2	15	552.3	5.8
6	547.5	5.8	16	552.9	1.6
7	550.9	0.7	17	562.9	2.7
8	552.0	5.9	18	559.4	5.4
9	553.7	9.5	19	555.8	1.7
10	557.3	1.9	20	547.6	6.7

Calculate the control chart limits for the Shewhart charts and plot the values on the charts. Interpret the results. Given a specification of 540 mm ± 5 mm, comment on the capability of the process. Use the above data to investigate the variations using cusum charts. Comment on any additional findings and, on the combined use of Shewhart and cusum charts for process management.

28 Consider a packaging process with short-term, hour to hour, variations of packed quantity, represented by a standard deviation, σ_s, of 5.2 grams. Irregular and apparently random fluctuations in the average level of the process on a much shorter time scale, can similarly be represented by a standard deviation of, σ_r, of 1.5 g. To formulate a quantity control system, estimate the combined medium-term effect of these two sources of variation and express the answer as a standard deviation, σ_m.

29 In a batch manufacturing process the viscosity of the compound increases during the reaction cycle and determines the end-point of the reaction. Samples of the compound are taken throughout the whole period of the reaction and sent to the laboratory for viscosity assessment. The laboratory tests cannot be completed in less than three hours. The delay during testing is a major source of underutilization of both equipment and operators. Records have been kept of the laboratory measurements of viscosity and the power taken by the stirrer in the reactor during several operating cycles. When plotted as two separate moving mean and moving range charts this reveals the following data.

Date and time		Moving mean viscosity	Moving mean stirrer power
07/04	07.30	1020	21
	09.30	2250	27
	11.30	3240	28
	13.30	4810	35
	BATCH COMPLETED AND DISCHARGED		
	18.00	1230	22
	21.00	2680	22
08/04	00.00	3710	28
	03.00	3980	33
	06.00	5980	36
	BATCH COMPLETED AND DISCHARGED		
	13.00	2240	22
	16.00	3320	30
	19.00	3800	35
	22.00	5040	31
	BATCH COMPLETED AND DISCHARGED		
09/04	04.00	1510	25
	07.00	2680	27
	10.00	3240	28
	13.00	4220	30
	16.00	5410	37
	BATCH COMPLETED AND DISCHARGED		
	23.00	1880	19
10/04	02.00	3410	24
	05.00	4190	26
	08.00	4990	32
	BATCH COMPLETED AND DISCHARGED		

Standard error of the means – viscosity – 490
Standard error of the means – stirrer power – 9

Is there a significant correlation between these two measured parameters? If the specification for viscosity is 4500 to 6000, could the measure of stirrer power be used for effective control of the process?

Handling attribute data, process capability and control charts

Worked examples based on attribute data are included in Chapters 11 and 12.

30   In the context of quality control explain what is meant by a 'number of defectives' control chart and a 'number of defects' control chart. Discuss the use of such charts for both constant and variable sample sizes.

31   The following record shows the number of non-conforming items found in a sample of 100 taken twice per day.

Sample number	Number of defectives	Sample number	Number of defectives
1	4	21	2
2	2	22	1
3	4	23	0
4	3	24	3
5	2	25	2
6	6	26	2
7	3	27	0
8	1	28	1
9	1	29	3
10	5	30	0
11	4	31	0
12	4	32	2
13	1	33	1
14	2	34	1
15	1	35	4
16	4	36	0
17	1	37	2
18	0	38	3
19	3	39	2
20	4	40	1

Set up a Shewhart '*np*' chart, plot the above data and comment on the results. Set up a cusum chart, plot the above data and comment on the results. Discuss the use of both charts for process control and management.

32   A control chart for a new plastic sheet product is to be initiated. After commissioning, twenty-five samples of plastic sheets were taken from the production facility. The examination of the sheets for flaws yielded the following results. Set up the control chart and comment on the capability of the process to meet a specification of a maximum of 10 flaws per sheet.

Sample number	1	2	3	4	5	6	7	8	9	10	11	12	13
Number of flaws/sheet	2	3	0	2	4	2	8	4	5	8	3	5	2

Sample number	14	15	16	17	18	19	20	21	22	23	24	25
Number of flaws/sheet	3	1	2	3	4	1	0	3	2	4	2	1

Plot the above data on a cusum chart. Does this reveal any additional information, and, if so, what action is required? Assuming that the average number of flaws in a sheet increases by 2, estimate the average run length to the detection of this change when using both the Shewhart and the cusum charts.

Appendix M   Some worked examples and answers to numerical problems in Appendix L

Chapters 5, 6, 7, 8, 9 and 10 include worked examples of variable data handling. The following examples give a more complete treatment of the data.

Lathe operation

A component used as a part of a power transmission unit is manufactured using a lathe. Twenty samples, each of five components, are taken at half-hourly intervals. For the most critical dimension, the process mean ($\overline{\overline{X}}$) is found to be 3.500″, with a normal distribution of the results about the mean, and a mean sample range ($\overline{R}$) of 0.0007″.

(a) Use this information to set up suitable control charts.
(b) If the specified tolerance is 3.498″ to 3.502″, what is your reaction? Would you consider any action necessary?
(c) The following table shows the operator's results over the day. The measurements were taken using a comparitor set to 3.5000″ and are shown in units of 0.001″. The means and ranges have been added to the results. What is your interpretation of these results? Do you have any comments on the process and/or the operator?

Record of results recorded from the lathe operation

Time	1	2	3	4	6	Mean	Range
7.30	0.2	0.5	0.4	0.3	0.2	0.32	0.3
7.35	0.2	0.1	0.3	0.2	0.2	0.20	0.2
8.00	0.2	−0.2	−0.3	−0.1	0.1	−0.06	0.5
8.30	−0.2	0.3	0.4	−0.2	−0.2	0.02	0.6
9.00	−0.3	0.1	−0.4	−0.6	−0.1	−0.26	0.7
9.05	−0.1	−0.5	−0.5	−0.2	−0.5	−0.36	0.4

Machine stopped – tool clamp readjusted

Time	1	2	3	4	6	Mean	Range
10.30	−0.2	−0.2	0.4	−0.6	−0.2	−0.16	1.0
11.00	0.6	0.2	−0.2	0.0	0.1	0.14	0.8
11.30	0.4	0.1	−0.2	0.5	0.3	0.22	0.7
12.00	0.3	−0.1	−0.3	0.2	0.0	0.02	0.6

Lunch

12.45	−0.5	−0.1	0.6	0.2	0.3	0.10	1.1
13.15	0.3	0.4	−0.1	−0.2	0.0	0.08	0.6

Re-set tool by 0.15″

13.20	−0.6	0.2	−0.2	0.1	−0.2	−0.14	0.8
13.50	0.4	−0.1	−0.5	−0.1	−0.2	−0.10	0.9
14.20	0.0	−0.3	0.2	0.2	0.4	0.10	0.7
14.35	**Batch finished – machine re-set**						
16.15	1.3	1.7	2.1	1.4	1.6	1.62	0.8

Solution
(a)   Since the distribution is known and the process is in statistical control with:

Process mean $\overline{\overline{X}} = 3.500''$

Mean sample range $\overline{R} = 0.0007''$

Sample size $n = 5$

Mean chart
From Appendix B for $n = 5$, $A_2 = 0.58$ and $2/3A_2 = 0.39$
   Mean control chart is set up with:
   Upper action limit $\overline{\overline{X}} + A_2\overline{R} = 3.50041''$

   Upper warning limit $\overline{\overline{X}} + 2/3A_2\overline{R} = 3.50027''$

   Mean $\overline{\overline{X}} = 3.50000''$

   Lower warning limit $\overline{\overline{X}} - 2/3A_2\overline{R} = 3.49973''$

   Lower action limit $\overline{\overline{X}} - A_2\overline{R} = 3.49959''$

Range chart
From Appendix C $D'_{.999} = 0.16$ $D'_{.975} = 0.37$
                          $D'_{.025} = 1.81$ $D'_{.001} = 2.34$

   Range control chart is set up with:
   Upper action limit $D'_{.001}\overline{R} = 0.0016''$

   Upper warning limit $D'_{.025}\overline{R} = 0.0013''$

   Lower warning limit $D'_{.975}\overline{R} = 0.0003''$

   Lower action limit $D'_{.999}\overline{R} = 0.0001''$

(b) The process is correctly centred so:

From Appendix B $\qquad d_n = 2.326$

$$\sigma = \bar{R}/d_n = 0.0007/2.326 = 0.0003$$

$$Cp = Cpk = \frac{(USL - \bar{\bar{X}})}{3\sigma} = \frac{(\bar{\bar{X}} - LSL)}{3\sigma} = \frac{0.002}{0.0009} = 2.22$$

The process is in statistical control and capable. If mean and range charts are used for its control, significant changes should be detected by the first sample taken after the change. No further immediate action is suggested.

(c) The means and ranges of the results are given in the table above and are plotted on control charts in the following figure.

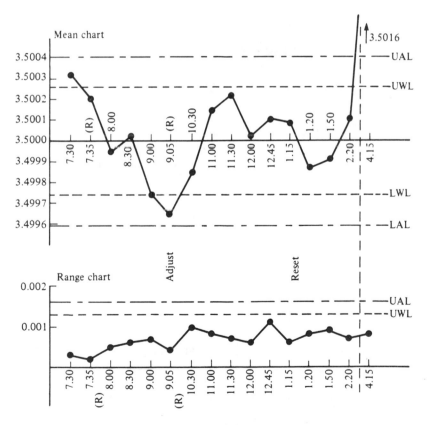

Control charts for lathe operation

Observations on the control charts

1   The 7.30 sample required a repeat sample to be taken to check the mean. The repeat sample at 7.35 showed that no adjustment was necessary.
2   The 9.00 sample mean was within the warning limits but was the fifth result in a downwards trend. The operator correctly decided to take a repeat sample. The 9.05 mean result constituted a double warning since it remained in the downward trend and also fell in the warning zone. Adjustment of the mean was, therefore, justified.
3   The mean of the 13.15 sample was the fifth in a series above the mean and should have signalled the need for a repeat sample and *not* an adjustment. The adjustment, however, did not adversely effect control.
4   With $n = 5$ and a *Cpk* of 2.2, the whole of the batch completed at 14.35 was within specification and suitable for despatch.
5   At 16.15 the machine was incorrectly reset.

General conclusions

There was a downward drift of the process mean during the manufacture of this batch. The drift was limited to the early period and appears to have stopped following the adjustment at 9.05. The assignable cause should be investigated.

The range remained in control throughout the whole period when it averaged 0.0007″, as in the original process capability study.

The operator's actions were correct on all but one occasion (the reset at 13.15); a good operator who may need a little more training, guidance, or experience.

Purchase of pins

Steel pins are purchased in lot sizes of 10,000 and used in the manufacture of textile machinery. The agreed specification for the pin lengths is 30 to 32 mm. Following problems with the lengths of the pins, the supplier was persuaded to carry out a process capability study from which it was concluded that the process was in statistical control with a *Cpk* of 2.1 (sample size $n = 4$). The manufacturer further agreed to maintain control of the pin length using mean and range control charts, but was unwilling to supply the details of either the process capability study or the control charts. In addition, there were still occasional problems with both short and long pins. A random sample of fifty pins was taken from one delivered batch and the lengths measured. These gave a mean value of 31.2 mm and a standard deviation of 0.4 mm.

1   How do you interpret the batch results?
2   What percentage, if any, of this batch would lie outside the tolerance band?
3   What options are open to both the customer and the supplier?

Solution
(a)   Let us assume that during the capability study the process was centred (this will be the easiest way for the manufacturer to achieve a $Cpk_{(potential)}$ of 2.1 with $n = 4$). Then:

$$2.1 = Cpk_{(potential)} = (USL - \overline{\overline{X}})/3\sigma = (32 - 31)/3\sigma$$

$$so\ \sigma = (32 - 31)/(2.1 \times 3) = 0.16\ mm$$

Had the process not been properly centred during the capability study this estimate of the standard deviation would be greater than that observed. If control charts were being used for product classification, and all pins produced prior to an action signal were eliminated from the despatches, the probability of finding any pins outside the specification limits would be extremely small – rising to a probability of 1 in 1000 on infrequent occasions.

Of course, the standard deviation of the delivered batch is greater than the standard deviation determined during the process capability study. If we assume that the distribution of the pins in the batch examined was normal, from the properties of the normal distribution we would expect to find only 0.3 per cent of pins outside the plus and minus three standard deviations zone. The batch results indicate that this zone is from:

$$31.2 - (3 \times 0.4) = 30\ mm\ to\ 31.2 + (3 \times 0.4) = 32.4\ mm$$

So, at the lower end of the specification (30 mm) there is about a 1 in 1000 chance of finding non-conforming pins and at the upper end (32 mm) there is a greater probability. If the distribution was non-normal, one of the tails would be longer than the normal distribution tail and, in either case, there would be even more non-conforming product at that side of the distribution. None of these probabilities of non-conformance are consistent with the declared $Cpk_{(potential)}$ and the use of mean and range charts for the control of both the process and the product.

Conclusion
The supplier's claims to have a process in statistical control with a $Cpk_{(potential)}$ of 2.1, and to be using mean and range charts for process control and product classification are suspect.

(b)   In order to estimate the proportion of product outside the tolerance zone, we shall be obliged, either to assume that the distribution is normal, or to carry out a test for normality (see Appendix A). Let us assume that the test for normality was carried out on the data from the batch sampled and that it showed the distribution to be normal. We can then use the table in Appendix A to estimate the proportion of non-conforming product on each side of the tolerance zone.

Lower tolerance limit

$$Z = (\bar{\bar{X}} - \text{LSL})/\sigma = (31.2 - 30)/0.4 = 3.0$$

At $Z = 3$ the proportion is 0.0013

Upper tolerance limit

$$Z = (\text{USL} - \bar{\bar{X}})/\sigma = (32 - 31.2)/0.4 = 2.0$$

At $Z = 2$ the proportion is 0.0228, so the total estimated proportion outside specification is $(0.0013 + 0.0228) = 0.0241$ or 2.41 per cent.

A skewed distribution would have resulted in either an increase or a decrease in this estimate depending on whether the skew was towards the lower or upper specification limit.

(c)   The *first option* is either to *approach the supplier*, or to *do nothing*. Since there is room for doubt concerning the supplier's claims, the option to do nothing should be dismissed. *So plan to approach the supplier.*

The existing evidence for suspicion is:

(i)   The reluctance of the supplier to make process data available.
(ii)   Having tested for normality, the estimated percentage of non-conforming product is 2.4.
(iii)   Occasional incidents of short and long pins are still occurring.

The *next option* is to *approach the supplier immediately* or to *collect further evidence*.

The supplier may be reluctant to make process data available for a number of reasons which could include:

● Uncertainty about whether the data and its analysis is fully understood by the supplier.
● The supplier's uncertainty about the customer's understanding of the data and its analysis.
● The possibility that in 'revealing his hand' to the customer, the process knowledge may be abused.
● The potential for leakage to competitors, etc.

Either before meeting formally to discuss the problem or during the early stages of such a discussion it will be useful to try and ascertain the existence and reasons for any reluctance.

The estimates of the mean and the standard deviation based on a random sample of fifty pins taken at random from one batch will be adequate to describe the characteristics of that batch. A similar sampling of several other batches will reveal whether the problem is general or limited to the one batch. If the discussion with the supplier is centred around the characteristics of the only batch studied, the problem may not have been well defined. Either there is a general problem which can be justified by examining several batches selected at random, or the problem relates to isolated batches. These may be different problems arising from different causes and requiring different solutions.

While any non-conforming pins should be tabled during the discussion remember that these only provide a convincing demonstration of the effect. The object of the discussion is to persuade the supplier to locate the *causes* of the effect and eliminate them in order to eliminate the effect.

If similar or identical pins are purchased from other suppliers, repeat the same analysis on several batches of their product, or, much better, seek their permission to make use of their process capability and process control chart data. If their capability is good, why should they wish to be shy about it.

The reader will be able to think of other reasons why the preferred option is to prepare for the discussion with the supplier, collect further evidence, and enter the discussions with a coherent and reasonably complete story.

Bottle filling

Bottles are being weighed gross after filling and a sample of empty bottles is also weighed to permit an estimate of the net weight of the contents. The results are:

Gross weight   Average 350 g Standard deviation 4.5 g
Tare weight   Average 60 g Standard deviation 1.5 g

What are the mean and standard deviation of the contents of the bottles?

The relationship between the average gross, net and tare weights may be expressed as:

$$\mu_{(net)} = \mu_{(gross)} - \mu_{(tare)}$$
$$\text{so } \mu_{(net)} = 350 - 60 = 290 \text{ grams}$$

The variance in the contents must be the sum of the variance associated with the weighing of the empty bottles and the variance associated with the weighing of the filled bottles. There are two sources of variation which are influencing the estimate of the net weight of the contents. So unlike the averages, which are subtracted above, we write for the variances:

$$\sigma_{(net)}^2 = \sigma_{(gross)}^2 + \sigma_{(tare)}^2$$

$$\text{or } \sigma_{(net)}^2 = (4.5)^2 + (1.5)^2$$

$$= 20.25 + 2.25 = 22.5$$

$$\text{so } \sigma_{(net)} = \sqrt{22.5} = 4.74g$$

The result of the standard deviations adding through their squares (the variances) is that the standard deviation of the contents is closer to the larger standard deviation of the gross weight than to the smaller standard deviation of the tare. Any improvement in the standard deviation of the tare would have only a minor effect on the standard deviation of the contents, whereas, an improvement in the standard deviation of the gross weight would reflect a significant improvement in the standard deviation of the contents.

Answers to numerical questions in Appendix L

6 $N = 60$; from Sturgess rule – 7 intervals
Highest value 3.16   Lowest value 2.85

Interval	Frequency	
2.825–2.874	1	
2.875–2.924	10	Distribution not properly
2.925–2.974	17	centred on 3.00 mm and some
2.975–3.024	20	material outside specification
3.025–3.074	8	limits.
3.075–3.124	2	$\bar{\bar{X}} = 2.98$ mm $\sigma = 0.05$ mm
3.125–3.174	2	

7   The distribution is bimodal and positively skewed.

(a)   Change in operator, material, maintenace, etc.?
(b)   Machines have different settings/operating regimes.

9   Pareto analysis:
Most frequent occurrences: cellophane wrap breakage (50 per cent) and undercooked biscuits (25 per cent).
Largest production losses: cartonner failure (57 per cent) and cellophane wrap breakage (21 per cent).

10 Pareto analysis:
Most frequent cause of claims: door lock mechanism (40 per cent), casing (28 per cent) and power additive unit (16 per cent).
Greatest costs of warranty: switching mechanism (24 per cent), electric motor (23 per cent) and electric control unit (20 per cent).

11 Pareto analysis:
Most frequent accident causes: handling goods (27 per cent), machinery (16 per cent) and falls (16 per cent)
Causes of greatest lost production: handling goods (41 per cent), machinery (25 per cent) and fall (13 per cent).

12 Pareto analysis:
Most frequently found information missing: age (20 per cent), bank account sorting code (19 per cent), income (11 per cent). *Note:* 'others' account for 10 per cent and need to be broken down.

13 Scatter diagram with weeks 1–4 on x axis and invoice numbers 0–500 on y axis gives 4 'quadrants' with $N = 6$, $n = 2$. Hence, N is $> 2 + (0.6 \times 2) = 3.2$, and some correlation is indicated.

17 (a) A high degree of confidence will exist if the process can be shown to be in statistical control and capable. A measure of statistical control can be established if the sample means are plotted and lie within plus and minus 3SEs of the grand or process mean, i.e. within the interval 0.564″ to 0.584″. To establish the capability of the process, the specification must be known. This information is not given.
(b) Set up a mean chart with both warning and action limits.
(c) Also use a range chart.

18 Grand or process mean $(\overline{\overline{X}})$ = 10.8 ppm   Mean range $(\overline{R})$ = 7.5 ppm

Mean chart		Range chart	
UAL	15.2 ppm	UAL	17.6 ppm
UWL	13.7 ppm	UWL	13.6 ppm
LWL	7.9 ppm		
LAL	6.5 ppm		

Mean and range charts show that process is in statistical control. Standard deviation (σ) = 3.2. If *Cpk* of 2 required the process would be capable of meeting a maximum specification of $\overline{\overline{X}} + 6SE + 3\sigma$ = 29 ppm.

19 Grand mean ($\overline{\overline{X}}$) = 1335.6   Mean range ($\overline{R}$) = 4.6

Mean chart		Range chart	
UAL	1338.3	UAL	10.6
UWL	1337.4	UWL	8.2
LWL	1333.9		
LAL	1333.0		

The mean and range charts show that:

- The range is in statistical control (confirmed by the Cusum plot of the ranges).
- The mean is not in statistical control. Inspection of this chart allows a run below the mean to be identified as an assignable cause and also an assignable movement of the exchange rate to a mean value above the UAL.

21

		Machine 1		Machine 2
Grand mean	($\overline{\overline{X}}_1$)	= 60.05 mm	($\overline{\overline{X}}_2$)	= 60.06 mm
Mean range	($\overline{R}_1$)	= 2.39 mm	($\overline{R}_2$)	= 0.40 mm
Standard deviation	(σ_1)	= 1.03 mm	(σ_2)	= 0.17 mm
Mean chart	UAL	61.43 mm		60.28 mm
	UWL	60.98 mm		60.21 mm
	LWL	59.12 mm		59.90 mm
	LAL	58.67 mm		59.83 mm
Range chart	UAL	5.58 mm		0.92 mm
	UWL	4.32 mm		0.71 mm
		In statistical control		Not in statistical control but in control on range
Cp		= 0.48		= 2.94
Cpk		= 0.47		= 2.39

Scatter diagram of machine 1 -v- machine 2 shows no significant correlations.

Action:

- Random variations on machine 1 are much greater than those on machine 2, investigate.
- Investigate assignable causes of the variations of the means on machine 2.
- Anticipate that with a proper control procedure machine 2 will be highly capable of meeting the requirements.

23 (a) 0.0228 or 2.28 per cent.
  (b) The Weights and Measures Act will not be contravened since the proportion below 14.6 ounces will be 0.0062 or 1 in 161.

24   *Mean chart*   *Range chart*
    UAL   43.66   UAL   16.8
    UWL   42.44   UWL   13.1
    LWL   37.56   LWL   3.2
    LAL   36.34   LAL   1.6

    (a)   ARL when mean is 40 is equivalent to proportion under the tails at 3 standard deviations, i.e. 0.0013×2 or an ARL of 385.
    (b)   When mean is 36, the proportion under the tail of the normal distribution is 0.3897 so the ARL is 2.57.

25 *Shewhart mean chart*
- At 1σ, which equals 2SEs, the proportion under the tail of the normal distribution is 0.1597, so the ARL is 6.3.
- At 2σ, ie. 4SEs, the proportion is $0.5 + 0.1587 = 0.6787$, so the ARL is 1.47.

Cusum charts
- At 1σ, ARL = 4.
- At 2σ, ARL = 2.

26 Grand or process mean = 60.8   Mean range = 3.94

Standard deviation = 1.91

    Mean chart   *Range chart*
    UAL   63.7   UAL   10.1
    UWL   62.7   UWL   7.6
    LWL   58.9
    LAL   57.9

Charts show that the ranges are in statistical control but the means are not. *Cp* is 0.34 and *Cpk* worse, since the process is not centred at 60. The random variations would appear to be excessive by comparison with the requirements. The cusum plot of the means shows a below target peformance from sample 1 to sample 27 followed by a marked deterioration. Similarly the cusum plot of the ranges shows a below target performance from sample 1 to about sample 20. (Neither of these are clear from the mean and range charts.) Re-analysing samples 1 to 20 only, shows a process which is in statistical control with a standard deviation of 1.46 and a *Cp* of only 0.46, so the random variations are still unacceptably large by comparison with the requirements. The cusum plots of the mean and ranges for this limited period show no trends.

27 Grand or process mean = 554.0 mm   Mean range = 4.2 mm

Standard deviation =   1.8 mm

Mean chart		Range chart	
UAL	551.5 mm	UAL	9.9 mm
UWL	555.6 mm	UWL	7.6 mm
LWL	552.3 mm		
LAL	551.5 mm		

Charts show the ranges are in statistical control and the means are not. $Cp = 0.915$. The Grand or process mean is outside the tolerance band so the Cpk is negative. The cusum plot of the means reflects the highly erratic behaviour of the means. The cusum of the ranges is also erratic and show no significant trends.

Action
Investigate the assignable causes of the changes in the means, eliminate them and then reassess the capability of the process.

28 $\sigma_m = 5.4$ grams

29   Any correlation is only significant at below the 95 per cent confidence limit, so the stirrer amps may not be used to control the viscosity.

31 $n\bar{p} = 2.1$

Control chart
UAL 6.5
UWL 5.5

Chart shows process is in control. Cusum plot shows that up to about sample 14 the process was running above the target and that after this it first became erratic and then more stable in a below target performance. $n\bar{p}$ for the last 20 samples is 1.5.

32 $\bar{c} = 2.96$

Control chart
UAL 8.5
UWL 6.5

Charts show process in control and, since UAL is 8.5 and specification is 10, it is capable. Cusum plot shows a below target performance from sample 12 to 25; during this period $\bar{c}$ is 2.35. If $\bar{c}$ increased to 4.96, say 5, the probability of finding 8.5 or more flaws would be 0.0681 (see Appendix H), i.e. an ARL of 14.7. The use of a V-mask with a 2σ decision interval would have an ARL of 2.

Appendix N Further reading

Besterfield, D.H. (1990). *Quality Control*, 3rd edition. New Jersey: Prentice-Hall Inc.

British Standards Institution (1980). *Handbook 23 – Quality Control*. London: BSI.

Caplen, R.H. (1988). *Practical Approach to Quality Control*, 5th edition. London: Business Books.

Crosby, P.B. (1979). *Quality is Free*. New York: McGraw-Hill.

Crosby, P.B. (1984). *Quality Without Tears*. New York: McGraw-Hill.

Cullen, J. and Hollingham, J. (1988). *Implementing Total Quality*. London: IFS (Publications) Ltd.

Davies, O.L. and Goldsmith, P.L. (eds), (1972). *Statistical Methods in Research and Production*, 4th edition. Edinburgh: Oliver and Boyd (for ICI).

Deming, W.E. (1982). *Quality, Productivity, and Competitive Position*. Cambridge, Mass: MIT Center for Advanced Engineering Study.

Deming, W.E. (1986). *Out of the Crisis*. Cambridge, Mass: MIT Center for Advanced Engineering Study.

Duncan, A.J. (1986). *Quality Control and Industrial Statistics*, 5th edition. Illinois: Richard D. Irwin.

Feigenbaum, A.V. (1983). *Total Quality Control*, 3rd edition. New York: McGraw-Hill.

Grant, E.L. and Leavenworth, R.S. (1988). *Statistical Quality Control*, 6th edition. New York: McGraw-Hill.

Gitlow, H.S. and Gitlow, S.J. (1987). *The Deming Guide to Quality and Competitive Position*. Englewood Cliffs, NJ: Prentice-Hall Inc.

Harrington, H.H. (1987). *The Improvement Process: How America's Leading Companies Improve Quality*. New York: McGraw-Hill.

Ishikawa, K. (translated by David J. Lu) (1985). *What is Total Quality Control? – the Japanese Way*. Englewood Cliffs, NJ: Prentice-Hall.

Ishikawa, K. (1982). *Guide to Quality Control*. Tokyo: Asian Productivity Association.

Jamieson, A. (1982). *Introduction to Quality Control*. Reston, A Prentice-Hall Co.

Juran, J.M. (ed) (1988). *Quality Control Handbook*, 4th edition. New York: McGraw-Hill.

Juran, J.M. and Gryna, E.M. (1980). *Quality Planning and Analysis*, 2nd edition. New York: McGraw-Hill.

Mann, N.R. (1985). *The Keys to Excellence: The Story of the Deming Philosophy*. Santa Monica, California: Prestwick Books.

Montgomery, D.C. (1985). *Introduction to Statistical Quality Control*. New York: John Wiley & Sons.

Moroney, M.J. (1978). *Facts from Figures*, 2nd edition. Harmondsworth: Penguin.

Murdoch, J. and Barnes, J.A. (1986). *Statistical Tables for Science, Engineering, Management and Business Studies*, 3rd edition. London: Macmillan.

Murphy, J.A. (1986). *Quality in Practice*. Dublin: Gill and Macmillan.

Oakland, J.S. (1989). *Total Quality Management*. Oxford: Heinemann Professional Publishing.

Ott, E.R. (1975). *Process Quality Control – Troubleshooting and Interpretation of Data*. New York: McGraw-Hill, Kogakusha.

Owen, M. (1989). *SPC and Continuous Improvement*. Bedford: IFS Publications.

Price, F. (1985). *Right First Time*. London: Gower.

Price, F. (1989). *Right Every Time*. London: Gower.

Scherkenbach, W.D. (1986). *The Deming Route to Quality and Productivity: Roadmaps and Roadblocks*. Washington, DC: Cee Press Books.

Society of Motor Manufacturers & Traders (1986). *Guidelines to Statistical Process Control*. London: SMMT.

Shewhart, W.A. (1931). *Economic Control of Quality of Manufactured Product*. New York: Van Nostrand.

Shingo, S. (1986). *Zero Quality Control: Source Inspection and the Poka-Yoke System*. Stamford, Conn: Productivity Press.

Shingo, S. (1987). *Key Strategies for Plant Improvement*. Cambridge, Conn: Productivity Press.

Taguchi, G. (1979). *Introduction to Off-Line Quality Control*. Magaya: Central Japan Quality Control Association.

Taguchi, G. (1981). *On-Line Quality Control during Production*. Tokyo: Japanese Standards Association.

Taguchi, G. (1986). *Introduction to Quality Engineering*. Tokyo: Asian Productivity Organisation.

Townsend, P.L. with Gebhart, J.E. (1986). *Commit to Quality*. New York: John Wiley & Sons Inc.

Index

A_1 to A_4 constants, 348, 350
  definition, 385
ABC analysis, 62
Absenteeism, 316
Accuracy, 87, 88
  definition, 385
  and process capability, 183
Action limit (line), 109, 348
  definition 385
Action zone, 77, 111
  definition, 385
Activity sampling, 315
Addition law of probability, 226
Addition of:
  dependent random variations, 191
  independent random variations, 185
  ranges, 185
  standard deviations, 185
  tolerances, 185
  variances, 185, 189
Adept Ltd UK, 347
Analysis of:
  existing data, 206
  failure, 9
  variance, 182, 191
Analysis:
  cause and effect, 16, 62, 289, 309
  Pareto, 14, 305
Answers to numerical problems, 407
Appraisal costs, 8
Approximations to:
  binomial, 380
  normal, 382
  Poisson, 380
ARL (average run length), 126, 136, 233
  curves, 373
  definition, 385

Assessing state of control, 118
Assignable causes, 261, 332
  definition, 385
  types, 219
Assignable causes of variability, 80, 151
Assurance, quality, 5, 8, 13
Attribute charts, 223, 256
  definition, 385
Attribute data summary, 251, 34
  definition 385
Attributes:
  approximations in control of, 380
  defined, 34
  process capability, 225
  process control, 223, 251
Auditing, measuring devices, 190
Audits, quality, 8
Autocorrelation, 378
Average, 88, *see also* Mean
  definition, 385
  fill level, 101
  moving, 161
  run length (ARL), 126, 136, 233, 373
Averages, and sampling, 98

B and B' constants, 351
  definition, 385
Bar:
  definition, 385
Bar charts, 36, 38, 207
Barriers to implementation of SPC, 328
Basic concepts, 1
Basic principles, 325
Basic tools, 14
Batch processes, 201
Bell-shaped curve, 85
Benefits of SPC, 327

Bimodal distribution, 39
Binary classification, 34
Binomial distribution, 227
    definition 385
Binomial expression, 227
Blending, 196
Blending, assessing its effectiveness, 198
Blending to reduce variation:
    example, 197
Bottle filling:
    example, 413
Boundaries, process, 18
Brainstorming 14, 62, 64, 305
    definition, 385

c chart, 224, 241, 318, 320
    definition, 385
c_n constants, 351
Calculation form, for $\overline{X}$ and R chart, 117
Capability 5, 13, 39, 120, 225, 306
    components, 183
    indices, 348
Capable:
    definition, 385
Cause and effect analysis, 14, 62, 289, 309
    action plan, 68
    applications, 69
    example, 64
    procedure, 64
Cause and effect diagram, 62
    definition, 385
Causes of process variability, 80, 151
Causes of random variability, 184
Cell intervals, 41
Central limits theorem, 102, 192
Central tendency, 38, 88
    definition, 386
Central value, measures, 88
Centre line:
    definition, 386
Chance, 69
Change of mean:
    detection, 136
Characteristic, operating, 127
Chart:
    c, 224, 241, 318, 320
    Cp, 348
    Cpk, 348
    income rank/contribution rank, 314
    for individuals, 152
    mean, 106, 107, 151

median, 157
mid-range, 157
moving mean, 161
moving range, 161
multi-vari, 157
np, 224, 225
p, 224, 237
range, 106, 108, 115, 151
rank, 314
standard deviation, 174
u, 224, 246
Charts:
    control, *see* Control charts
    tally 14
Cheating, 324
Checksheets, 14
Checklist:
    definition, 386
'Chumbo' chart, 40
CL (control limit):
    definition, 386
Classification of product from control
        charts, 142
Collection of data, 296
Combined mean charts and cusum charts,
        264
    attribute data, 261
    combined Shewhart and cusum charts,
        264
Commitment, 329
Common causes of variability, 80
    definition, 386
Complaints, 9
Computer methods, 347
Computers:
    use of in SPC, 205
Concepts, basic, 1
Confidence in decision making, 76
Confidence limits, 363
Conformance, 5
Conforming units, 223
    definition, 386
Constants:
    mean chart, 348
    median chart, 350
    range chart, 349
    standard deviation chart, 351
Continuous data:
    definition, 386
Continuous processes, 202
Continuously recorded data, 141

Contribution rank/income rank chart, 314
Control, 1, 5, 13
  definition, 386
Control charting, 76
Control charts, 14, 74, 107, 305
  decision zones, 77
  definition, 386
  performance, 126
  trouble shooting, 212
  reflecting drift of process mean, 214
  reflecting drift of process scatter, 216
  reflecting frequent irregular shifts of
    mean, 214
  reflecting frequent irregular shifts of
    scatter, 217
  reflecting shifts of mean and scatter, 219
  reflecting sustained shifts of process
    mean, 213
  reflecting sustained shifts of process
    scatter, 215
  for variables, international, 124
Control limits (lines), 109, 124
  definition, 386
Control, process, 106, 120
Control, process in, 107, 118
Correlation during unsteady operation, 196
Correlation test:
  with scatter diagrams, 74
Correlations between dependent variables
    from SPC data, 191
Costs:
  appraisal, 8
  failure, 8
  of quality, 5, 7
  prevention, 7
Countable data:
  definition, 386
Cp:
  definition, 386
  delivery, 210
  production, 210
  meaning and use, 133
Cpk:
  delivery or Csk, 145
  potential, 141–7
  production or Ppk, 144
  definition, 386
  limiting value, 138
  meaning and use, 133
  operating characteristic curves, 140
  and sample size, 139

Csk, 145
Cumulative distributions, 57
Cumulative Poisson probability curves, 362
Cumulative Poisson probability tables, 352
Cumulative sum charts, 256
Curves:
  ARL (average run length), 373
  OC (operating characteristics), 373
Customer demands for SPC, 325
Customer requirements, 1
Customer/supplier relationships, 18, 132,
  146
Cusum charts, 256, 310, 312
  attributes, 258
  decision procedures, 268
  definition, 386
  variables, 262
Cusum score, 258
  definition, 386

D_2, D_4, D^m, and D' constants, 349
  definition, 386
Data:
  analysis, 296
  autocorrelated, 378
  collected or observed, 107
  infrequent, 173
  injury, 320
  and the process, 305
  rational subgrouping of, 104
  statistical, 107
Data collection:
  formats, 35
  design of formats, 36
  planning, 296
Data presentation:
  use of pictures, 36
Decision intervals for V-masks:
  attributes, 269
  variables, 273
Decision procedures:
  cusum charts, 268
Decision zones:
  control charts, 77
Defect, 223
  definition, 386
Defective:
  definition, 386
  product, 8
  product, proportion – estimating, 96
  proportion, 224, 237

Defectives, 223
Defects, chart (c), 241
Defining customers/suppliers, 18
Degree of confidence in decision making, 76
Deming cycle, 294
Deming, W. Edwards, 11
Describing the process, 19
Design, 5, 284, 298
    cusum chart for means of variables, example, 262
    of experiments, 182, 191
    for quality, 331
    system, 299
Design of V-masks:
    attributes, 269
    variables, 272
Designing processes, 20
Detection, 2
    definition, 386
Detection of change of mean:
    attributes, 256
Detection of trends:
    variables, 117
    attributes, 256
Development, 284
Development of the process, 28
Deviation, standard, 92
    definition, 386
Difference between two means, 363
Discrete data, 34
    definition, 387
Dispersion, 38
    definition, 387
Distribution, 284
    binomial, 227
    normal, 85, 94
    Poisson, 241
    rectangular, 102
    triangular, 103
d_n or d_2 constant, 348
    definition, 386
Dot plot, 39
Downgrading, 9
Drifts of mean:
    on control charts, 214
Drifts of process scatter:
    on control charts, 216

Economic design, 287
Eighty-Twenty Rule, 52

Effect and cause analysis, 62
Equipment, inspection, 8
Error, standard of means, 100
Errors on invoices, 318
Estimated:
    mean, $(\overline{X})$, 93
    proportion defective, 96
    standard deviation, (σ), 94
Estimating:
    process capability with existing data, 203
    random effect of measuring, 185
    random effect of sampling, 185
European Centre for TQM, 324
EWMA, 172
Examination, process, 25
Examples:
    addition of variances, 186
    analysis of existing variable data – in-process, 210
    analysis of existing variable data – raw materials, 206
    blending to reduce variation, 197
    cause and effect analysis, 64
    combined mean charts and cusum charts, 264
    control charts for variables, 142
    cusum chart for attributes, 258
    cusum and Shewhart charts, 278
    design of a variable cusum chart, 262
    histograms, 44
    scatter diagrams, 70
    worked, 407
Existing data analysis:
    example, 206
Experiments, statistically planned, 300
Exponential constant, 241
Exponentially weighted moving average (EWMA), 172
External failure costs, 9

F test for variances, 367
Facts from Figures, 241
Failure analysis, 9
Failure costs, 8
Firefighting, 61
Fishbone diagram, 63
Flow processes, 201
Flowchart of process capability studies, 203
Flowcharting, 14, 18, 21
    for implementation of SPC, 334
    symbols, 24

Forecasting, 327
  income, 311
  moving average, 168
Frequency:
  definition, 387
  polygon, 84
  of sampling, 125
Frequency distributions, 37
  definition, 387

Gaussian distribution, 85
Goodwill, loss of, 9
Grand mean, $\overline{\overline{X}}$, 89
Graphs, 14, 46
  choice of scale, 48
Green-field, 333

Hartley's Constant (d_n or d_2), 113, 348, 386
Hawthorne Effect, 69
Hidden plant/operation, 10
Histograms, 14, 36, 84, 41, 305
  definition, 387
  examples, 44
Historic data, 332
Hunting, 99

i-chart, 152
Identification, product, 107
Implementation of SPC, 324
  barriers, 328
  improvement, 295
Improvement, 13, 21, 29, 225
  implementing, 295
  never-ending cycle, 294
  planning, 295
  process, 290
  teams, 289
In control (process), 80, 107
In process data analysis:
  example, 210
In statistical control, 80
Income:
  forecasting, 311
  rank/contribution rank chart, 314
  ranking, 312
Independent random variations:
  addition, 186
Indirect quality costs, 9
Individuals:
  charts, 148, 152

definition, 387
Industrial Quality Control, 127
Information, process, 18
Infrequent data, 173
Initial steps to SPC, 333
Injury data, 318
Inspection equipment, 8
Installation, 284
Internal failure costs, 8
International control charts – variables, 124
Invoice errors, 318
Ishikawa diagram, 63
ISO 9000 (International Standards Organisation), 330

Japan, 325
Job processes, 201
Juran, J. J., 2
Just-in-time, 327

Kaisen teams, 289
Key:
  slopes on a cusum chart for variables, 263
Kurtosis, 345

LAL (lower action line or limit), *see* Limits; Lines
Large sample size, 342
Lathe operation:
  example, 407
Law, Pareto, 52
Laws of probability, 226
LCL (lower control limit), *see* Limits; Lines
Liability, 9
Limit theorem, central, 102
Limits:
  confidence, 363
  control, 109, 124, 232, 385
  definition, 387
  see also Lines
Line graphs, 46
Lines:
  action, 109, 113, 115, 155, 157, 173, 176, 232, 239, 243, 249
  warning, 109, 113, 115, 155, 157, 176, 232, 239, 243, 249
Lockyer, Keith G., 316
Lockyer's Five Ps, 65

Log normal distribution, 345
Logistics, 284
Long term:
    objectives, 334
    problem solving, 61
Loss:
    function, total, 297
    of goodwill, 9
Lower action line (LAL), *see* Limits; Lines
Lower control limit (LCL), *see* Limits;
    Lines
Lower warning line (LWL), *see* Limits;
    Lines
LSL (lower specification limit):
    definition, 387
LWL (lower warning line or limit), *see*
    Limits; Lines

Magic and SPC, 324
Maintenance, 284
Management commitment, 326
Market research, 284
Marketing, 284, 387
Mean:
    grand, 89
    process, 89
    sample, 107
Mean charts, 107, 151
    ARL curves, 373
    constants, 348
    definition, 387
    moving, 161
    OC curves, 373
    sensitivity to detection of changes of
        mean, 136
    with varying sample size, 155
    zones, 111
Mean range, 113
    definition, 387
Mean sample size, 238
    definition, 387
Means:
    scatter diagrams, 193
Measures of central value, 88
Measures of spread of values, 92
Median, 90
    chart, 157
    chart constants, 350
    definition, 387
Meeting customer requirements, 1

Methodology:
    SPC implementation, 329
Mid-range chart, 157
Mixing, 196
Mode, 38, 91
    definition, 387
Modified control charts, 147
Modifying processes, 20
Monitoring:
    process, 331
    quality, 331
Moroney, M. J., 241
Motivation for SPC, 325
Moving average, exponentially weighted,
    172
Moving mean:
    chart, 161
    definition, 387
Moving range:
    chart, 161
    definition, 387
Mu (μ), 89
    definition, 387
Muhlemann, Alan P., 316
Multiplication law of probability, 226
Multi-vari chart, 157

n – (sample size), 100
    definition, 388
Narrowing down problems, 32
Never-ending improvement, 61, 69, 182,
    294, 326
Noise:
    components, 132
Non-conformance, 202
Non-conforming units, 223
    definition, 388
Non-conformities, 223
    definition, 388
Non-manufacturing, SPC, 305
Non-normal distributions, 131
Non-normality, 338, 343
Normal approximation to binomial, 382
Normal distribution, 85, 94, 131, 338
    definition, 388
    using, 96
Normal probability paper, 338
North West Analytical, USA, 347
Number defective np chart, 224, 225, 257
    for absenteeism, 316
    definition, 388

Number of defects (*c*) chart, 224, 241, 318, 320
Number of defects per unit (*u*) chart, 224, 246

Oakland, John S., 10, 289, 316
OC (operating characteristic), 127
  for *Cpk*s, 140
  curve, 373
  Cusum charts, 277
  Shewhart charts, 277
One-sided specifications, 134
Operating characteristic (OC), 127
Operating characteristics of:
  *Cpk*s, 140
  cusum charts, 277
  Shewhart control charts, 277
Out of control (process), 80
Out of control processes:
  management, 201
Out of statistical control, 80
Ovality, 157

p chart, 224, 237
  definition, 388
Packaging, 284
Parameter design, 299
Patterns on control charts, 212
Pareto Analysis, 14, 305
  definition, 388
  procedure, 52
Pareto curve, 58, 59
Performance of control charts, 126
Pictorial graphs, 47
Pie charts, 48
Piece to piece variation, 157
Planning:
  for quality, 331
  improvement, 295
  quality, 8
Poisson approximation to binomial, 380
Poisson distribution, 241
  definition, 388
  tables, 352
Policy, quality, 2, 329
Polygon, frequency, 84
Population:
  definition, 388
Potential capability, 133
Ppk, 144

Precision, 81, 92
  definition, 388
  and process capability, 183
  required in measurement, 187
Predicted variability, 88
Presentation of data:
  its impact, 49
Presentation of results, 335
Prevention, 3, 326
  costs, 7
  definition, 388
Probability:
  binomial, 227
  definition, 388
  laws of, 226
  paper, 338
Problem solving, 51
  short and long term, 61
Problems:
  SPC implementation, 324
Procedures, standardizing, 20
Process, 1, 3
  boundaries, 18
  and data, 305
  definition, 388
  description, 19
  design, 20
  development, 28
  development and assessment, 284
  examination, 25
  flowcharting, 14
  improvement, 29, 290
  in control, 107, 118
  information, 18
  modification, 20
  operation, 284
  planning, 284
  stability, 81
  types, 201
  understanding, 18
  variation, 80
Process capability, 120, 127, 306
  attributes, 225
  definition, 388
  and control, 332
Process capability for variables:
  measurement, 129
Process capability index:
  definition, 388
Process capability indices, 133
  interpretation, 135

process capability studies using a
  flowchart, 204
Process changes:
  shown on control charts, 213–19
Process control, 120, 127, 284
  attributes, 223
  basic principles, 202
  definition, 388
  procedures, 203
  and teamwork, 281
  variables, 106
Process control charts and improvement,
  292
Process management, 184
Process mean, 89
  definition, 388
Process operator's responsibility, 324
Process stability, assessing, 116
Process variability, causes, 80
Procurement, 284
Product:
  defective, 8
  identification, 107
  requirements, 7
  specification, 107
  testing/checking, 284
Product classification:
  from control charts, 142
Production and Operations Management,
  316
Profit on sales, 308
Progress, 326
Project processes, 201
Proportion defective:
  definition, 388
  estimating, 96
  (*p*) chart, 224, 237
Proportions under tail of normal
  distribution, 340
Purchased goods:
  example, 410
Purpose, fitness for, 1

Quality, 1
  audits, 8
  component parts, 32
  of conformance, 6
  definition, 389
  of design, 6
  planning, 8
  policy, 2, 329

'Quality Analyst' (computer package), 347
Quality assurance, 5, 8, 13
Quality circles, 62
Quality costs, 7
Quality improvement teams, 289
Quality management:
  short and long term, 184
Quality organization, 331
Quality systems, 284, 332
Questioning technique, 27
Quick and rough methods of assessment,
  190

R (range), 92, 106
  definition, 389
R chart, 108, 115, 151, 178
Random causes:
  definition, 389
  of variability, 80, 151
Random variation:
  management, 182
Randomness:
  definition, 389
Range, 92, 106
  definition, 389
  mean, 113
  sample, 107
  variability, 184
Range chart, 108, 115, 151, 178
  ARL curves, 373
  constants, 349, 350
  moving, 161
  OC curves, 373
Rank:
  chart, 314
  plotting, 62
Ranking, 55, 312
Rating, vendor, 8
Rational subgrouping, 104
Raw material input data – analysis:
  example, 206
Rectangular distribution, 103
Rectification, 8
Reduction in variation, 293, 299
Reinspection, 9
Relaxed control charts, 147
Remedial action, 326
Repair, 9
Requirements:
  meeting them, 1, 129
  product, 7

service, 7
Research, market, 284
Responsibility, 325
Returns, 9
Rework, 8
Right First Time, 9, 13
Risk of erroneous decisions, 335
Rough and quick methods of assessment, 190
Rules:
  for assessing state of control, 118, 124
  interpretation of cusum charts, 261
Run:
  chart, 152
  definition, 389
Run length, average, 126
Runs in attributes:
  detection, 256
Runs on mean chart, 112

s (standard deviation estimate), 94
σ (standard deviation), 92
Sales, 284, 306, 308
  forecasting, 256
Sample:
  definition, 389
  frequency, 125
  means, 107
  ranges, 107, 115
Sample size, 104, 125, 342
  definition, 389
  varying-mean chart, 155
Sampling, 75
  activity, 315
  and averages, 98
  definition, 389
Scatter diagrams, 14, 70, 191
  definition, 389
  examples, 70
  of means, 193
  procedure, 73
Scatter plot, autocorrelation, 378
Scheffe, Henry, 127
Scrap, 8
Senior management's role in SPC, 326
  using mean charts, 136
Sensitivity to change:
  cusums and V-masks, 275
Service requirements, 7, 284
Servicing, 9

Setting targets, 97
Shewhart charts, 256
  definition, 389
Shewhart control charts and cusums in combination, 275
Shewhart, Walter, 11
Shifts of mean:
  on control charts, 213, 214
Shifts of process scatter:
  on control charts, 215, 217
Short-term objectives, 334
Short-term problem solving, 61
Sigma (σ), 93
  definition, 389
Sigma (σ) chart, 174
Significance, tests of, 363
Size, sample, 104, 125
Skew distributions, 45, 343
  definition, 389
Small sample size, 342
SPC:
  benefits, 327
  definition, 389
  implementation, 324
  in non-manufacturing, 305
  successful users, 326
  for variables, summary, 127
SPC data:
  its use in seeking correlations between variables, 191
SPC implementation:
  the steps, 329
SPC system, 11, 284
Special causes:
  definition, 389
  of variability, 80
Specification:
  attributes, 224
  definition, 389
  one-sided, 134
  and process capability, 183
  product, 107
Specifications, subjective, 250
Specifying, 284
Spread, 38
  definition, 389
Spread of values, measures, 92
Spur to action:
  using cause and effect analysis, 66
  using presentation, 49
  using Pareto Analysis, 52

SQC:
  definition, 389
Stable:
  definition, 389
Stable process, 81
  assessing, 116
Stable zone, 77
  definition, 389
Standard deviation, 92
  addition of, 185
  chart, 174
  definition, 389
  of means, 100
  variability, 184
Standardizing procedures, 20
Statistical control, 80
  definition, 390
Statistical data, 107
*Statistical Method from the Viewpoint of
  Quality Control*, 11
Statistical process control, 11
  definition, 390
Statistical quality control (SQC), 325
Statistically planned experiments, 300
Statistics, 76
  definition, 32, 390
Steps in assessing process stability, 116
Storage, 284
Sturgess Rule, 42
Subgrouping, rational, 104
Subjective assessments, 250
Successful users of SPC, 326
Summary of SPC with $\overline{X}$ and R charts, 127
Supplier, customer relationships, 18
Symbols, 385
  flowcharting, 24
System, process control, 284
System, SPC, 11, 284
Systematic approach, 32

t (student's), 364
  definition, 390
T (tolerance):
  definition, 390
t test for means, 364
Tables, constants, 338–71
Taguchi, 132
  methods, 182, 297
Tally charts, 14, 37
  definition, 390

Target:
  definition, 390
Target values for cusums:
  attributes, 258
  variables, 262
Targets, setting, 96
Teamwork, 287, 333
Technical service, 284
Terms, glossary, 385
Test:
  F, 367
  t, 364
Tests of significance, 363
Theorem, central limit, 102
Third party:
  SPC implementation, 327
Tolerance, 131
  definition, 390
Tolerance band:
  increased, 131
  narrowed, 132
  specification, 132
Tolerance design, 300
Tools, basic, 14
Total loss function, 297
Total quality costs, 10
Total Quality Management, 10, 289
Total Quality Management (TQM), 11
Training, 8, 326
  process operators, 328
  for quality, 328
Trend, 112
  definition, 390
Trends in attributes:
  their detection, 256
Trends on mean chart, 112
Triangular distribution, 103
Trouble shooting with control charts, 212
Truth, 136
Type 1 and 11 errors, 131

u chart, 224, 246
  definition, 390
UAL (upper action line or limits), *see*
  Limits; Lines
UCL (upper control limit), *see* Limits;
  Lines
Understanding, 326, 329
  the process, 18
  variation, 94

Unimodal distribution, 40
Universe:
  definition, 390
Unsteady operation:
  correlations, 196
Upper action line (UAL), *see* Limits; Lines
Upper control limit, *see* Limits; Lines
Upper warning line (UWL), *see* Limits;
  Lines
USA control charts for variables, 124
Use of control charts for product
    classification:
  example, 142
Using the normal distribution, 96
USL (upper specification limit):
  definition, 390
UWL (upper warning line or limit), *see*
    Limits; Lines

V-masks, 268
  attributes, 269
  definition, 390
  variability, 76
  variables, 273
Variability:
  components, 184
  range estimates, 184
  standard deviation, 184
Variable data:
  definition, 390
Variables, 80
  defined, 34
  international control charts, 124
  process control, 106

Variances, 185
  additions, 186
  definition, 390
  F test for, 367
Variation:
  assessment, 132
  definition, 390
  process, 80
  reduction in, 293, 299
  understanding, 94
Vendor rating, 8
Verification, 8
Vertical axis scale:
  cusums, 262
Vital few, 60

Warning limit (line), *see* Limits; Lines
  warning zone, 77
  definition, 391
Waste, 9
Weibull distribution, 345

x, definition, 391
$\overline{X}$ chart, 107
X-bar ($\overline{X}$), 89
  definition, 391
X-bar-bar ($\overline{\overline{X}}$), 102
  definition, 391

Z value, 96
Zero defects, 192
Zone:
  action, 111, 385
  stable, 111
  warning, 111
Zones of control, 111, 121